THE BODY

THE BODY

A COMPLETE USER'S GUIDE

HOW YOUR BODY FUNCTIONS ⟩⟩⟩
AND HOW TO KEEP IT HEALTHY AND STRONG

REVISED EDITION

Foreword by RICHARD RESTAK, M.D.

NATIONAL GEOGRAPHIC

WASHINGTON, D.C.

Contents *Foreword 8*

BOARD OF ADVISERS

WILLIAM R. AYERS, M.D.
Professor of Medicine, Emeritus
Georgetown University
School of Medicine
Washington, D.C.

ROBERT M. BRIGGS, M.D.
Chairman, Department of Plastic
and Reconstructive Surgery (retired)
St. Barnabas Medical Center
Livingston, New Jersey

ELIZABETH BROWNELL, M.D.
Family Practice Physician
Waukesha Memorial Hospital,
Division of Family and
Emergency Medicine
Waukesha, Wisconsin

TED M. BURNS, M.D.
Professor of Clinical Neurology
University of Virginia,
School of Medicine
Charlottesville, Virginia

ROBERT A. CHASE, M.D.
Emile Holman Professor of Surgery
and Professor of Anatomy, Emeritus
Stanford University,
School of Medicine
Stanford, California

D. ROBERT DUFOUR, M.D.
Consultant Pathologist and
Attending Physician, Veterans
Affairs Medical Center, Liver Clinic
Washington, D.C.

DAVID E. FLEISCHER, M.D.
Professor of Medicine
Mayo Clinic College of Medicine
Scottsdale, Arizona

**JEAN L. FOURCROY, M.D.,
PH.D., M.P.H.**
Medical Officer, Center for Devices
and Radiological Health (retired)
United States Food and
Drug Administration
Washington, D.C.

DONALD S. KARCHER, M.D.
Chair, Department of Pathology,
and Chief, Flow Cytometry Service
George Washington University
Medical Center
Washington, D.C.

JOHN E. MORLEY, M.B., B.CH.
Dammert Professor of Gerontology
and Director, Division of Geriatric
Medicine and Endocrinology
Saint Louis University Medical Center
St. Louis, Missouri

POLLY E. PARSONS, M.D.
E. L. Amidon Professor and Chair
of the Department of Medicine
University of Vermont,
College of Medicine
Burlington, Vermont

ROBERT S. ZOHLMAN, M.D.
Clinical Professor of
Medicine (retired)
University of California, San Francisco
San Francisco, California

AUTHORS

RICHARD RESTAK, M.D.
Foreword

PATRICIA DANIELS
Chapters One, Five, Six,
and Nine

TRISHA GURA, PH.D.
Chapters Ten and Eleven

SUSAN TYLER HITCHCOCK
Chapter Four

LISA STEIN
Chapters Two and Three

JOHN THOMPSON
Chapters Seven, Eight, and Twelve

ACKNOWLEDGMENTS

Theodore Abraham, M.D.,
Director, Johns Hopkins
Hypertrophic Cardiomyopathy
Center of Excellence; Arthur
Burnett, M.D., Professor of
Urology, Johns Hopkins University
School of Medicine; Manus
Patten, Ph.D., Assistant Professor
of Biology, Georgetown
University; David Perlmutter,
M.D., Medical Director,
Perlmutter Health Center.

The medical illustrations in this
book were created by Scientific
Publishing Ltd., a research and
development publishing company
specializing in high-quality medical and scientific illustrations and
educational products.

(Opposite) Blood cells are critically
important to the body. Red blood
cells supply oxygen; round, white
blood cells fight disease; and
platelets help in healing.

(Page 8) This visualization shows
how the skeleton supports and
shapes the body.

FOREWORD *by Richard Restak, M.D.*

The Body: A Complete User's Guide successfully meets three challenging tasks. First, it provides new scientific information about the human body in ways that can be understood by the general public. Second, it illustrates how this knowledge relates to everyday life. Finally, it frankly acknowledges that our knowledge about the body is far from complete and will continue to evolve.

The Body meets these challenges through the ingenious interweaving of the main text—a model of clarity, precision, and readability—with relevant sidebars and fact boxes that provide practical examples and tips for improving health. As a result readers are rewarded with that rarest of publishing creations: a book that integrates science and humanism.

The chapter on bones documents how minute electrical currents and ultrasound waves can reduce healing time in arm and shinbone fractures by 35 to 45 percent. A survey of the bones in the skull segues gracefully into why most blows to the head do not result in concussion. For fans of murder mysteries comes this tidbit about the hyoid bone, a tiny U-shaped structure at the base of the tongue: "In forensic medicine the hyoid is examined to detect whether the deceased has been murdered—it sometimes snaps when a person is strangled or choked."

The section on the spine provides the clearest explanation I've ever encountered outside of a medical textbook about disks in the neck and back, how injuries occur, and what treatment options are available. And thanks to the entry on the knee joint, sports fans will understand why this powerful joint is injured so frequently by athletes and what can be done to repair such potentially career-ending injuries.

Readers will also learn about the kidney's extraordinary daily filtering capacity (more than 300 times the body's blood volume!) and discover the best ways to avoid a painful kidney stone. In the discussion about the skin, the largest organ of the body, we discover why there are no known ways of avoiding age-related skin changes. Given this inevitability, the book suggests that we should perhaps take "a kinder view of wrinkles, accepting them as well-earned character lines, each with a story to tell, rather than as reminders of passing time." (See what I mean about humanism?)

Thanks to its breadth of coverage, clarity of writing, and marvelous integration of information—ranging from Renaissance art (Did you know Leonardo da Vinci spent years studying the human body, dissecting more than 30 cadavers and producing hundreds of anatomical drawings?) to gymnastics (women's hips have a wider range of movement than men's, resulting in greater flexibility for splits and other maneuvers)—*The Body* will prove accessible and enjoyable to a broad spectrum of readers.

Discover this book's excellence for yourself by opening it to any chapter that interests you. I'm confident that after reading a page or two you'll agree that a lot of hard work and dedication has gone into the preparation of *The Body*. Its overall excellence is likely to remain unsurpassed for a very long time to come.

Chapter 1
BEGINNINGS

THE CELL

Life in Miniature

Just as an atom is the basic unit of physics, so the cell is the basic unit of biology. All living things are made of cells; some organisms are no more than a single cell, while others contain trillions in hundreds of specialized forms. The human body is one of these complex, multicellular organisms. To understand its functions—from a sneeze to the birth of a baby—one must understand its cells.

Human cells have numerous shapes and tasks but share fundamental ingredients and structures. All consist mainly of carbon, hydrogen, oxygen, and nitrogen. Every cell has an outer boundary, the plasma membrane.

Inside, the cell is divided into two main regions, the cytoplasm and the nucleus. The cytoplasm contains fluid and a variety of organelles, "little organs," that perform cell metabolism. The cytoplasm surrounds the nucleus, the largest organelle, which is protected by its own membrane. The nucleus is the control center and location of chromosomes, the genetic material.

The adult body contains 50 to 100 trillion cells in about 200 specialized forms. They range in size from the sperm, a few micrometers wide, to the egg cell, about 100 micrometers, the size of the period that ends this sentence. (Including the axon, the part of the nerve cell that conducts impulses, the nerve cells that run from the spine to the toes take the prize, at more than three feet long.) Cells of similar types usually group together as tissues and organs, and their distinctive shapes and internal workings are tailored to their uses. For instance, smooth muscle cells have a long, spindle shape and durable microfilaments that allow them to contract; cells of the intestines are covered with microvilli, tiny fingerlike extensions that increase intestinal surface area; and

DID YOU KNOW?

MICROSCOPIC CELLS

Most cells and cell structures are measured in micrometers (μm, also called microns) or nanometers (nm).

+ 1 μm = one-millionth of a meter

+ 1 nm = one-billionth of a meter

+ Typical animal cells measure 10–30 μm long. Bacterium are 2–8 μm long.

+ A human DNA molecule is about 2 nm wide but can measure several centimeters long.

Organelles in a typical cell (seen in cross section here), such as the nucleus and the oval mitochondria, work together in a range of specialized jobs.

cells of the eye possess photoreceptors that can detect light.

Each kind of cell has its own life cycle. All are "born" from the division of preexisting cells; most will grow, divide, and eventually die, to be replaced by new cells. In some tissues, such as those of the skin, this cycle runs almost continuously. In others, such as those of the liver, cells reproduce slowly unless damaged; then their division speeds up, regenerating the tissue. In yet other tissues, such as those of the heart, mature cells cannot divide, and damage to the organ is repaired with scar tissue. The processes that control cell birth, death, and specialization are still not well understood, and solving these mysteries is a major goal of biologists today.

Cell Parts

Although cells come in a veritable zoo of forms and functions, almost all have a similar basic anatomy. All, for instance, are contained by a plasma membrane—not a skin, but an oily fluid whose molecules are constantly shifting about on the cell's surface. This membrane quickly seals itself around a puncture—a handy trait for scientists who want to inject or remove something from a cell.

Consisting of a double layer of lipids—organic molecules of carbon, hydrogen, and oxygen—interspersed with proteins, the membrane is selectively permeable. Some molecules, like oxygen or alcohol, diffuse directly through them. Many others—glucose, for example, or amino acids—travel through protein channels or are transported by protein carriers.

DID YOU KNOW?

GOLGI CELLS

Camillo Golgi (1843–1926) was an Italian physician who won the 1906 Nobel Prize for the identification of nerve cells. Structures bearing his name include:

+ **Golgi cell:** a type of nerve cell.

+ **Golgi tendon organ:** a sensor that detects changes in muscle tension.

+ **Golgi complex (above):** a cell organelle concerned with modifying and packaging proteins.

Proteins in the membrane also serve as receptors that bind to particular molecules (such as insulin) as identity markers that distinguish familiar from foreign cells, and as linking molecules that hold cells together.

Most cells have similar lipids in their membranes, but proteins vary dramatically among cell types.

Beyond the membrane lies the cytoplasm, which includes all the structures of the cell, excluding the nucleus, and the fluid that contains them. The fluid, called cytosol, is made largely of water and has nutrients, wastes, and other chemicals dissolved or suspended in it. Packed into the cytosol are the organelles, specialized structures that carry out specific jobs within the cell in a smooth-running process that would be the envy of any industrial manager.

The organelles include the cytoskeleton, a network of tubules and filaments that supports the cell and gives it its shape. Microtubules surrounded by plasma membrane are at the core of the cilia (hairlike extensions) and flagella (whiplike tails) that can be found on some cells. Another organelle, the centrosome, contains tubelike centrioles that help to organize the stages of cell division.

Extending from the cell's nuclear membrane is a network of folded membranes called the endoplasmic reticulum (ER). The "rough" part of the ER is studded with little, granular ribosomes, which make proteins to be used by or secreted from the cell. (More ribosomes are scattered throughout the cytoplasm.) The smooth ER, lacking ribosomes, makes fatty acids and steroids, detoxifies drugs, releases glucose into the blood, and performs other tasks depending on the kind of cell it occupies.

The mitochondria are the cell's power stations. Anywhere from a few to thousands of these large, wiggling, oblong organelles dot the cytoplasm, with the largest number found in the most active cells, such as in the kidney or liver. In the process of aerobic respiration, mitochondria convert oxygen and nutrients into adenosine triphosphate (ATP). This important molecule releases energy that the cell uses for its own activities, including transporting substances around the cell, muscular contractions, and the movement of chromosomes during cell division.

Interestingly, mitochondria contain their own separate DNA and RNA; when the cell needs more energy, they can even reproduce. These organelles were once two separate organisms that joined each other and eventually evolved into animals and plants. Also intriguing to geneticists is that mitochondria appear to be inherited almost exclusively through the maternal line, undergoing change over time, a useful trait in the study of evolution (see "Genetic Histories," page 26).

The Golgi complex (also called the Golgi apparatus) is an organelle made of flattened, disklike sacs. It modifies and packages proteins made in the endoplasmic reticulum before passing them on to other parts of the cell; some, for instance, may be incorporated into the cell membrane, while others may leave the cell altogether.

EPITHELIAL CELLS
A color-enhanced micrograph of cells from the pancreas reveals DNA (blue) and mitochondria (green).

NERVE CELLS
These cells form the brain's support system, feeding the neurons that transmit messages.

MUSCLE CELLS
A cross section of heart muscle cells shows large mitochondria sandwiched between muscle fibers.

Organelles can destroy as well as produce; this job falls largely to the lysosomes, spherical bodies containing digestive enzymes. Lysosomes digest invaders "eaten" by the cells, such as bacteria and viruses; they also break down worn-out organelles and recycle their components. Similar, tiny organelles called proteasomes destroy damaged or unneeded proteins, while peroxisomes contain enzymes that neutralize free radicals.

Largest of the organelles, the nucleus is both the cell's boss and its archivist. It contains a complete copy of the body's genetic material and coordinates the cell's activities, including growth, protein production, and cell division.

Almost all human cells have a nucleus, typically a spherical or oblong body about five micrometers wide. A few kinds of large cells, such as those in skeletal muscles, have more than one. Only mature red blood cells lack a nucleus; therefore, they can't reproduce or repair themselves and die within a few months.

The nucleus is enclosed by a fluid-filled double membrane known as the nuclear envelope. Minute pores penetrate both layers

Mitochondria are the cell's power stations, converting oxygen and nutrients into energy for activities like cell division.

of the envelope to connect the inside of the nucleus to the rest of the cell; these portals allow proteins from the cytoplasm to enter the nucleus and RNA to leave it. Clinging to the outer membrane of the nucleus is the rough endoplasmic reticulum, pocked with little ribosomes. Like the cell itself, the nucleus is filled with a gel-like fluid, here called nucleoplasm. The fluid holds one or more little spherical bodies, nucleoli, that look like the nucleus's nucleus. These membraneless clumps of protein, DNA, and RNA assemble and then export the ribosomal RNA.

Weaving its way through the nucleoplasm is the cell's motherlode: its DNA (see "Inheritance," page 22). In a mature, nondividing cell,

this takes the form of chromatin. Chromatin is made of both DNA and little balls of proteins called histones. Each long strand of DNA coils tightly around the histones like string around sticky beads, so that the approximately six feet of DNA in the human nucleus fits compactly into its microscopic space. Some of this chromatin is genetically active, working constantly to dictate the cell's protein production; other portions are inactive and so are more condensed.

The coiled DNA strands of the chromatin are usually compacted by a factor of 40 (compared with the same strand stretched out straight). Then, as the cell prepares to divide, the chromatin coils in on itself far more, by a factor of 10,000, forming the dense little bundles of information known as chromosomes (see "Inheritance," page 22).

A Cell's Life

A human cell's life cycle has two main portions: interphase, which is the period from formation to division; and the mitotic phase, when it reproduces by dividing into two identical daughter cells.

Interphase is the time of cell growth and productive activity. During the first part of interphase, after a cell is "born" from the division of a parent cell, the new cell gets busy making proteins, carbohydrates, lipids, and the like, as well as organelles. These materials accumulate in the cytoplasm, and the cell gets bigger, its membrane expanding. As this stage comes to an end, the centrioles near the cell's nucleus begin to replicate in preparation for cell division.

The DNA within the cell's nucleus replicates during the second stage of interphase. First, the coiled threads of DNA in the chromatin begin to unwind from the histones. Then an enzyme unzips the two strands of the DNA molecule (see "Inheritance," page 22). The bases that form the teeth of the DNA zipper—adenine connecting to thymine, and guanine connecting to cytosine—pull apart as the enzyme breaks the hydrogen bonds that held them together. Free-floating DNA nucleotides in the nucleoplasm are guided to the separate DNA strands, forming an exact duplicate of the missing strand, to create two identical DNA molecules. The new molecules coil around newly made histones and condense to form a pair of identical chromatids, joined by a buttonlike centromere.

This middle part of interphase lasts about six to eight hours in a typical mammalian cell. When it is over, the cell is committed to reproduction. It enters the final, short period of interphase, continuing to grow and make proteins for perhaps four to six hours until cell division begins.

DID YOU KNOW?

WATSON AND CRICK

+ In 1944, Canadian biologist Oswald Avery showed that DNA carried the genetic information within bacteria—but just how it did it, no one knew. In 1953, **American James Watson and Englishman Francis Crick** figured out the answer. They saw that the DNA molecule was a double helix, above, composed of two sugar-phosphate strands linked by four repeating bases. The discovery began a revolution in human biology and medicine.

When a cell divides, its nucleus splits first, in the process of mitosis, followed by the cytoplasm, in cytokinesis. (Reproductive cells—the sperm and egg—reproduce by meiosis; see "Process of Meiosis," opposite.) The amazing dance of the chromosomes that takes place during mitosis is clearly visible through the microscope; biologists divide it into four steps known as prophase, metaphase, anaphase, and telophase.

In prophase, the chromatids condense into tightly coiled packages, each one made of two identical DNA molecules joined by a centromere. Around them, in the nucleus, the nucleoli disappear. The tubular centrioles in the nucleus, which had replicated themselves in interphase, separate into two identical pairs. From them grows a set of tiny tubules called the mitotic spindle, which stretches from one end of the cell to the other.

During the next step, metaphase, the paired chromatids line up neatly along the midpoint of the mitotic spindle. Then, during anaphase, the centromeres that held the pair of identical chromatids together split; the chromatids, now called chromosomes, are pulled apart and quickly dragged by their centromeres

PROPHASE
A cell undergoing mitosis enters prophase; microtubules appear in green, chromosomes in blue.

METAPHASE
During metaphase, chromosomes (blue) line up along the mitotic spindle (green).

CYTOKINESIS
Microfilaments cinch a cell in half during the stage in which a cell splits in two.

along the mitotic spindle's tubules toward opposite poles of the cell.

Once the two groups of chromosomes have reached their destinations, telophase begins. The chromosomes uncoil and become, once again, loose threads of chromatin. A new nuclear membrane forms around the chromatin, nucleoli reappear, and the mitotic spindle disappears.

Throughout the earlier phases of mitosis, the cell remains intact. During late anaphase, cytokinesis begins. An indentation forms in the middle of the membrane, where microfilaments tighten around the cell like a belt. Soon the microfilaments pinch the cell in half, creating two new cells with identical sets of chromosomes in their new nuclei. Despite its amazing mechanics, mitosis in mammalian cells typically takes less than an hour; the entire cycle from cell formation to division might last 20 to 24 hours, although some cells last a lifetime.

What prompts a cell to divide? Biologists don't know for sure. Cell size might be one trigger: When a cell reaches a certain ratio of surface area to volume—when it gets fat—chemical signals might prompt division. Normal cells also stop dividing when they begin to touch each other. Scientists are eager to decipher the signals that turn cell division on and off, since cancer cells notoriously ignore these normal controls.

Cell Death

Scientists would like to uncover the secrets governing cell division, but they may be even more eager to find out what causes cell aging and death. Most human cells eventually stop dividing and all of them die. Biologist Leonard Hayflick, working in the 1960s, found that cultured human cells divide about 50 times, on average, before they no longer reproduce and eventually perish: This upper boundary to a cell's division is now known as the "Hayflick limit."

Cell deaths can be planned or unplanned. Planned, or programmed, cell death is called

apoptosis. Apoptosis prunes away unneeded tissues in the developing body, such as the webbing between fingers and toes in a fetus or unused neurons in the growing brain. Outside the womb, the process also kills cells that have been invaded by certain viruses. When the body determines that cells must go—a mechanism

DID YOU KNOW?

MEIOSIS

+ **Sperm and egg** cells divide by meiosis. First, the 23 pairs of chromosomes replicate. Then, chromosomes exchange genetic material; the cell divides into two cells with 23 chromosomes, each with two chromatids. Division occurs again, producing four daughter cells with 23 chromatids. Each one is distinct from the parent cell and each other.

PROCESS OF MEIOSIS

EARLY PROPHASE
CHROMOSOMES
CENTROMERE
CENTRIOLE

LATE PROPHASE

METAPHASE

ANAPHASE

TELOPHASE

SECOND TELOPHASE

KEEPING HEALTHY

THE TRUTH ABOUT TELOMERES

How do you know if your cells are healthy? One indication may be the length of your telomeres—protective strands of DNA that cap your chromosomes, similar to the plastic tips on the ends of shoelaces.

If telomeres are too short, they fail to protect cells from aging and disease. Although scientists can't use telomere length to diagnose specific conditions, they have linked short telomeres to an increased risk of diabetes, cancer, heart disease, and hypertension.

This field of research is still in the early stages, but genetics, diet, and lifestyle all appear to be major factors in telomere length. So how can you keep your telomeres long and healthy?

+ **Eat foods rich in antioxidants, fiber, and healthy fats:** Studies have found that people who eat fruits, vegetables, whole grains, fish, and olive oil have longer telomeres.

+ **Find healthy stress busters:** Trauma, depression, and anxiety are linked with shorter telomeres. Try healthy ways to deal with stress—such as exercise or meditation.

+ **Quit smoking:** Tobacco use appears to shorten telomeres.

To see how these healthy habits would play out in real life, a study followed 35 men for five years. One group followed a vegetarian diet, got regular exercise, reduced stress, and attended support group meetings. The other group did not make changes. At the end of the study, the men who made the healthy changes saw an increase in telomere length. The study provided more persuasive evidence that a healthy lifestyle leads to longer telomeres, and hopefully a longer life.

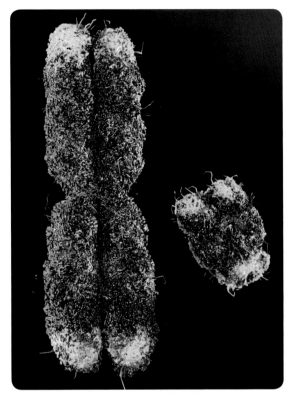

Computer rendering of X and Y chromosomes shows telomere regions in yellow.

that is not fully understood—enzymes called caspases destroy the cell's DNA and internal structures. The cell shrinks, detaches from its neighbors, and is promptly gobbled up by other cells.

Unplanned cell death—necrosis—comes about as a result of disease or injury. Cells that lose their oxygen supply, for example, or are frozen or burned may swell and burst. The immune response this triggers can cause inflammation (swelling and redness) in the surrounding tissues.

Cell aging is still a mystery. Simple damage over time may contribute to it as environmental toxins accumulate in the body. Free radicals may also play a part. These are molecules that are missing a single electron; when they enter a cell, they pull electrons from the cell's molecules, damaging them. Some substances, such as vitamins C and E, are antioxidants that may help prevent this damage.

Gradual genetic damage may promote aging, as well. Every chromosome is capped by a telomere, a long string of bases that carries no genetic information. Each time a cell reproduces, a piece of the telomere cap is snipped off. The erosion and loss of the telomeres may signal the cell to stop dividing.

This theory picked up support when, in the 1980s, biologists discovered an enzyme that protects telomeres. Dubbed telomerase, it can repair and extend the telomere string. Telomerase is virtually absent in most adult cells, but it is found in germ cells (the cells that produce sperm and eggs) and in cancer cells, which reproduce without aging.

Most likely, many factors working together cause cell aging and death, and those factors may vary from one cell type to another. Learning more about them could lead to cures for diseases ranging from cancer to Alzheimer's disease.

Radiologists use specialized scanners to monitor the progress of cancer treatments.

INHERITANCE

Passing On Traits to the Next Generation

Two key molecules within the cell's nucleus, DNA and RNA, hold the key to the body's central mysteries: its birth and death, its diseases, its very identity. DNA is deoxyribonucleic acid, an immensely long organic molecule. It's built like a ladder: The sides are chains of sugars called deoxyribose that alternate with phosphate groups. Four bases attached to the sugars pair up, each pair forming a rung: Adenine links to thymine, and guanine to cytosine. The DNA molecule twists around itself in the spiral called the double helix. Within the nucleus of almost every cell are 46 of these spiraling molecules in 23 matched pairs—46 human chromosomes.

The basic unit of each chromosome is a nucleotide formed of a sugar, a phosphate group, and a base: a piece of the side of the ladder attached to half a rung. Genes are simply groups of nucleotides, and the order of the nucleotides in the gene—for instance, ATTGCCA—dictates the gene's purpose. Human DNA contains nearly 3.2 billion nucleotides; the average gene consists of thousands of nucleotides, but some genes have more than one million. An organism's complete set of DNA is called its genome and contains 20,000 to 25,000 genes. A large portion of DNA seems to consist of noncoding strands of nucleotides. These might be "extinct" genes, traits discarded in the course of evolution, or they might be strands whose meaning has not yet been deciphered.

3-D computer rendering of DNA strand

Transcription

What do genes do, exactly? These molecular snippets loom large in the popular imagination, depicted as dictators of everything from criminal behavior to cold sores. Yet a gene is nothing more than a set of instructions for making a particular protein or RNA in a cell. Its molecular code—the sequence of A, T, C, G nucleotides that gives it its identity—spells out the order in which amino acids (the building blocks of proteins) must be strung together.

The gene doesn't even transmit the instructions. That job falls to ribonucleic acid, or RNA, and consists of two steps: transcription, when information from the gene is copied onto a strand of RNA, and translation, when the information in the RNA is used to make the protein in the cell's ribosomes.

During transcription, the DNA helix pulls apart. One of its strands serves as the template for a new strand of RNA, whose complementary bases match those on the DNA—except that RNA contains uracil (U) instead of thymine. A strand of DNA with TTAACC would therefore produce the RNA sequence AAUUGG. Three kinds of RNA are made in this way: messenger RNA (mRNA), which carries the protein instructions out of the nucleus to the ribosomes; ribosomal RNA (rRNA), which helps the ribosomes build the proteins; and transfer RNA (tRNA), which brings amino acids, dissolved in the cytosol, to the ribosomes.

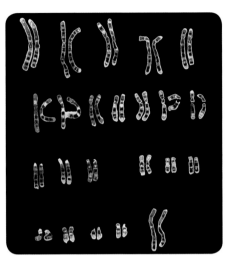

Twenty-three pairs of matching chromosomes end with two X chromosomes, showing that this set belongs to a woman.

So all that genes do is direct a cell's protein production. Still, this is an extraordinarily powerful job. Proteins form the basic material of the human body, accounting for up to 30 percent of the mass of each cell. They include enzymes that carry out a cell's chemical reactions; they make up the hemoglobin in the blood, the collagen in the skin, hormones that regulate growth, antibodies that fight disease, and innumerable other vital parts and functions. In controlling proteins, genes control the growth and workings of every cell in the body. They also dictate the growth, specialization, and development of every cell in a human embryo.

Virtually all cells contain an identical, complete set of the genetic code in their nuclei. Why, then, does a kidney cell look and act differently than a brain cell? It's because different genes are active in different cells. In any given cell, only a small percentage of the genes are at work, most performing functions that are basic to any cell. Just

BREAKTHROUGH DECODING THE GENOME ⊢

In 1990, the Human Genome Project began work on an ambitious program: mapping the entire human genome by identifying the sequence of all three billion base pairs on our chromosomes. Research centers in the United States, Europe, and Asia collaborated, completing the work in April 2003, ahead of schedule. As well as identifying the order of every nucleotide in human DNA, the project pinpointed the location of the genome's 20,000 to 25,000 genes and created maps to link inherited traits to certain genes. Today, the entire sequence is now available online, for free.

A scientist reviews a display of DNA.

Members of the same family share some chromosomes, but only identical twins share a complete set of genetic material.

a few are activated for the tasks that identify that particular cell or tissue. Thus, all cells contain the genes for making insulin, for example, but only in the pancreas are those genes active.

All kinds of signals can activate genes, and this is where the environment and even emotions interact with the subtlest workings of the body. Molecules within the cell may switch on genes, but genes are also turned on or off by hormones or other chemical signals that enter the cells from outside. Fear or stress, for instance, can signal genes to produce proteins in response and to stimulate or inhibit the activity of neurons in the brain. The interplay of genes and environment is one of the more exciting new areas of biology.

Heredity

Genes are inherited. Because they are produced by meiosis, the gametes (or sperm cells and egg cells) are the only cells to contain just 23 chromosomes (see "Did You Know?," page 19). When a sperm merges with an egg, their chromosomes combine to form the usual 46—half from the mother, half from the father—that will be duplicated again and again as the embryo grows and develops.

These inherited chromosomes are not, however, exact copies of those in the parents' cells. Unlike cell division elsewhere in the body, every time sperm cells and egg cells divide, their genes cross over and are randomly reassigned between matching chromosomes. As a result of this process, every sperm cell and egg cell is slightly different genetically.

The genes on each of the linked chromosomes in every cell code for the same proteins and the same traits. This redundancy provides critical backup for the cell: If one gene is damaged, the matching gene on the linked chromosome will usually pick up the slack. The matched genes—called alleles—aren't always identical, however. If they do code for the same form of a trait, the individual is called homozygous for that trait. If they code for a different version, the individual is heterozygous for that trait.

Sometimes, one functional allele is enough to produce the trait. Such an allele is called dominant, while its partner is called recessive. Geneticists typically depict this pairing with capital and lowercase letters. For example, brown-eyed man #1 may have inherited dominant alleles for brown eyes (B) from both of his parents, resulting in a BB genotype. Brown-eyed man #2 may have inherited one dominant allele for brown eyes and one recessive allele for blue eyes, giving him a genotype of Bb.

The mix-and-match game of inherited traits begins when sperm meets egg. Let's say brown-eyed man #1 (BB) marries blue-eyed woman #1 (bb). Each of their children will have a Bb pair of genes, and will therefore have brown eyes.

Now let's say brown-eyed man #2 with a recessive blue-eyed gene (Bb) marries a similar brown-eyed woman (Bb). The odds are 3 out of 4 that a child will have brown eyes (BB or Bb) and one in four that it will have blue (bb). Note that these are probabilities only. It is possible for the second couple to have four blue-eyed children.

Few traits are expressed by one gene: Almost all result from several genes working in tandem. Other traits—such as ABO blood types—are

KEEPING HEALTHY

FAMILY AILMENTS

Almost everyone can identify an ailment that "runs in the family": heart disease, perhaps, or asthma. It is true that many diseases have a genetic component, with family members sharing an increased risk of developing the same illness.

Compiling a family medical history can give a clearer understanding of your own health risks and help you and your doctor plan your own health care, including lifestyle changes, further medical tests, or even genetic testing.

A family medical history should include parents and siblings, and preferably grandparents, aunts, uncles, and half-siblings. Try (tactfully) to assemble the following information for each person:

+ **Birth date**

+ **Death date (if applicable)**

+ **Major medical conditions, such as diabetes, coronary artery disease, cancer, stroke, depression, or Alzheimer's disease**

+ **Lifestyle factors and choices, such as smoking, diet, exercise, and alcohol consumption**

+ **Miscarriages or stillbirths, and causes (if known)**

+ **Dominant racial and ethnic background (some diseases, such as sickle-cell anemia, are linked to race or ethnicity)**

+ **Other health issues, like allergies, asthma, arthritis, migraines**

+ **Cause of death (if applicable)**

GENETIC HISTORIES

Every human alive today is descended, DNA studies suggest, from hunter-gatherers who migrated from Africa roughly 60,000 years ago. They made their way past the shores of the Red Sea and into western Asia. They then spread out to Australia and East Asia, followed by Europe and the Americas.

RESEARCH

The story of their epic journey is written in our genes. Genetic markers in DNA present a picture of our ancestors and their migrations, allowing us to trace the broad pathways of ancient human exploration. Using techniques that have become available only in the past 20 years, some geneticists are now mapping ancient human history.

One of the leading inquiries into human migration is the Genographic Project, an international study sponsored by National Geographic and IBM, with funding for field research provided by the Waitt Family Foundation. The project is led by population geneticist Spencer Wells, a National Geographic explorer-in-residence. At ten research centers around the world—in sub-Saharan Africa, North Africa/Middle East, India, East/Southeast Asia, Australia/Pacific Ocean, North Eurasia, Central/Western Europe, North America, and South America—geneticists are collecting blood samples and cheek swabs from living people in order to study their DNA. At an 11th center, in Australia, researchers analyze DNA from ancient human remains. Researchers hope that comparing samples from ancient and modern populations will reveal new insight into how humans populated the world.

MARKERS

The genetic markers that make this possible are found in two places: the Y chromosome and the DNA of mitochondria, the energy-producing organelles found in each cell (see "The Cell," page 12). Although virtually all of our DNA is recombined with every generation, these two parts of the genome remain for the most part unshuffled from parent to child.

Only men possess a Y chromosome (their sex chromosomes are XY, whereas women's are XX), so that chromosome is passed down virtually unchanged from father to son over many generations. Mitochondrial DNA, on the other hand, is passed down only through the maternal line, from mothers to children of both genders, because the mitochondria in a man's sperm cells are concentrated in the sperm's tail, which drops off before fertilization of the egg. Therefore this DNA, too, escapes reshuffling.

Very rarely, but apparently at a steady rate over time, a harmless mutation will occur in the DNA of the Y chromosome or the mitochondria. This mutation, called a genetic marker, will then be passed down through the generations.

By comparing shared and distinct markers within populations, geneticists can determine when certain groups of people split off from others. They can even trace the markers back through time to the most recent pair of ancestors common to living humans, known as "Y-chromosome Adam" and "Mitochondrial Eve," two Africans who lived about 140,000 years ago and 125,000 years ago, respectively.

FINDINGS

Though much remains to be learned, the broad outlines of human migration have now been drawn. Modern humans arose about 200,000 years ago in Africa (a finding reinforced by fossil evidence). These hunter-gatherers spread across the continent until about 70,000 to 50,000 years ago, when they moved onto the Arabian Peninsula and then throughout the rest of world.

Some of them then moved rapidly across Asia and into Australia some 50,000 years ago. Because of an ice age, sea levels were remarkably low at the time, so that just 62 miles of open water separated the Australian landmass from Asia; the newcomers may have navigated by sight from island to island. Others traveled into Central Asia (about 40,000 to 50,000 years ago) and some into Europe (about 40,000 years ago).

One question with regard to Europe that the Genographic Project's findings could possibly answer is the contribution, if any, of the Neanderthals to the genetic makeup of modern humans. European descendants of prehumans who left Africa around 500,000 years ago, Neanderthals mysteriously died out about 30,000 years ago.

One theory holds that the newcomers from Africa gradually pushed the Neanderthals aside. Citing new genetic evidence, however, some scientists have recently suggested that there may have been interbreeding between the two groups, or even that commingling caused the Neanderthals to disappear, as they were absorbed into the other group. Although mitochondrial DNA analysis shows that the two

Researchers used genetic markers to trace human history back to Africa.

groups had distinct maternal lineages, other DNA markers indicate that some degree of interbreeding took place. In fact, individuals from most non-African populations exhibit some Neanderthal DNA markers, which were apparently acquired as the modern human population expanded out of Africa.

The exact arrival date of humans in the Americas is in dispute, but DNA evidence suggests that sometime between 15,000 and 20,000 years ago immigrants from Asia crossed from Siberia to Alaska in a single major expansion event. It is estimated that they then spread into South America, 12,000 to 15,000 years ago.

Today, the Genographic Project is working to collect and analyze DNA from around the globe, particularly from isolated, indigenous populations before they assimilate into the broader genetic pool. Project researchers obtained some 85,000 DNA samples during the first five years of the study and continue to expand this sample set while analyzing a larger set of DNA markers to better characterize patterns of human migration and genetic diversification.

expressed by multiple allele forms, meaning the individual can display characteristics of more than one allele, as in AB blood type.

Nor are dominant traits necessarily more common, or desirable, than recessive ones. Huntington's disease and polydactyly (extra fingers or toes) are caused by dominant genes; normal vision and normal cartilage formation are found on recessive alleles.

In the game of heredity, two chromosomes stand out from the rest: the X and Y sex chromosomes. The Y chromosome contains the genes that determine maleness; every man's sex chromosomes are an X and Y set, while every woman's are XX.

The X chromosome is much larger and richer in genetic information, containing about 1,300 genes versus dozens on the Y. This means that most characteristics on the X chromosome will be expressed in men, because they are not countered by matching genes on the Y. Traits that are linked to genes on the X or Y chromosomes are called sex-linked; muscular dystrophy and hemophilia are examples of X-linked traits (or more precisely, damage to genes on the X chromosome).

KEEPING HEALTHY
COLOR BLINDNESS

Red-green color blindness is an X-linked trait. Women who carry the mutated gene on one X chromosome rarely have the disorder because they usually have a normal gene on the other X chromosome to offset it. A man who inherits the mutated gene from his mother will find the trait expressed, because he has no matching normal gene on the Y chromosome.

Inheriting Illness

Genetic diseases can be linked to a single damaged gene, to entire misplaced chromosomes, or to many genes working together. Those linked to a single gene—say, hemophilia—are the easiest to track down, but many disorders are probably caused by the complex and difficult-to-trace interaction of genes and environment.

Normal genes do not code for disease. When we say that someone has the gene for muscular dystrophy or breast cancer, what we mean is that the person has a mutated version of a particular gene, and therefore the protein encoded by the gene is changed or missing in that person's cells.

A mutation is a permanent change in the structure of a gene. Mutations may be inherited, passed down from parent to child in every cell; they can also be triggered in selected cells by a host of environmental factors, including malnutrition and exposure to solar radiation. Possible mutations include a swapped base in the genetic code—adenine in the place of cytosine, say—or missing or repeated stretches of DNA.

Many mutations appear meaningless genetically. Some produce harmless variations in a population: different hair color or blood types, for example. The mutations of most concern are the destroyers. Thousands of diseases result from genetic damage. In some, the location of the mutation determines the severity of the disorder. Hundreds of different mutations can affect the genes that control mucus production, for example. Some have no effect; some result in a mild case of cystic fibrosis and others in a severe one.

A young woman with Down syndrome most likely has an extra copy of chromosome 21.

Inherited genetic disorders typically follow the standard rules of inheritance. For example, if one parent has a dominant mutated gene and the other parent's gene is normal, each child has a 50-50 chance of inheriting the condition. People who have one recessive mutated gene are called carriers. Carriers do not usually have the disease, but each child of two carriers has a one-in-four chance of having the disease, a similar chance of being healthy, and a two-in-four chance of being a carrier himself.

In recent years, geneticists have begun to pinpoint the location of some genetic diseases. The neurological disorder Huntington's disease, for instance, is caused by excessive repetitions of the nucleotide triplet CAG on chromosome 4.

Most people average about 18 of these triplets in a gene that codes for the protein huntingtin. In people with Huntington's disease, the triplets keep repeating 40, 50, or more times, resulting in an abnormal protein that negatively affects the brain. Although symptoms of the disease typically manifest around middle age, in some cases the strings of triplets increase as they are inherited, leading to an earlier onset of the disease in succeeding generations.

Because scientists can identify the malformed gene, they can tell people whose parents have Huntington's disease whether they, too, will develop it. The disease currently has no cure, and the decision to be tested can be agonizing.

Some of the most prevalent genetic disorders are caused not by one mutated gene, but by the replication of all or part of an entire chromosome. An extra copy of chromosome 21, created in error during meiosis of a parent's sperm or egg cell, causes most cases of Down syndrome. Found in 1 out of every 800 to 1,000 babies, Down syndrome brings with it a host of medical problems including mental retardation, poor muscle tone, and heart and digestive defects.

Other trisomy ("three body") disorders include trisomy 18 (marked by three copies of chromosome 18) and trisomy 13 (three copies of 13), both linked to mental retardation and other problems. One or more extra X chromosome in boys (giving them XXY sex chromosomes) causes Klinefelter syndrome, a disorder whose onset (typically during puberty) leads to abnormally low levels of testosterone and development of some female characteristics.

THE STORIES IN OUR GENES

Researchers have identified disorder-causing genes on each of the 23 chromosomes. Most carry a few defective genes with no signs of disease, and many genes only contribute to susceptibility. Lifestyle choices and environmental factors can raise or lower disease risk.

CHROMOSOME	DISORDER	CHROMOSOME	DISORDER
1	Retinoblastoma, pancreatic cancer	13	Breast cancer, Alzheimer's disease
2	Goiter, leukemia / T-cell lymphoma	14	Early-onset glaucoma, congenital hypothyroidism
3	Tay-Sachs disease, colorectal cancer	15	Small-cell lung cancer, susceptibility to Asperger's syndrome
4	Endometrial carcinoma, Crohn's disease	16	Huntington's disease, polycystic kidney disease
5	Dementia, ovarian cancer	17	Allergic rhinitis, diabetes mellitus
6	Leukemia / B-cell lymphoma, Paget's disease of the bone	18	Estrogen resistance, epilepsy
7	Myotonic dystrophy, cystinuria	19	Cystic fibrosis, growth-hormone-deficient dwarfism
8	Fatal familial insomnia, Creutzfeldt-Jakob disease	20	Prostate cancer, myopia
9	Autoimmune polyglandular disease, amyotrophic lateral sclerosis	21	Fructose intolerance, dilated cardiomyopathy
10	Ewing's sarcoma, giant-cell fibroblastoma	22	Congenital cataracts, leukemia
11	Color blindness, hemophilia	X	Sickle-cell anemia, albinism
12	Turner's syndrome, gonadal dysgenesis	Y	Lung cancer, morbid obesity

Genetic disorders may also arise when a single cell in an early embryo mutates. As the embryo divides and grows, some cells will contain the mutation and others will not, a condition called mosaicism.

Rarely, two unrelated chromosomes will break and exchange chromosomes during cell division, a process called translocation. When this occurs, for instance, a piece of chromosome 21 might switch places with a piece of chromosome 14. Outwardly, the person carrying these swapped genes will have no symptoms, but when he or she produces sperm or egg cells, some of them will have the extra chromosome 21 material. The unsuspecting carrier could then parent a child with Down syndrome.

Genetic disorders range widely in their severity. Some mutations are fatal from the outset; perhaps 50 percent of fertilized eggs fail to develop normally for this reason and are silently miscarried in the first days of pregnancy. Other traits, like male-pattern baldness, are not life-threatening.

Many common diseases—such as cancers and heart disease—probably involve a combination of genetic and environmental factors. Scientists are beginning to pinpoint the location of genes connected with some diseases, but mutations on these genes typically indicate only a tendency to develop a disease in certain circumstances.

Gene Therapy

Few areas of genetic research hold more promise than gene therapy. With the mapping of the human genome and advances in the manipulation of genetic material come the possibility that doctors could cure a host of genetic diseases using genes themselves.

This could be done in one of three ways: Scientists could replace a damaged gene with a healthy copy, introduce new genes into the body to fight disease, or deactivate or repair a misbehaving gene.

New genes usually won't incorporate into a cell's genetic material if just injected into the cell. They need to be delivered by a carrier, called a vector. Most researchers use a deactivated virus, such as an adenovirus (cold virus), that will infect the target cell without causing illness. The researcher splices the gene into the DNA of the vector and injects the vector into the affected tissue. There, it infects the cell and incorporates the healthy gene into the cell's DNA. The cell and its descendants then produce functional proteins.

A deactivated cold virus can act as a vector in gene therapy.

Gene therapy is still being researched and is available only in experimental trials. Since the first test in 1990 there have been hundreds of trials with limited success and some tragic failures. In 1999, an 18-year-old patient with a liver disorder died when the treatment resulted in massive organ failure. In 2002, two French children who had been successfully treated for immune system disorders developed a leukemia-like condition.

Right now, gene therapy is not long-lasting and must be repeated periodically in the patients, but this can stimulate a harmful immune response. Disabled viruses make useful vectors, but they could possibly regain their ability to cause illness once in the patient.

Ethical questions also surround the procedure: What will happen if a parent thinks that short stature, below-average IQ, or impulsive behavior is a disorder to be cured? Who will decide? What should the guidelines be?

On the other hand, gene therapy's potential is undeniable, especially for single-gene diseases. The same experiment that resulted in two cases of leukemia also successfully treated several other children. With more study, today's risky business could be tomorrow's spectacular cure.

DEVELOPMENT

Single Cells to Human Beings

Transforming a single fertilized cell into a human infant is one of the most extraordinary processes in biology. Even today, with our understanding of cell biology advancing on a daily basis, many of the complex mechanisms that govern embryonic growth and human development remain enigmatic, but promising, subjects for further research and study.

For the human egg, the literal journey toward fertilization begins when the female's egg pops out of an ovary and drifts into the nearby fallopian tube. At this point, the egg is midway through the process of cell division by meiosis; it carries two sets of 23 chromosomes, one set in the main body of the egg (called a secondary oocyte) and the other encapsulated in a little polar body, which is later discarded. As it moves down the fallopian tube, propelled by hairlike cilia in the tube's lining, the egg is chaperoned by a cloud of encircling smaller cells called the corona radiata.

Meanwhile, some 250 million sperm are furiously swimming up the reproductive tract. Consisting of little more than heads, filled with 23 chromosomes, and tails (or flagella) Packed with mitochondria to produce energy, the sperm die by the millions as they traverse the uterus and enter the fallopian tube where the egg is slowly making its progress down the tube. Only a few hundred sperm will survive to meet the egg. The first to arrive won't even get to fertilize it. Instead, as they begin to batter the outside of the egg, they release enzymes that disperse the corona radiata cells surrounding it and gradually expose its surface.

Finally, one sperm penetrates the egg's outer covering, the zona pellucida, and the egg internalizes it. As soon as this event happens, calcium ions flood the cell and its charge changes from negative to positive, repelling

DID YOU KNOW?

TWINS

+ **About 12 out of every 1,000 births** in the United States result in twins, although multiple births are increasing due to fertility treatments.

+ **Fraternal, or dizygotic, twins** are produced when two eggs, released simultaneously in the mother, are fertilized by different sperm. These twins will be genetically distinct, no more alike than any other siblings.

+ **Identical, or monozygotic, twins** result when a single fertilized egg splits into two embryos, almost always within eight days of fertilization. Monozygotic twins are genetically identical and always the same sex.

A two-cell zygote is the first product of fertilization. From these identical cells will come all the structures of the human body.

all other sperm. The egg completes its process of meiosis and ejects the polar body. The nuclei of the sperm and egg, each containing 23 chromosomes, then merge.

The egg cell is now officially fertilized. Almost immediately, its chromosomes duplicate and the cell divides by mitosis, creating two identical daughter cells. This tiny entity, still packed within the confines of the egg's tough membrane, is called a zygote.

Proceeding slowly down the fallopian tube, the zygote divides again, into four cells, and then eight—each new cell just half the size of its predecessor because the zygote is still packed within the egg's hardened shell. About three days after fertilization, the zygote is a lumpy little ball of 16 or more cells, all identical. At this point, owing to its appearance, it is known as a morula, from the Latin word for "mulberry."

After four or five days, the morula finally arrives at its destination, the uterus. There, it begins to change. Its outer shell breaks down, allowing fluids from the uterus to enter the

KEEPING HEALTHY
PRENATAL CARE

From 1957 to 1961, pregnant women across Europe, and some in the United States, took a new drug called Thalidomide to ease morning sickness and sleeplessness. Almost 10,000 children born of those pregnancies had birth defects, many in the form of shortened or missing limbs. Too late, doctors realized that Thalidomide was a teratogen—an agent that causes developmental damage to an embryo or fetus.

Although the placenta serves as a filter between the maternal and fetal bloodstreams, as a rule whatever the mother ingests, the fetus also takes in. Many substances that barely affect the mother can hurt the fetus. Teratogens include alcohol, which can cause the irreversible brain damage and malformations of fetal alcohol syndrome; environmental chemicals such as mercury or herbicides; radiation; infectious agents such as rubella or toxoplasma; some prescription and over-the-counter medications; and illegal drugs. The most delicate period usually is during the first trimester of pregnancy, when the embryo is growing rapidly.

ball of cells. The cells, still duplicating, reorganize into a hollow sphere with a lump of cells, called the inner cell mass, at one end. The tiny mass—still no bigger than the original egg—is now called a blastocyst. The inner cell mass will eventually develop into the embryo, while the other cells, called the trophoblast, will become the fetal portion of the placenta. The inner cell mass is also one possible source of stem cells.

Implantation

About six or seven days after fertilization, the blastocyst "implants" as it attaches to the wall of the uterus. The inner cell mass is closest to the wall. Producing enzymes that eat away at the uterine tissue, it extends its outer layer and embeds itself in the uterine lining. Its outer cells also begin to produce the hormone human chorionic gonadotropin (hCG), which helps to prepare the uterine lining and prompts the empty egg follicle back in the ovary, which has turned into a little gland called the corpus luteum, to continue its hormonal secretions.

For the next few months, the corpus luteum will secrete the hormones estrogen and progesterone, which sustain the embryo as it grows.

A week after fertilization, the blastocyst is still no larger than a speck of dust. In its second week of development it will begin the spectacular process of differentiation and growth that will transform it into an embryo, then a fetus, and eventually an infant.

The process of implantation takes about a week. During this period, structures begin to form that will protect and nourish the developing embryo and, as it is called at a later stage, the fetus: the placenta, the amniotic sac, and the umbilical cord. The outermost cells in the blastocyst multiply and extend fingerlike projections called chorionic villi into the uterus, where they develop blood vessels. These blood vessels will eventually give the embryo access to the mother's blood supply through the umbilical cord. Cells of the uterus around the chorionic villi form a layer called the decidua basalis. Together, the villi and the decidua basalis will become the placenta.

ORGANOGENESIS

In an embryo, three layers of cells give rise to the body's different structures.

	ECTODERM	MESODERM	ENDODERM
STRUCTURES	All nervous tissue	Kidneys, gonads, reproductive ducts	Epithelium of digestive tract
	Epidermis, hair, sweat glands, nails	Skeletal, smooth, and cardiac muscle	Thyroid, parathyroid, and thymus glands
	Melanocytes	Cartilage, bone, connective tissue	Liver and pancreas
	Cornea and lens of eye	Blood, bone marrow, lymphoid tissues	Epithelium of respiratory tract, auditory tube, tonsils
	Tooth enamel	Synovial membranes of joints	Epithelium of urethra and bladder

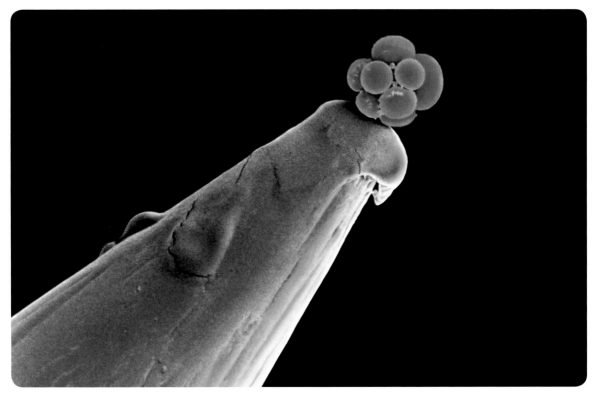

Perched on the head of a pin, this ten-cell embryo—known as a morula—is about three days old.

Meanwhile, the cells of the inner cell mass are shuffling about, reorganizing themselves into two small, hollow balls pressed together inside the blastocyst. One of the balls will become the amnion, the sac that will enclose the growing embryo. The other ball will develop into a yolk sac (yes, even human eggs have yolks). This yolk sac is the site of red blood cell production in the early weeks of pregnancy; later it will be incorporated into the growing embryo, forming part of the gut, blood vessels, and gonads (ovaries or testes).

The most significant action takes place where the yolk sac and amnion are pressed together, forming a double layer of cells called the embryonic disk. During the third week after fertilization, in a process called gastrulation, the two layers of the embryonic disk become three. In response to some unknown signal, a raised ridge with a groove down the middle—the so-called primitive streak—forms on the embryonic disk.

Cells from opposite sides of the disk swim toward the streak in the middle. Other cells then go into motion. Surface cells scoot over to the primitive streak and dive down into the groove, emerging underneath to form a bottom layer of cells called the endoderm. Other cells follow, forming a mesoderm layer on top of the endoderm. The cells that remain on top are called the ectoderm.

From these three layers will arise all the structures of the human body in a process called organogenesis (see "Organogenesis," opposite). Exactly how this process is controlled—how the cells know where to go, and how they decide to turn on and off the genes that regulate their development, directing them to specialize into different tissues—is still one of the key mysteries of biology.

The three-layered, ridged disk is now an embryo, and it develops at a furious pace. Some of the mesodermal cells collect into a lengthwise supporting rod called the notochord. The notochord releases chemicals that cause the cells of the ectoderm to thicken into the neural plate.

The plate then curls up at the edges until it forms an enclosed pipe, the neural tube. The top end of this tube will become the brain, and the rest the spinal cord. (If this process goes awry and the tube does not close completely, the baby may be born with a birth defect such as spina bifida or anencephaly.) Cells nearby move about the embryo to become, eventually, such diverse features as sensory nerves, bones of the face, and connective tissue.

The heart and blood vessels also develop early. In the third week after fertilization of the egg, blood vessels form and link up throughout the embryo. A primitive heart tube arises from the mesoderm, bends into an S shape, and starts to beat. It connects to the blood vessels

5-6 WEEKS
A few weeks into its development, an embryo's upper limb buds begin to look like arms and hands.

6 WEEKS
The heart and liver show up as dark masses within the $1/2$-inch-long embryo.

8 WEEKS
Though only the size of a walnut, the fetus now has all its organs, and its brain is growing rapidly.

to create a rudimentary cardio-vascular system.

During the fourth week, the embryo triples its size (although it is still well under an inch in length). The formerly flat embryonic disk curls into an irregular cylinder with a recognizable head and tail. Part of the yolk sac now merges with the embryo's midsection to form a primitive gut. Eyes and ears begin to appear on the head, while upper and lower limb buds form, eventually to be transformed into arms and legs.

From the fifth through the eighth week after fertilization, almost all remaining structures and organs, such as the bones, liver, and lungs, begin to appear. The brain grows rapidly, and with it the head. Arms and legs lengthen, and fingers and toes appear; at first the hands and feet are webbed, but soon the webbing disappears. Muscles begin to twitch weakly. At the end of the eighth week, the embryo is about 1.2 inches long and weighs approximately 0.03–0.06 ounces (2 grams). It is ready to move into the next stage of its life, as a fetus.

Fetal Development

At nine weeks, the foundations of all of a fetus's major organ systems have been laid down. From that time until 38 weeks after fertilization, its body begins a campaign of growth, development, and consolidation. From week to week, and from month to month, the fetus comes closer to being able to survive outside the womb.

From 9 to 12 weeks, the tiny body elongates to about 3.5 inches from the top of its head to the end of its rump, but the head still makes up about half its total length. The face is formed and the eyelids are developed, but closed. The prominent liver takes over as the primary producer of red blood cells until late in pregnancy. The brain and spinal cord grow as well—the brain alone is adding up to 250,000 neurons a minute—and the fetus's sex can be determined from the genitals.

During weeks 13 to 16, more muscle and bones develop; the fetus can move vigorously. The kidneys and other gastrointestinal organs become more mature. The fetus's skin is almost transparent, and a fine hair, called lanugo, appears on its head. Growing rapidly, the fetus reaches a length (head to rump) of 5.5 to 7 inches.

From weeks 17 to 20 after fertilization, more muscle development and overall growth mean that the mother can now feel the fetus move. Fine lanugo hair covers its entire body. During

12 WEEKS
Fingers appear as the cells between them die in the programmed cell death known as apoptosis.

16 WEEKS
Visible in a translucent hand are the beginnings of a fetus's bones.

20 WEEKS
This ten-inch-long girl looks well-developed, but she is not yet ready for life outside the womb.

the next month, in weeks 21 to 24, fine details appear: Footprints and fingerprints form, as well as eyebrows and eyelashes; the eyes are well developed, and air sacs (alveoli) take shape in the lungs. The fetus starts to gain weight quickly.

In the seventh month (weeks 25 to 28), the brain grows rapidly and the nervous system matures enough to be able to control some body functions. The bone marrow takes over production of red blood cells, a role it will maintain after birth and through adulthood. By the end of this period, a baby born prematurely might be able to survive, but only with a high risk of complications.

From weeks 29 to 36, the fetus grows from about 15 to 19 inches head to toe and gains approximately 4 pounds. The bones are fully developed, eyes are open, and fingernails and toenails appear. Fat is laid down under the skin. Babies born toward the end of this time have a good chance of survival.

Beginning sometime from 36 to 38 weeks, thicker hair starts to grow on the head. The fetus is considered full term now and ready for birth, reaching an average size of 19 to 21 inches and weighing between 6 and 10 pounds.

Throughout this period of growth, of course, the fetus has been immersed in amniotic fluid. It does not breathe, nor does it eat in any conventional sense. It survives during the long months of development because of remarkable adaptations of its cardiovascular system. The blood that flows through a fetus's veins and arteries picks up all its oxygen and nutrients from the mother's blood as it is filtered through the placenta. Having already traveled in the mother's body, the maternal blood is somewhat depleted of oxygen. To make up for this, fetal blood carries a special form of hemoglobin that is better than adult hemoglobin at binding to oxygen (and which, incidentally, never carries sickle-cell anemia).

The fetus's blood, returning from the placenta, enters its body through the umbilical vein. It circulates through the heart and body but largely bypasses the lungs and liver, using three special blood vessels as shunts around these organs. Just enough blood flows through the lungs and liver to keep those organs alive and growing.

While in the womb, this 32-week human fetus is connected to the mother's blood supply through the umbilical cord.

Eventually, the deoxygenated blood, now laden with wastes, leaves the fetus's body via two umbilical arteries. All of these shunts will shut themselves off neatly at birth, when the fetus emerges from the mother's body as a newborn baby.

Birth

When the fetus is full term, hormones in the mother's bloodstream interact with hormones released by the fetus itself to initiate labor. The mother's uterus begins to contract, pushing the baby toward the dilating cervix.

Squashed and stressed by the contractions, the baby's body responds by releasing epinephrine and norepinephrine, hormones that begin to clear the baby's lungs of fluid and pump a greater volume of blood to the heart and brain.

As it is exiting the mother's body, the baby no longer receives oxygen from the mother's placenta. Rising levels of carbon dioxide in its blood signal respiratory centers in its brain to jump-start the process of breathing. Outside the womb, the baby takes a breath, filling its lungs with air for the very first time.

With the lungs working, the baby's cardiovascular system must also kick in fully to transport the oxygen. The foramen ovale—an opening between the atria of the fetal heart—closes, sending blood to the lungs. Blood vessels that bypassed the lungs and liver also close off, as do the umbilical arteries.

After a baby is born, its systems still continue to grow and develop outside the womb. This is particularly true of its nervous system and brain. Nerve fibers throughout the baby's body are still acquiring the myelin sheaths that ease the transmission of nerve impulses. "Lower" brain functions, such as breathing, circulation, and swallowing, are working, but "higher" functions, such as sophisticated thinking and coordination, are still developing as the brain continues to add cells and connections.

In the first three months, a baby's synapses—the connections necessary for learning to occur—multiply twentyfold. A young infant's brain development is marked by great advances in motor control, social awareness, and language. At birth, a baby cannot hold up her head, see clearly past about 12 inches, or communicate except through cries or coos. By one year, she may well be able to walk, play games, and speak a few words. She is a recognizable member of the human family— all from the union of two single cells.

APGAR SCORING SYSTEM

The Apgar score is used to evaluate the health of newborns. Healthy newborns typically score an 8 or 9.

	0	1	2
HEART RATE	Absent	Less than 100 beats per minute	More than 100 beats per minute
RESPIRATION	Absent	Slow; irregular weak cry	Good; strong cry
MUSCLE TONE	Limp	Some flexing of arms and legs	Active motion
REFLEX	Absent	Grimace	Grimace and cough or sneeze
COLOR	Blue or pale	Body pink; hands and feet blue	Completely pink

Chapter 2
BODY ARMOR

THE SKIN

The Largest Organ of the Body

Beauty may be only skin-deep, but how far down does the skin go? The skin is made up of a series of layers, each with its own tasks and goals. An outer layer forms a waterproof barrier, and the deepest layer acts as a cushion for internal organs. The layer between is packed with nerve endings, sweat glands, and hair follicles.

In contrast to the body's other organs, the skin is nearly flat with varying thicknesses. The thinnest skin (from 0.004 to 0.006 inch) is found on the eyelids, while the thickest (about 0.18 inches) is on the palms of the hands and the heels of the feet. That narrow expanse contains two kinds of tissue. The moisture-proof epidermis is the protective outermost layer, which can be seen and touched. Below it, the thicker dermis provides support and gives the skin both strength and suppleness.

The epidermis is multitiered. Its outer surface, the stratum corneum, or horny layer, is a sheet of dead skin cells that serves as a barrier against the outside world. These squamous (scale-like) cells originate in the epidermis's corrugated lower region—the stratum basale, or basal layer—where new basal cells are generated through mitosis. Neighboring Langerhans' cells scout for disease-causing intruders and other foreign substances. The epidermis also contains melanocytes, cells that make melanin, a radiation-absorbing pigment, which gives skin its color and protects it from the sun's ultraviolet (UV) rays. Occasional Merkel cells, found in the lowest regions of the epidermis, function as sensory receptors for touch.

As the basal cells move up through the epidermis, they fill with keratin—a fibrous, waterproof protein that also makes up nails and hair—and bind together in tightly packed sheets. Cut off from the blood supply of the lower epidermis, the keratinized cells, called keratinocytes, die and rise to the skin's surface,

DID YOU KNOW?

SKIN STATS

+ **The skin of an average adult**, if laid flat, would cover about 18 to 22 square feet.

+ **The skin accounts** for as much as 15 percent of body weight.

+ **Some 30,000 to 40,000** dead skin cells are sloughed off every minute, or about 40 pounds (18 kilograms) in an average lifetime.

+ **Fresh skin cells** continuously replace shed cells, resulting in a new epidermis about once a month.

THE SKIN

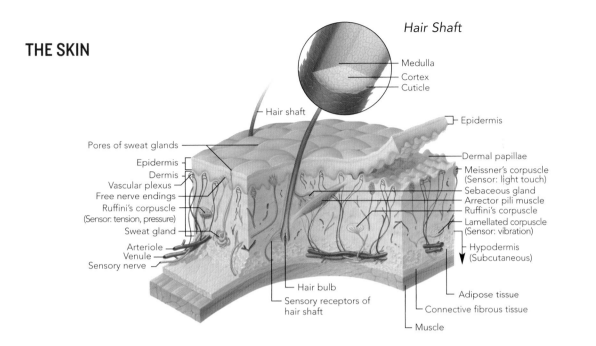

Hair Shaft

Medulla
Cortex
Cuticle

Hair shaft

Epidermis

Pores of sweat glands

Epidermis
Dermis
Vascular plexus
Free nerve endings
Ruffini's corpuscle
(Sensor: tension, pressure)
Sweat gland
Arteriole
Venule
Sensory nerve

Dermal papillae
Meissner's corpuscle
(Sensor: light touch)
Sebaceous gland
Arrector pili muscle
Ruffini's corpuscle
Lamellated corpuscle
(Sensor: vibration)
Hypodermis
(Subcutaneous)

Hair bulb
Sensory receptors of
hair shaft

Adipose tissue
Connective fibrous tissue
Muscle

where they make up the visible, outermost layer of skin. The body sheds millions of old skin cells daily, making way for a fresh group of cells.

Going Deeper

Just beneath the epidermis is the dermis, a tough sheath of connective tissue that makes up the bulk of the skin. The dermis is often referred to as the "true skin" because, unlike the continually renewing epidermis, it stays the same throughout life. The dermis contains blood and lymph vessels, nerve endings, hair follicles, sweat and sebaceous glands, and the proteins collagen and elastin.

Where the dermis and epidermis meet are bumpy features, called dermal papillae, that project up into the epidermis. These contain different structures; some papillae contain blood vessels and others contain nerve endings and touch receptors. Blood vessels nourish skin cells by bringing them oxygen and nutrients and by removing waste produced by cell

metabolism. Sensory receptors in the dermis detect pressure, vibration, and light touch; bare nerve endings sense painful stimuli such as irritating chemicals and extreme heat and cold. Collagen in the dermis gives skin its flexible strength; elastin enables the skin to return to its original shape when tugged and twisted.

Beneath the dermis lies the subcutaneous tissue—a separate region, not technically part of the skin, that anchors it to underlying structures. Generally much thicker than the dermis, the subcutaneous tissue—also known as the hypodermis (a region "below the dermis")—varies in thickness depending on the person, because this is where fat cells are located. Fat is deposited into the cells of adipose tissue, which makes up most of the subcutaneous. Despite its poor reputation, fat plays an important role in the body: It holds in heat, helps to cushion organs and tissues from injury, and serves as a fuel reserve when the body uses more energy than it consumes.

The subcutaneous tissue also serves as a bridge between the skin and the rest of the

CELLS OF THE EPIDERMIS

CELLS	FUNCTION
Keratinocytes	Produce keratin, the protein that endows skin with its protective properties. The most numerous cells in the epidermis
Melanocytes	Spider-shaped cells that produce melanin, the pigment that gives skin its color. Found in the lower level of the epidermis
Langerhans' cells	Protective cells that work with the immune system to fight invaders. Found throughout the epidermis
Merkel cells	Sensory receptors for touch. Found in the lowest levels of the epidermis

body. It hooks onto underlying muscles and, in spots where there are no muscles, such as the knuckles, attaches directly to bones. While skin seems to fit tightly, it's actually loosely hinged to allow the body to move freely.

Skin Types

The skin is divided into two major types, based on function and structure: thick skin, which covers the soles of the feet and palms of the hands, and thin skin, which covers the rest of the body.

Thick skin isn't all the same thickness. It ranges from 0.024 inches (0.6 mm) to 0.18 inches (4.5 mm). Thick skin contains more epidermal cells than thin skin and features five tiers of epidermis (from the inside out): basal layer, or stratum germinativum (where cells are made); stratum spinosum (spiny, interlocked cells that support the skin); stratum granulosum (thin middle layer where keratinization takes place and cells begin to die); stratum lucidum (a translucent sliver of flattened, dying cells); and the stratum corneum (the outermost layer). Thick skin contains many more sweat glands than thin skin, but it does not have sebaceous glands, hair follicles, or arrector pili

(the smooth muscles attached to hair follicles that make the hairs lie flat or stand erect).

In thin skin—with an epidermis from 0.004 inches (0.10 mm) to 0.006 inches (0.15 mm) thick—the epidermal layers are less dense, and there is no stratum lucidum. Thin skin is softer and more flexible than thick skin, has fewer sweat glands, and contains hair follicles, sebaceous glands, and arrector pili.

The sensitivity of thick and thin skin varies, depending on the number and type of sensory receptors for touch, pressure, heat, cold, and pain. In general, nerve endings are more densely clustered in thick skin. Fingertips, for example, have many receptors and

DID YOU KNOW?

STRETCH MARKS

+ **Skin's elasticity** is a marvelous ability, but it is possible to stretch it too far and too fast. Growth spurts, pregnancy, and other rapid fluctuations in weight can all damage the skin and tear the dermis. The streaks left behind, commonly called "stretch marks," are scars created by the dermal tearing. At first, stretch marks may appear reddish or purple, but they eventually lighten and may disappear over time.

are very sensitive to touch. The thin skin of the upper arm, on the other hand, is less sensitive, because it has very few touch receptors.

The thin-skinned lips are an exception; they have many nerve endings, which makes them extremely sensitive to touch and temperature. The skin of the lips is composed of a very thin epidermis and dermis; there are no sweat glands or sebaceous glands, which leaves them susceptible to chapping. The lips also have no hair follicles or arrector pili.

The lips contain many dermal papillae, which nourish cells in the basal layer. The proximity of these capillaries to the skin's surface gives the lips their color. In lighter-skinned people, lips may appear pinker because the cells contain less pigment (which also makes them vulnerable to damage from the sun's UV rays). Darker-skinned individuals may have some melanin in their cells, making the pink color less prominent and the skin less vulnerable to UV damage.

Barrier

Acting like armor, the skin is an effective defense system, keeping invading bacterium from penetrating it (unless it is broken, as through a bite, cut, or puncture). Because it is water resistant yet also a powerful excreter of wastes and sweat, it helps to keep body fluids in balance and regulate temperature. The skin cushions the organs and tissue against injury and, by detecting touch, pain, pressure, and degrees of temperature, tells the body about the world.

To protect the human "fortress" the skin employs multiple defenses. Most would-be invaders never get past the acid mantle of sweat and sebum that coats the skin and contains antibacterial and antifungal substances, including human defensin, which literally punches holes in bacteria. The ongoing action of shedding the outermost cells deters other microbial invaders from the skin's surface. Normal flora do colonize the skin and attack any invaders that manage to land. Because these "home" bacteria have already occupied most of

Keratinocytes are the most numerous cells in the outer, protective layer of the skin.

the available colonization sites, their sheer numbers deter newcomers. They also produce proteins called bacteriocins that kill specific species and make other substances, including fatty acids and peroxides, that inhibit nonspecific invaders. In addition to these defenses, the epidermis's multilayered sheets of tightly packed keratinocytes form a dense barricade against microbes trying to penetrate the interior. When microorganisms gain entry through damaged skin, another army of cells rushes to the body's defense (see "Healing," page 50).

The layers-thick armor of keratinized cells, together with sebum on the skin's surface, also forms a moisture barrier, regulating the flow of fluids in and out of the body and helping to maintain the proper chemical balance. Some substances can penetrate the skin in limited amounts. Among them are: water (which is absorbed by the epidermis, as when fingers plump up after a long bath); gases such as oxygen and carbon dioxide; fat-soluble vitamins A, D, E, and K, which are not repelled by the skin's oily coating but dissolve into it, and from there into the epidermis and dermis; and certain drugs such as steroids.

Pigmentation

Excessive exposure to the sun's UV rays can be harmful to the body, but some sunlight exposure is necessary to help the body produce vitamin D, essential to healthy bones. To protect the body from the harmful effects of too much sun, the skin produces melanin, a radiation-absorbing pigment that also determines the wide variety of colors of the human skin (and hair). Melanocytes in the basal layer of the epidermis produce melanin in tiny cell structures called melanosomes. There, the enzyme tyrosinase interacts with the amino acid tyrosine to create pigment. When melanosomes are packed with melanin, the melanocytes pass them to the keratinocytes, the cells in the upper layers of the epidermis.

Skin color depends primarily on the concentration of melanosomes and the amount of melanin they contain. Fair skin has much less melanin, fewer melanosomes, and lower levels of eumelanin than darker skin. The quantity and kind of melanin a person has is genetically determined, but other factors play a role in skin color, including UV exposure (tanning is the result of the skin's producing more and bigger melanin granules after being stimulated by UV rays), stress, hormones, and some medications.

Increased melanin production can result in temporary or permanent hyperpigmentation, such as brown splotches called liver spots (or age spots) caused by years of cumulative sun exposure. In melasma, a typically temporary condition, dark spots appear on the cheeks, forehead, and upper lip, triggered by hormonal changes from menopause or pregnancy (as well as by oral contraceptives and hormone replacement therapy).

Another common change in skin color—freckling—is an accumulation of pigment in the bottom layer of the epidermis. Freckles first develop in childhood from repeated exposure to the sun. Freckling may be a protective mechanism, since it is most common in those with fairer complexions. Freckles are not associated with skin cancers or other illnesses.

Decreased melanin production may cause hypopigmentation, or loss of skin color. Scars from cuts or burns, for instance, commonly lack pigment. In albinism, a rare, inherited condition, some or all melanocytes are missing tyrosinase, the enzyme needed to make melanin. People with albinism do not have pigment in the skin, hair, and eyes; they may have extreme sensitivity to the sun. A condition called vitiligo creates white patches of skin. It is believed

Students illustrate the wide range of variation in human pigmentation.

to be an autoimmune disorder in which the immune system attacks some melanocytes.

Changes in skin tone are common and usually harmless, but sometimes they signal problems elsewhere in the body. A bluish purple cast to the skin and nail beds may indicate cyanosis (poorly oxygenated blood), which can occur during heart failure and severe respiratory disorders. Transient reddening of the face and neck, ears, and upper chest is associated with embarrassment or anxiety (blushing), but it may indicate an underlying condition. Hypertension can cause flushing, as can allergies to some food additives and medications.

DID YOU KNOW?

BIRTHMARKS

+ **Newborn skin** may have distinctive red patches called hemangiomas (clusters of blood vessels), below. Some, like strawberry marks (red, raised spots) or stork bites (flat, pink to red), may fade with time. Others, like port-wine stains (flat, maroon, or dark red) or cavernous hemangiomas (bluish-red marks that extend into the dermis) often persist and must be removed surgically.

Pale skin can indicate distress as well. There are numerous causes of pallor, including stress and illness. Unusual paleness may indicate blood supply problems. Blood may be diverted away from the skin when blood pressure is low or, if there is an infection someplace else in the body, to vital organs. The most common cause of paleness is iron-deficiency anemia (especially in young women), resulting in a decrease in red blood cells.

Other disorders can cause the skin to look yellow or bronze. A yellow cast usually indicates liver disease in which bilirubin (yellow bile pigment, a component of bile secreted by liver cells) builds up in the blood and is deposited in body tissues. In infants, jaundice can be a normal, transient condition. A bronze appearance to the skin is a sign of Addison's disease, a disorder of the adrenal glands.

Regulator

The skin plays a major role in maintaining a healthy body temperature, which typically hovers around 98.6° Fahrenheit (F) or 37° Celsius (C). Slight variations are typical; normal body temperature can fluctuate a degree in either direction. Extreme changes, however, can be deadly, as they create an unlivable environment for organs and tissues.

When the body is too cold, blood vessels in the dermis constrict, decreasing the flow of warm blood to the skin and conserving it around the central organs. To cool down, blood vessels dilate, bringing warm blood closer to the skin's surface, which is why exercise may make the face redden. At the same time, the hypothalamus, the brain's heat regulator, signals the eccrine sweat glands to produce more sweat—a fluid containing mostly water with small amounts of ammonia and urea, sugar, and salts. The hotter the body becomes, the more

A highly magnified sweat gland (in purple) opens on the skin's surface. Skin cells flake off around the pore's opening.

it perspires. In contact with the air, the sweat evaporates, cooling the skin's surface. (It's also common to sweat when nervous, because sweat glands respond to overstimulated muscles and overstimulated nerves.)

After the water in sweat evaporates, it leaves behind the salts sodium chloride and potassium. The rate of evaporation depends on the humidity of the surrounding air. In muggy weather, the air is already saturated with water vapor, evaporation slows, and the body takes longer to cool down. The body needs fluids to replace both water and electrolytes (mineral salts, such as sodium and potassium) to avoid dehydration.

Sweat glands also aid the body in eliminating waste. Sweat glands excrete waste products such as urea, a by-product of protein metabolism. If urea in the cells is not disposed of regularly, headaches, nausea, and, in extreme cases, death can result. Sweat also helps to flush out toxic metals absorbed from pollution and to eliminate lactic acid, which causes stiff muscles and contributes to fatigue.

KEEPING HEALTHY
PUT YOUR SKIN ON A DIET

The key to healthier skin may be found in your refrigerator. Certain foods can help counteract the effects of sun, aging, and other skin damage.

Two natural processes cause skin to become damaged and less supple over time. Glycation occurs when sugar molecules bind with proteins in the skin, causing damage to collagen and elastin fibers. Oxidation is a natural process that occurs as skin cells are exposed to oxygen, but too much exposure to the sun can cause oxidative damage. You can't stop these processes, but by making the food choices below, you can help nourish and protect your skin:

+ Vitamin E protects the skin from cell damage. Munch on leafy green vegetables, whole grains, nuts, and vegetable oils.

+ Vitamin C promotes wound healing. Eat red peppers, broccoli, citrus fruits, strawberries, and kiwi.

+ Vitamin A encourages cell growth. Sources include low-fat dairy products, fish, meat, and orange and red fruits and vegetables.

+ Resveratrol helps clear out sugar-protein combos. Find it in blueberries, red grapes, grape juice, and wine.

+ Omega-3 fatty acids (in salmon, tuna, flaxseed, walnuts, canola oil, and soybeans) help build healthy cell membranes that hold moisture in the skin.

+ Cut down on sugar, white bread and rice, pasta, and potatoes. Too many sweets and simple carbohydrates can speed up glycation. Check ingredient labels for common sugars, such as cane juice, fruit juice concentrate, corn syrup, fructose, and glucose.

+ Cinnamon, cloves, allspice, sage, tarragon, and rosemary all might help prevent glycation.

HEALING

Repairing the Skin, Protecting the Body

I n the course of a lifetime the skin suffers countless injuries, from nicks and scrapes and more serious wounds, to irritations, infections, and infestations, to burns, small and large. Although the body's internal systems do most of the work of repairing the skin, treatment ranging from minor first aid to a doctor's care may be needed, depending on the problem, to help ease discomfort, minimize damage, and speed the healing process.

HOW A CUT HEALS

Cut blood vessel

Wound

Blood

Constricted blood vessel

Blood clot

Exudate

Granulation tissue:
• Fibroblast • Lymphocyte
• Macrophage

Scab

Scab falls off

Regenerated
epidermal
tissue

Scar

Wound Healing

When superficially injured, the epidermis speeds up cell replacement. The new cells surround the wound and migrate in sheets across it. In a few days the damaged area is completely resurfaced, and the abrasion is gone. Wounds that penetrate the dermis and the subcutaneous tissue, where destroyed cells do not regenerate, are another matter. Because multiple tissue layers must be mended, the healing process is more complex, involving a series of cellular and vascular activities that occur in three phases: inflammatory, proliferative, and remodeling.

In the inflammatory phase, blood from severed blood vessels floods the wound, washing away microbes and debris. To control the bleeding (a process called hemostasis), the

When skin is cut, blood cells spring into action to begin the healing process. Produced by platelets, strands of fibrin bind blood cells to form a scab.

broken vessel quickly constricts to slow blood flow. Platelets, a component of blood, line the wound and clump together to patch punctures in blood-vessel walls. The platelets interact with blood proteins (called clotting factors) to make fibrin, a protein whose strands fan out from the platelets, creating a mesh that traps red blood cells and other platelets to form a blood clot. The clot stops further bleeding, holds the edges of the wound loosely together, and effectively walls in the injured area; this prevents bacteria, debris, and other harmful substances from spreading to surrounding healthy tissue. The part of the clot exposed to air hardens, forming a reddish brown scab.

Meanwhile, the damaged tissue releases inflammatory chemicals that signal the body to respond to the injury. Local blood vessels dilate, increasing the blood supply to the area, and become more permeable, allowing fluid and infection-fighting white blood cells (neutrophils and monocytes) to pass into the injury site within minutes. The neutrophils and monocytes devour invading bacteria and debris.

After several hours, the number of neutrophils (which are short-lived) declines. Macrophages now take over. These are monocytes that have grown larger and produced granules, which contain enzymes to digest bacteria and other invading cells. The macrophages engulf and consume remaining invaders and debris, dead and damaged cells, and old neutrophils. Swelling occurs around the wound, caused by the buildup of fluid in the tissues. The area becomes red and warm, from the increased blood supply, and pain results from inflammatory substances that stimulate the nerves. Inflammation, a sign of healing, usually lasts 24 to 48 hours but can last up to two weeks.

Rebuilding

The proliferative phase of healing is characterized by robust growth underneath the scab. Damaged blood vessels regenerate. In the dermis, cells called fibroblasts produce growth factors and collagen fibers to create new tissue in the injured areas. Some of the fibroblasts have contractile properties that enable them to pull the separated margins of the wounds together.

During this stage, delicate pink granulation tissue fills the wound beneath the scab, taking the place of the clot. Macrophages eat the original blood clot and destroy remaining dead and dying cells. The granulation tissue creates a new capillary bed. (The tissue is actually named for the capillaries, which jut out from its surface, giving it a granular appearance.) These capillaries are fragile and bleed easily (as demonstrated when someone picks a scab). Epithelial cells proliferate and then migrate over the granulation tissue.

During the remodeling phase, the epidermis returns to its normal thickness, and eventually the scab falls off. Damaged tissue has been replaced, and there's underlying scar tissue. The scar (a permanent fibrous patch) may be visible or invisible, depending on the severity of the wound. Raised scars within the borders of the original wound are known as hypertrophic scars; those that extend beyond the boundaries into normal tissue are called keloids. Scars give extra strength to the skin in deeply wounded areas but may lack flexibility and elasticity. Collagen fibers in scar tissue are more tightly packed than in normal skin, and there are fewer blood vessels, which may cause scars to be lighter in color than the surrounding skin.

Keeping the wound covered with a bandage can speed healing. Sutures (external or buried), butterfly bandages, and wound closure tapes can also quicken the pace by bringing divided edges closer, decreasing the distance that cells have to travel to bridge the wound. Debridement, the surgical removal of dead or diseased tissue in wounds, also speeds healing and lowers the risk of infection.

KEEPING HEALTHY

HEALING POWER

Minor wounds should heal in about five days with proper care. Here are some dos and don'ts for faster healing:

+ DO clean the wound with soap and water.

+ DON'T use alcohol, hydrogen peroxide, or Betadine, which may damage skin.

+ DO use petroleum jelly twice a day under the bandage. This can replace antibiotic ointments, unless the wound is infected.

+ DON'T leave the wound exposed to air.

+ DO apply a fresh bandage daily.

+ DO seek medical attention if a wound is very long or deep, was caused by a bite, or becomes infected.

+ DO make sure your tetanus booster is up to date.

Irritation

Skin can be irritated by many things: bacteria and viruses, mites and mosquitoes, poison ivy and poison oak, and many other potential causes. Genetic or other underlying disorders may also reveal themselves through skin irritation. The most common symptom is pruritus, or itching, which can be triggered by a primary skin condition, allergies, or systemic diseases including liver disease, kidney failure, diabetes, and cancer. Sometimes the skin itches when there are no visible lesions. Generalized itching at night can be a symptom of Hodgkin's disease, a lymphoma.

At times, rashes accompany itching. They may come in a variety of shapes and sizes, depending on the trigger, and may make the skin warm, tender, lumpy, cracked, and dry. The pattern and evolution of a rash is key to pinpointing its source. The red, slightly bumpy rash present with the measles virus, for example, almost always starts on the head (after a fever) and spreads down the body. While most rashes are minor, some may be signs of serious diseases, such as Lyme disease, typically caused by deer ticks infected with *Borrelia burgdorferi,* a bacterium contracted from rodents. The rash begins as a tiny red spot at the tick's site and expands over days or weeks. If the rash looks

Eczema is common during childhood.

DID YOU KNOW?

WHY WE ITCH

+ **An itch activates** the neural pathways of the movement-controlling part of the brain, which signals for a scratch. Scratching may create a slight pain to override the itch (pain impulses travel more quickly). Things like poison ivy, allergies, bug bites, and dryness or heat can trigger itches, but the cause of itching absent other symptoms remains unclear.

like a bull's-eye with a clear center, it should be examined by a physician.

Dermatitis and eczema are interchangeable terms for several conditions that cause skin inflammation. They have different causes but similar symptoms: patches of skin with raised red lumps and/or blisters. Contact dermatitis occurs after the skin comes in contact with a substance that either irritates it or causes an allergic reaction. The skin itches and may even blister, depending on length of exposure to the substance. Common irritants include laundry detergents, metals from jewelry or clothes fasteners, some rubber products, some cosmetics, plants such as poison ivy, and some medications.

Urticaria, or hives, can be triggered by many different factors, including stress, allergies, and reactions to drugs or food additives (such as monosodium glutamate). Some people break out in hives after perspiring, after being in the sun, or in extremely cold temperatures. After exposure to one of these factors, the skin responds by producing the chemical histamine, which sets in motion a chain of immune system reactions that result in skin inflammation. Outbreaks usually begin with itching followed by a rash that consists of pale bumps (wheals) surrounded by a clearly defined area of redness. In some cases hives may indicate a serious allergic reaction and might require prompt medical attention.

Burns

Burns occur when extreme heat damages tissue. Fire is a common cause, but many other things—like sunlight, friction, steam, hot liquids, electricity, radiation, and chemicals—can burn the skin. A burn can range from mild to life-threatening, depending on the depth and the amount of affected surface area. Superficial burns respond to simple hygiene and are usually gone in a few days without any blistering or scarring. Critical burns can melt away the skin's layers and structures in a moment and, in horrific instances, go even deeper to damage muscle and bone. At the surface, severely damaged skin can no longer carry out crucial functions, leaving burn victims vulnerable to disease, environmental assaults, extremes of temperature, dehydration, and death.

The three types of burns are classified according to the damage they cause. First-degree burns, like a mild sunburn, are the least serious because they only damage the skin's top layer. Affected skin turns red and may sting, but the burn usually heals by itself in three to six days. Peeling or flaking may occur, but scarring is unlikely.

A second-degree burn penetrates the epidermis and parts of the dermis. Pain, swelling, redness, and blisters result. Severe sunburns with blisters are second-degree burns. Blisters (lymph fluid that collects between the epidermis and dermis after burning or friction, causing the layers to separate) are a sign of deep damage: The more that burned skin blisters, the deeper the burn has gone into the dermis. If there is no infection, second-degree burns usually heal in about three to four weeks, but there may be some scarring.

A third-degree burn destroys both skin layers, the fat layer, and their structures. Most skin function is lost, and nerve endings are destroyed. Third-degree burns can turn the skin white, bright red, or black. The slowed healing process leaves patients more vulnerable to infection because damaged blood vessels prevent the transport of disease-fighting cells to the burn site. Charred tissue becomes parched as sebaceous and sweat glands have been destroyed and are no longer pumping out lubricating sebum and sweat.

Burns are considered critical, according to the American Burn Association, if the patient has second-degree burns over 25 percent of the body, third-degree burns over 10 percent of the body, or if there are third-degree burns on the face, hands, feet, or anal and genital regions. Facial burns are extremely dangerous because of the possibility of burned respiratory passageways, which can swell and lead to suffocation. Burns over joints are also worrisome because scar tissue, less flexible and elastic than normal skin, can reduce mobility.

Exposure to UV rays from the sun destroys skin cells, causing them to slough off.

BREAKTHROUGH BAXTER BURN FORMULA

In the 1960s, more than 40 percent of severely burned patients died from their injuries. Today, that figure has dropped to 6 percent due in large part to the Baxter Burn Formula (also called the Parkland Burn Formula) developed by Dr. Charles Baxter. Because of the dehydrating effects of burns, administering massive amounts of fluids is critical to the patient's recovery. To avoid over- or underhydrating the patient, Baxter developed a way to calculate the proper volume of fluid to be given during the first 24 hours of resuscitation. A Texan by birth, Dr. Baxter was also director of the emergency room at Parkland Memorial Hospital in Dallas, Texas, on November 22, 1963, and one of the team of doctors who tried to save President John F. Kennedy. He said later that "there were probably more tears shed in that room than in the surrounding hundred miles."

Repair

The imminent threat from severe burns is dehydration, which can cause kidney failure and other serious problems. Burn patients are immediately given massive fluids intravenously, as well as nutrients to supply the thousands of calories needed to replace lost proteins and repair tissues. There is also a high risk that nearby healthy tissue will begin to die because of reduced blood flow to surrounding skin. Burn areas often are wrapped in bandages soaked in antibacterial medications to prevent infection. After a patient is stabilized, infection becomes the main threat and leading cause of death.

Skin grafts are often the next step to hasten healing. If the burn area is small, an autograft made from the patient's own skin can be done. In the case of more extensive burns, the area is flooded with antibiotics and then covered with a temporary graft to guard against infection, reduce pain and fluid loss, and allow the patient to continue to stabilize. Biological temporary grafts include human donor skin or fetal amniotic membrane (each is called an allograft).

Synthetic skin made of a silicone "epidermis" sewn onto a "dermis" composed of collagen and ground cartilage may also by used. The patient's body will reject this temporary graft once the immune system recognizes it as a foreign substance. Burn victims cannot tolerate immunosuppressants to fight the rejection process, as their immune systems are too weak.

When the patient is stable and the temporary graft removed, healthy skin from the patient is grafted onto the area. If a patient does not have enough healthy tissue for a graft, new skin can be grown from the patient's own body. Cultured epithelial autograft (CEA) uses bits of epidermal tissue taken from unburned parts of the body. Isolated skin cells are grown on a sheet of fabric that dissolves when the sheet is placed over a wound. In less than a month, enough epidermis to cover the entire body can be grown from a piece of tissue just three square centimeters in size.

PROTECTING THE SKIN

Spending hours in the sun can be hazardous to your health: Chronic exposure to ultraviolet (UV) radiation is the leading risk factor for developing skin cancer, an abnormal growth of cells that can turn deadly.

WAVELENGTHS

Changes in skin color, like tans and sunburns, indicate skin cell damage. They're caused by UV radiation and can increase the chances of developing cancerous growths. Ultraviolet radiation is broken into three wavelengths: UVA, UVB, and UVC. Nonthreatening UVC rays are completely absorbed by Earth's atmosphere and pose little threat to the skin, unlike UVA rays, which are not absorbed by the ozone layer and can penetrate the dermis and damage collagen and elastin fibers. UVB is partly absorbed by the ozone layer; it penetrates the skin less deeply but causes at least 75 percent of UV damage, including premature aging, burning, and skin cancer. UVB damages DNA—even if a sunburn does not occur—creating mutations that can cause cancer in epidermal skin cells.

The Food and Drug Administration (FDA) determines the risk of sun damage on a scale from 1 to 6 for different skin types. People with fair skin tend to burn easily, because they produce less radiation-absorbing melanin; those with higher-number skin types have darker skin; they can still burn but not as easily. The same people who are most likely to burn are also most vulnerable to skin cancer.

DIFFERENT CANCERS

There are three main forms of skin cancer—all on the increase. Basal cell carcinoma, the

This lifeguard has had over 600 melanomas removed. Every sticker represents a former lesion.

most common, is curable in 99 percent of cases because of its slow growth rate and tendency not to spread, or metastasize. It occurs in cells in the lowest layer of the epidermis. Lesions are usually shiny, round, reddish growths with raised edges but can vary greatly. Basal cell carcinomas often result from cumulative exposure to UV rays and develop on exposed areas such as the face, hands, and forearms.

Squamous cell carcinoma originates in keratinocytes of the epidermis, in the layer just above the basal layer. It may grow and spread rapidly, but it can be cured if diagnosed early and removed surgically or by radiation therapy. (The mortality rate is 1 in 100 diagnosed cases.) This kind of skin cancer generally begins as a flat, red, scaly area that may become raised and bumpy as it grows and develop into an open sore. Any suspicious lesions should be examined by a doctor.

The most dangerous form of skin cancer is melanoma, a malignancy of melanocytes, the pigment-producing cells. Melanoma occurs in the skin and can appear in the eye. Compared to other skin cancers, melanoma is more unpredictable, spreads more rapidly, and is more resistant to chemotherapy. Although melanoma occurs in the skin, some suggest that its behavior is closer to that of a sarcoma, a cancer of deeper, connective tissues (such as tendons and cartilage). Continuing research into the exact nature of melanoma could bring about earlier detection methods and greater cure rates.

A melanoma may be a flat brown patch with uneven edges and small black spots; a raised brown patch with black, blue, red, or white spots; or a black or gray lump. It may also develop in a mole. Normal moles are overgrowths of melanocytes, but most are noncancerous, or benign. A new mole or a mole that becomes bigger, changes color or shape, bleeds, itches, hurts, or becomes inflamed should be examined by a doctor.

Melanoma lesions that have penetrated the skin more than 0.03 inches (about 0.80 mm) are likely to have spread through lymphatic and blood vessels to other body areas, sometimes bypassing adjacent lymph nodes. Melanoma accounts for about 3 percent of skin cancers but the large majority of skin cancer deaths. Twenty percent of people with malignant melanoma die within five years of being diagnosed, so early detection is key to survival.

Researchers recently discovered that moles, which are harmless clumps of pigment cells, start the same way as malignant tumors—with a mutated gene that allows cells to divide and reproduce profusely. In moles, the process suddenly grinds to a halt, and the mole becomes a benign tumor whose cells likely won't ever divide again. Scientists are optimistic that this observation could lead to a new, highly effective way of stopping malignant growths.

PREVENTION

Skin cancer is largely preventable by taking some simple precautions, among them wearing sunscreen and protective clothing. Sunscreen protects the skin by absorbing and/or reflecting UVA and UVB rays. The FDA requires all sunscreens to carry a sun protection factor (SPF) label that indicates the amount of protection provided. Most dermatologists recommend sunscreens with an SPF of 30 for protection against sunburn. An SPF of 30 protects against about 97 percent of UVB radiation. It should be re-applied every two hours.

Although SPF ratings apply mainly to UVB rays, many sunscreens now include ingredients that protect the skin from some UVA rays. To get the full effect of a sunscreen, it should be applied 20 minutes before going outside. Between 60 and 80 percent of UV radiation passes through clouds, so sunscreen is needed on cloudy days, too. The FDA also requires that "water resistant" sunscreens be able to maintain their SPF for 40 minutes after submersion.

The American Cancer Society recommends limiting time in the sun when its the strongest (from 10 a.m. to 4 p.m.), staying in the shade as much as possible, wearing sunglasses with 99 to 100 percent UVA and UVB protection, and wearing a wide-brim hat. Snow, water, and sand increase sun exposure by reflecting incoming UV rays, making it especially important for boaters, beachcombers, and skiers to cover up with sunscreen and protective clothing to avoid damaging their skin.

CHANGES

From Childhood to Maturity

During a person's lifetime, the skin undergoes many changes as the body grows taller and larger. Some are structural and part of the body's natural maturing process, with stages through which everyone progresses. Others, often associated with aging, result from cumulative damage to the skin caused primarily by sun exposure, other environmental factors, and lifestyle choices (like smoking or diet). These aspects can vary greatly from one person to another. In these instances, heredity plays an important part too.

Infancy

Babies have a very thin outer layer of skin coupled with a much thicker layer of subcutaneous fat. The combination creates that enviable "baby soft" skin. The extra-thick fat layer serves not only as a cushion, however, but also as a barrier to keep toxins from reaching the rest of the body through the thin keratin layer, which is more permeable than in adult skin.

Newborns may exhibit various temporary skin conditions. At birth, the skin is reddish, and for a few hours the fingers and toes may appear bluish from poor blood circulation. Some babies have bruises and other marks from the birth process that will soon disappear. In the first week after birth, a mild rash often develops where clothing rubs the skin, and there may be dryness and peeling at wrists and ankles. Many normal newborns, particularly premature babies and those that are breastfed, develop jaundice, a yellowing of the skin and the whites of the eyes caused by a buildup of bilirubin in the blood (see "The Skin," page 42). As the spleen disposes of old red blood cells, it breaks down their hemoglobin, creating bilirubin, a yellow pigment, which is transported to the liver for eventual excretion. The condition usually clears up within a few weeks but should be monitored carefully by a doctor.

Puberty

As children approach puberty, their skin begins to undergo significant changes. A surge in hormones stimulates the sebaceous glands, which have been relatively inactive throughout childhood, to secrete more sebum. The eccrine glands produce more sweat, and the apocrine sweat glands secrete a thicker, milky perspiration. Acne and other adolescent skin concerns usually clear up by early adulthood as sebaceous and sweat glands calm down. During this stage, the skin is soft, supple, and strong, thanks to collagen, elastin, subcutaneous tissue, and lubricating oils.

Elastin and lubricating oils make young skin pliable. As skin ages, these substances decrease, which causes wrinkling.

Maturity

As the body ages, both the dermis and the epidermis thin. The skin becomes drier; collagen and elastin fibers in the dermis split, weaken, and decrease in number. The skin starts to sag and fold over on itself, forming wrinkles. As a person nears 40, creases may appear around the eyes, nose, and mouth, but the pronounced effects of aging are not usually seen until the late 40s or early 50s. Generally, women show signs of aging before men do because their skin tends to be thinner and dryer. In the elderly, the combined loss of subcutaneous fat, collagen, and elastin makes the skin look almost transparent and blood vessels more apparent.

With advancing age, the number of Langerhans' and other disease-fighting skin cells decreases and cell turnover in the epidermis slows, so wounds begin to take longer to heal. These changes, combined with diminished circulatory capacity, also make the elderly more vulnerable to infection.

Meanwhile, the sebaceous glands become smaller, causing dry and broken skin. The number and output of sweat glands declines, leaving the elderly more vulnerable to heatstroke; at the same time, the loss of insulating subcutaneous fat makes them more sensitive to the cold.

Older skin produces fewer melanocytes, too, and they tend to cluster in areas exposed to the sun, causing darker brown blotches that appear on the face, hands, and chest, commonly called liver or age spots. Lighter-skinned individuals, who have less melanin to start, usually show these age-related changes more rapidly.

Ultraviolet (UV) light markedly accelerates aging by activating enzymes that break down collagen and elastin; the sun's UV rays also make skin rough and can overstimulate pigment cells, causing unsightly age spots and other skin conditions, including skin cancers (see "Protecting the Skin," page 56). Smoking also can damage the skin, causing changes that mimic the effects of aging. Smoking causes narrowing of blood vessels in the outermost layers of the skin, which impairs blood flow, depleting the skin of oxygen and important nutrients such as vitamin A. Smoking also appears to speed the decline of collagen and elastin fibers, creating thick, leathery skin.

We can stretch and lift and fill in the creases, but there are no known ways to avoid true age-related changes to the skin. Given the inevitable, one could try taking a kinder view of wrinkles, accepting them as well-earned character lines, each with a story to tell, rather than as reminders of passing time.

DID YOU KNOW?

BOTOX

+ **Americans spend more** than $1.4 billion annually on injections of Botox, the trade name of a toxin produced by botulism-causing bacterium. Botox was first used to treat muscular disorders of the eyes. A Canadian ophthalmologist noticed it also softened the wrinkles around her patients' eyes, which led to its now popular cosmetic applications.

THE HAIR

The Long and the Short of It

Some five million hairs grow on the human body, most obviously on the head, although the typical scalp contains only about 100,000 hairs. Body hair, found everywhere except the lips, palms, soles of the feet, nipples, and external genitalia, accounts for the rest. The hair plays an important role in protection. It protects the scalp from the sun's ultraviolet rays and prevents heat loss. Eyelashes and eyebrows shield the eyes from sun, foreign particles, and perspiration. Hair in the nostrils and ears helps keep out dust and other irritants.

Hair Parts

Each hair shaft—the visible part of the hair—is just a flexible strand of dead tissue. Hair grows in follicles, tube-like structures that originate in the dermis of the skin. In the base of the follicle is the hair bulb, a region of cell creation.

A thicket of eyelash hairs (colored blue in this scanning electron micrograph) emerge from follicles in the underlying layer of skin.

A piece of dermal tissue, or papilla, packed with capillaries (tiny blood vessels) projects into the bottom of the bulb, providing nutrients to the undifferentiated cells, which are pushed upward as more cells are created. In the upper part of the bulb, the cells develop into distinct types and group themselves into concentric layers. The hair cells wither, die, and are compressed into a hard, dense shaft growing out of the skin. As new cells push the older sections of hair up, the overall shaft grows longer. Like the outermost layer of the skin, the hair shaft is made up almost entirely of keratin proteins. In the follicle, each hair bulb is surrounded by a hair root plexus, a network of sensory nerve endings that are sensitive to light touch and send out impulses if a hair is disturbed. That's why it hurts when hair is plucked or pulled and tickles when a bug alights. Haircuts are painless, however, because there are no sensory receptors in the hair shafts.

Types and Growth

There are three types of human hair that occur during different stages of life. In the womb, extremely fine, nonpigmented hair called lanugo covers nearly the entire fetus by the fifth month of pregnancy and sheds before birth, except on the scalp, eyebrows, and eyelashes. Short, downy "peach fuzz" called vellus replaces lanugo hair a few months after birth. Vellus hair grows in most places on the human body in both sexes. Terminal hair is fully developed hair that is generally longer, coarser, thicker, and darker than vellus hair, including eyelashes, eyebrows, scalp, armpit and pubic hair, and, in men, hair on the face, chest, abdomen, legs, and arms.

Some 95 percent of the hair on the body of an average man is terminal and the rest vellus. In contrast, vellus hair accounts for about 65 percent of the hair on a woman's body; the rest is terminal. The reason for the difference is that during puberty, boys produce massive amounts of androgens (male hormones), stimulating the growth of terminal hair over much of the body. In girls, the ovaries and adrenal glands produce only small amounts of androgens, which promote hair growth in the armpits and pubic region.

Hair growth is cyclical, alternating between periods of continuous growth and rest. In the anagen portion of the cycle, growth is continuous. (As much as 90 percent of the hair on a person's head is in this phase.) Next, in the catagen (regressive) period, growth abruptly stops and the hair bulb shrivels; only about 1 percent of hair is in this stage, which lasts two to three weeks, at any time. There is no activity during the telogen (resting) phase, which lasts about three months. Then the follicle sheds the hair, and a new hair begins growing.

Growth rates vary with sex, age, and body location. Hair that can grow long, such as scalp hair, has a prolonged growth phase; shorter hair—eyelashes and eyebrows, for instance—has a relatively short growth period. Eyelashes grow for only one to six months, and eyebrows remain active for three to four months, which is why they don't become as long as hair on the head. Each scalp hair, for example, grows about half an inch (12.5 mm) per month for two to six years, then rests for a few months, sheds (about 50 to 100 head hairs normally fall out daily), and is replaced. If it isn't cut, scalp hair can grow up to five feet (152.4 cm).

Hair Color

Melanocytes in the base of the follicle produce pigment, melanin, that is passed to cells in the hair shaft. Hair color results from the ratio of eumelanin (black/brown melanin) to pheomelanin (sometimes called red melanin). Black and brown hair contain a high amount of eumelanin; red and blond hair have more of the lighter pigment. Red hair is also colored by trichosiderin, an iron-containing pigment. Melanin varies in the hair of different parts of the body, which is why their colors may be different.

Melanocytes are genetically programmed to produce a certain amount of pigment at specific ages. Graying results from a progressive decline in the amount of tyrosinase, an enzyme needed to make melanin. Some people start graying as early as their teens; others, not until their 60s, but graying ordinarily begins in the 30s. The process is usually gradual; it may take more than ten years for a person to become completely gray. Gray hair retains some pigment but white hair has none.

THE NAILS

Useful Tools on the Fingers and Toes

In our everyday encounters with objects, the fingernails and toenails are often the first parts of the body to make contact. They absorb the force of impact while also acting as feelers, sending messages and conveying touch to the sensitive tissue in the tips of the fingers and toes. As a counterforce, they enhance that sense of touch. When we grasp an object, for instance, pressure on the fingertips causes resistance in the fingernails. The degree of resistance helps us gauge the object's texture and weight.

Fingernails are also practical tools. They make it easier to pick up small objects, unknot a shoelace, and peel an orange. Like the claws from which they have evolved, they can scratch an attacker or, more usually, an itch. The nails also give doctors a window into the body; changes in the nails' appearance can be an indicator of distress or disease in other systems.

THE NAIL

- Nail body
- Lunula
- Eponychium
- Proximal nail fold
- Nail root
- Hyponychium
- Phalanx *Bone of fingertip*
- Epidermis
- Dermis

Nail Parts

The nails are made up of several parts. The nail body is the most visible one; the hard part of the nail, it is a modification of the stratum corneum, or outer layer of epidermis. This translucent plate is composed of three tight layers of dead, keratinized cells that originate in a specialized epidermis called the nail matrix.

Nail growth begins in a part of the matrix called the nail root, an area of intense cell production. The lunula ("moon" in Latin)—the white crescent under the base of the nail plate, and most visible on the thumbnails—forms one end of the root; the other end passes below the thin flap of skin at the base of the nail plate, called the cuticle or eponychium.

After nail cells are produced in the root, they become heavily keratinized. The nail body

DID YOU KNOW?

+ **The embryo begins** growing nails in the ninth week. Fingernail length is sometimes used to estimate whether the baby is pre-, full-, or post-term.

+ **It takes 6 months** to grow a complete fingernail, 18 months for a toenail.

+ **Children's nails grow** fastest on the middle finger, in the daytime, and in summer.

+ **Typically, toenails are thicker** than fingernails. Men's nails are thicker than women's.

grows out over the nail bed, which extends from the lunula to the nail's free edge—the part of the nail that is trimmed in a manicure. Parallel ridges in the underside of the nail plate interlock with ridges in the nail bed, forming a tight bond. Beneath the free edge a thick ridge of skin, called the hyponychium, binds the nail plate to the tip of the finger (or toe).

Folds in the skin overlap the base and sides of the nail plate, holding it in place and guiding the direction of its growth. The proximal nail fold, at the base of the nail plate, produces a thin, tough outer epidermis, the cuticle. This skin adheres to the nail's surface, which creates a waterproof seal. The cuticle is renewed as skin cells flake off. Many people also be trim it as part of their grooming.

Nails grow continuously and don't shed. The nail plate consists of about 25 tightly packed layers of dead, keratinized cells organized into three distinct layers. The top layer is a few cells thick; the bottom layer, only one or two. A random, overlapping arrangement of keratin fibers in those layers—not unlike the hair shaft—gives the nail its flexibility, resiliency, and strength. In the center layer, however, which makes up about three-quarters of the nail plate, long keratin fibers are arranged parallel to the lunula. The combination allows the nails to tear easily across the width of the nail plate but not down its length, preventing the pain and infection that would result from tears that exposed the nail bed. This self-trimming mechanism would have kept early humans' nails neat and functional. Besides containing keratin, the nail plate includes trace elements (mainly iron, zinc, and calcium) and small amounts of lipids and water.

The normal nail has a smooth upper surface, but vertical ridges may appear with aging. The nail plate appears pink from dermal capillaries below, while clusters of new cells give the lunula its white color. Physicians can diagnose some autoimmune connective tissue conditions by examining the dermal capillaries, which form unique patterns that are altered by such disorders.

Damage to the cuticle, nail folds—hangnails, for instance—or hyponychium can allow germs to enter the nail bed and cause infection. Biting the cuticles and nails may introduce bacteria from the mouth into the numerous blood vessels alongside or underneath the nail. The fingertip bone, or phalanx, is more vulnerable to infection.

Illness Indicators

Many ailments not only affect nail growth, but can cause disfiguring changes in the nails, alerting doctors to the existence of the underlying condition. Brittle, concave, spoon-shaped, or ridged nails, for instance, may indicate iron-deficiency anemia. Nails separating from the nail bed could be a sign of a hyperthyroid disorder such as Graves' disease; black marks that look like splinters under the nail plate could indicate respiratory or heart disease; hard, curved, yellow nails could indicate serious bronchial problems or a fluid backup in the lymph glands.

Chapter 3
STRUCTURE

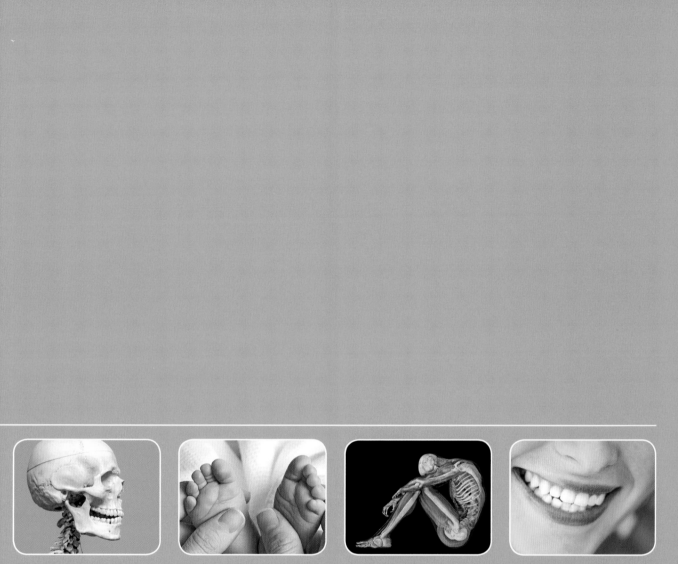

THE SKELETON

Support and Protection

The human skeleton is made up of 206 bones of different shapes and sizes. The skull, backbone, and ribs form the axial, or central, part of the skeleton. Extending symmetrically on either side, the appendicular skeleton includes the bones of the arms and hands, legs and feet, and the pectoral and pelvic girdles, which attach the appendages to the axial skeleton.

Function

The skeleton's major tasks are to support the body, provide a framework for the muscles to enable movement, and shelter the internal organs. The 12 pairs of ribs that curve around the chest cavity from the spine (also called the backbone) protect the heart, lungs, and organs of the upper abdomen; the spine protects the delicate spinal cord; the pelvis protects the uterus and bladder; and the skull safeguards the brain. The brain may be a mighty thinker, but its soft tissue would be extremely vulnerable to injury were it not for the thin yet sturdy skull encasing it (see "The Skull," page 86).

The skeleton enables humans to stand upright and to resist the constant pull of gravity. The spine provides some flexibility (hence, our ability to bend, curl, sit, twist, and tumble), but it also enables humans to stand tall and, together with cervical muscles, to hold their hefty heads high (the brain and skull together weigh approximately 15 pounds).

Bones, although relatively light—the average adult skeleton weighs about 20 pounds (9 kg)—can bear an enormous amount of weight and withstand forces of movement with the strength of cast iron. They are said to be "twice as tough as granite" in resisting compression. When a person takes a leisurely stroll, for instance, each foot strikes the ground with a force of about three times that of his or her weight. Increase the tempo to a brisk walk, jog, or run, and the impact increases to five or six times the person's weight. This means that in someone who weighs, say,

DID YOU KNOW?

BONE FACTS

+ From head to toe, there are 206 bones in the adult human body.

+ The femur, or thighbone, is the longest bone in the body. In an average adult, it measures 18 inches (46 cm) in length.

+ The shortest bones are the ossicles, "little bones" in the ear that are named for their shape. Smaller than the hammer (malleus) or anvil (incus), the stirrup (stapes) is just 0.1–0.13 inches (2.6–3.3 mm) long.

THE SKELETON

Frontal bone
Parietal bone
Temporal bone
Orbit
Maxilla
Zygomatic bone
Mandible
Hyoid bone
Clavicle
Scapula
Sternum
Humerus
Sacrum
Ilium
Radius
Ulna
Carpal bones
Phalanges
Metacarpals
Pubic symphysis
Coccyx
Femur
Patella
Fibula
Tibia
Metatarsals
Phalanges

Parietal bone
Occipital bone
Temporal bone
Mandible
Atlas (C1)
Axis (C2)
Vertebral column
Clavicle
Scapula
Humerus
Ilium
Sacrum
Ischium
Coccyx
Radius
Ulna
Carpal bones
Metacarpals
Phalanges
Femur
Tibia
Fibula
Calcaneus

150 pounds, his or her lower limbs are subjected to a walloping 450 to 900 pounds of force during normal physical activity.

Development

When the skeleton first forms, it consists of rubbery cartilage. Bone-making cells called osteoblasts deposit mineral salts in this soft template, gradually converting most of it to bone. Ossification, or bone building, continues after a baby is born and lasts throughout life. For instance, the xiphoid process, a flap of cartilage just below the sternum (breastbone) often doesn't ossify, or harden, until a person reaches 40 years of age or even older.

Bones are living tissues that grow with the rest of the body and can repair themselves. They undergo continuous renewal, even in adulthood, during a process called remodeling. This dynamic process allows bones to develop proper proportions, to gain strength, and to mend when they break. If the remodeling process is interrupted, it can result in potentially debilitating conditions such as the bone-thinning disease osteoporosis (see "Keeping Healthy," page 75).

Initially, bones are only partially ossified and their ends consist of cartilage. As the cartilage grows, the ossified areas also expand until the adult bone is complete. Most growth (in length) occurs in the rounded area at the end of the bone, called the epiphysis. A baby's skeleton consists of about 300 "soft" bones (composed mostly of cartilage) that eventually fuse to form the 206 bones in the adult skeleton.

During childhood, the cartilage grows and is slowly replaced by solid bone in a process that may continue until a person reaches the mid-20s. The cartilage almost completely disappears in an adult skeleton; it remains only in certain parts of the ear, nose, mouth, trachea and bronchi, the front of the ribs, and on the surface of the joints. Fetal cartilage serves as a model for the bones; mineral salts, such as calcium, are deposited onto the template, and the cartilage becomes calcified, or hard, a process called endochondral ossification. Most bones, called endochondral bones, are formed this way.

Bone Growth

Ossification starts in the middle of the bone in the medullary cavity. Osteoblasts continue to ossify the original cartilage cells. As a child ages, the osteoblasts that convert cartilage to bone become mature bone cells called osteocytes and retire from the bone-making process. During childhood and adolescence, the rounded ends of

Bone formation starts in the womb and continues throughout life.

MALE PELVIS

FEMALE PELVIS

Differences between the male and female pelvis:
A male's pelvic inlet (on the left) is much narrower than that of a female's (on the right).

the long bones in the arms and legs are separated from the shaft (or diaphysis) by a cartilage growth plate, also called the epiphyseal plate.

As the cartilage cells divide and reproduce, the bone lengthens; when the cartilage cells in the growth plate stop dividing, the plate ossifies, leaving only a thin epiphyseal line. The bone stops growing in length, and adult height is reached, usually between the late teens and early 20s. Even though bone growth halts, however, mass may still be added or lost. The body manufactures normal bones by balancing the operations of bone-building osteoblasts and other bone-destroying cells called osteoclasts. At the same time that osteoblasts deposit protein and mineral salts to make new bone, osteoclasts remove old and aging bone cells. Cell material that's no longer needed is excreted as waste, but calcium is recycled through the bloodstream and used by osteoblasts to create new bone.

Gender Gaps

There are several key differences between the male and female skeletons. In most individuals,

the female skeleton is smaller and lighter than the male skeleton. The easiest differences to see are that women tend to have broader hips and their so-called true pelvis—the ring formed by the pubic bones, ischium, lower part of the ilium, and the sacrum—is wider and rounder. The sacrum is tilted back more in a woman, and the coccyx (tailbone) moves more easily, because the opening has to be wide and flexible enough for a baby to pass through during birth.

The female skull is, on average, 20 percent smaller than the male skull, and a woman's facial frame is generally less angular—her chin is less square, her jawbone is smaller and less pronounced, and her forehead tends to be more rounded and longer from top to bottom. In addition, the ocular orbits, or eye sockets (cavities that house and protect the eyeballs), generally appear higher and rounder in women than in men.

Women generally have narrower shoulders and shorter arms than men do, which is why they typically have less throwing leverage. In

An x-ray shows how the upper arm "hangs" from the pectoral girdle.

cuts through the center of the body. Eighty bones make up this portion of the skeleton, including the spine, skull, ribs, and sternum (breastbone).

The axial skeleton also includes the hyoid bone, a tiny U-shaped structure at the base of the tongue (just above the larynx, or voice box), which anchors the tongue and muscles used during swallowing. (In forensic medicine, the hyoid is examined to determine whether the deceased has been murdered—it sometimes snaps when a person is strangled or hanged, choking the victim and causing death.) The hyoid is the only bone in the body that does not articulate with another bone; it is supported by ligaments attached to the styloid processes of the temporal bones and multiple muscles.

Appendicular

The appendicular skeleton consists of the 126 bones of the body's "hanging parts," or append-ages. Working with the muscles, this group of bones can manipulate objects and move around itself. Its bones protect the organs of digestion, excretion, and reproduction. The appendicular skeleton is organized into the upper extremities (arms and hands), the lower extremities (legs and feet), and the bones that attach them to the axial skeleton. The arms and legs have the same number of bones (30 in each limb) and share the same fundamental structure: a single long bone (upper arm, thigh) connected by a joint (elbow, knee) to a pair of long bones (forearm, lower leg), connected by another joint (wrist, ankle) to a collection of tiny hand or foot bones.

In the upper arm, the bone is called the humerus; the large bone in the forearm is the ulna and the other bone is the radius. In the leg, the femur is the huge thighbone, the tibia is the large shinbone, and the fibula aligns alongside.

addition, their broader hips increase the angle between the pelvis and thighbone, making it difficult for women to raise their knees as high or to push off the ground with as much force as men can while, for example, running or jumping. The trade-off: The female body is generally more stable, because women's lower hips and slender shoulders create a lower center of gravity. Women also have a wider range of movement in their thighs than men do because the acetabulum, a hollow area in the hip that the head of the femur (thigh bone) fits into, is shallower and the right and left acetabulum are farther apart in women, providing greater flexibility.

Axial Skeleton

The skeleton is divided into two parts: the axial and appendicular systems. The axial skeleton protects the major organs of the nervous, respiratory, and central circulatory systems. The term "axial" refers to an imaginary line, or axis, that

Girdles

The appendicular skeleton also includes two "girdles," the bones that connect the limbs to the axial skeleton. Whenever someone jumps, kicks a ball, swings a tennis racket, or writes a letter, he or she is using limbs and girdles (and, of course, muscles, tendons, and ligaments). The girdles are rings of bones ("girdle" means to encircle) that attach each limb pair to the axial skeleton. At each shoulder, the upper arm bone (specifically, the humerus) attaches to the pectoral girdle, consisting of the clavicle (collarbone) and scapula (shoulder blade). The thighbone (femur) connects to the hooplike pelvic girdle. When you place your hands on your hips, they rest on the sloping upper edges of the pelvic girdle.

The pectoral girdle is the weaker, and has only one true joint between it and the axial skeleton, the sternoclavicular joint. It is also the more flexible of the two girdles, which gives the arm bones great range of motion. Coupled with the flexibility of the hand bones, it makes the upper limbs perfect for manipulating objects. However, the intricate connections at the shoulder joint allow more motion, but leave the shoulder susceptible to dislocation if wrenched or twisted.

The leg bones are firmly attached to the stationary pelvic girdle, which enables them to support the body's weight. Two curved hip bones (each composed of three fused bones, the ilium, ischium, and pubic bone) make up the pelvic girdle. The pelvic bones meet at the front of the girdle, while at the rear they are firmly attached to the sacrum at the sacroiliac joint. The pelvic girdle and sacrum together form the pelvis, a bowl-shaped structure that supports and protects lower digestive, reproductive, and urinary organs. The opening in the center of the pelvis, the pelvic inlet, is much larger in diameter in females than in males, to provide sufficient space for a baby's head to squeeze through during birth.

Limbs

The bones in limbs each have to absorb and stand up to a great deal of pressure. Human leg bones (and hip, knee, and ankle joints) must be powerful enough to support the body in an upright position. To help meet this demand, the human thighbone has evolved into the strongest, largest bone in the body; both the femur and the muscles attached to it must be strong enough to withstand the intense stress that walking, running, jumping, and other movements place on them.

BREAKTHROUGH THE X-RAY

In 1895 German physicist Wilhelm Röntgen (1845–1923) discovered a form of radiation that could penetrate the body, the x-ray, right. X-rays are absorbed by the dense bones, which appear as shadows on film; they pass through softer tissues, which appear less distinctly. The x-ray allowed doctors to see inside the body without surgery, and it was quickly adopted as a diagnostic tool. X-rays are used to diagnose bone fractures and breaks, reveal dental cavities, and identify foreign objects. Many soft-tissue abnormalities can be detected with computed tomography (CT)—computer-enhanced x-rays.

THE BONES

The Multitaskers of the Body

Bones are strong and, because of their internal structure, surprisingly light. As living organs, they bustle with activity. Inside, blood cells are being created by bone marrow. Outside, competing cells continually build and break down bone in a process known as remodeling. Bone-making cells called osteoblasts build new bone by secreting a matrix of collagen, other proteins, and minerals such as calcium and phosphorus. At the same time, cells called osteoclasts absorb old bone cells, releasing their calcium into the blood, where it can be used again to make new bone. (Any material not reused is excreted as waste.)

Bone Growth

During childhood, bone formation outpaces bone breakdown, enabling the bones to become longer, denser, and heavier. Osteoblasts continue to calcify the original fetal cartilage, and growth takes place throughout the bone. As a child ages, osteoblasts become trapped in the matrix and mature into bone-maintaining cells called osteocytes. Eventually, the only areas of growth are near the ends of certain bones in two small pads of remaining cartilage called growth plates, or epiphyseal plates. In each plate, the inner edge of cartilage is gradually ossified; at the same time, new cartilage grows at the outer edge, lengthening the bone. Bones stop growing when these plates fully ossify.

Growth and maintenance of bone depends on nutrition and levels of certain hormones. Healthy bone growth requires large amounts of calcium and phosphorus and smaller amounts of fluoride, magnesium, iron, and manganese. Vitamin A helps keep osteoblasts functioning.

Osteoblasts play a key role in bone formation.

Vitamin C supports the manufacture of collagen—the main bone protein—and the development of osteoblasts into osteocytes. Vitamins K and B_{12} promote protein synthesis. Vitamin D promotes absorption of calcium and phosphorus and is essential to bone mineralization, growth, and repair.

When and how much the bones of a child grow depends on levels of insulin-like growth factors (IGFs) produced by bone tissue and the liver. IGFs promote cell division at the epiphyseal plate and in the periosteum (outer bone membrane) and enhance synthesis of the proteins needed to build new bone. Production of IGF-1 in the liver is stimulated by human growth hormone (HGH) secreted by the pituitary gland.

Hormones from the thyroid gland (thyroxin) and from the pancreas (insulin) are also needed for normal bone growth. At puberty, sex hormones (estrogen and androgens) stimulate increased osteoblast activity and synthesis of bone matrix, leading to the sudden growth spurts that occur during the teenage years. Ultimately, the sex hormones shut down growth at the epiphyseal plates. At this point, bones stop growing longer. By the time a person reaches his or her mid-20s, the bone-building and bone-destroying processes begin to occur at about the same rate; by the mid-30s, bone loss starts to exceed bone formation.

Bone Marrow

Bones consist of two kinds of bone tissue: compact bone, which forms the hard outer layer; and spongy bone, a lacy tissue in the interior. Gel-like tissue called marrow is found in the gaps of some spongy bone, such as in the long bones of the arms and legs.

Red marrow is the body's blood cell factory, churning out more than 100 billion new blood cells every day. In babies, all bone marrow is red, but in adults, red marrow is found only in the ends of the long bones of the limbs and in the breastbone, spine, ribs, shoulder blades, pelvis, and skull. Yellow bone marrow serves mainly as a storage bin for fat tissue; it can be found in the medullary cavity in the shaft of long bones.

BONE STRUCTURE

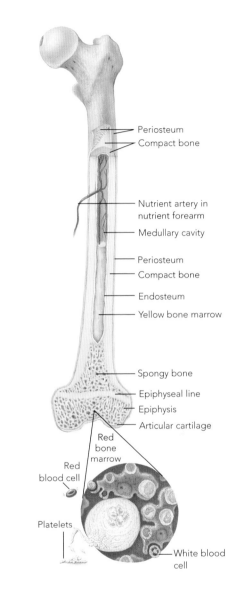

Periosteum
Compact bone
Nutrient artery in nutrient forearm
Medullary cavity
Periosteum
Compact bone
Endosteum
Yellow bone marrow
Spongy bone
Epiphyseal line
Epiphysis
Articular cartilage
Red bone marrow
Red blood cell
Platelets
White blood cell

Blood cells begin as stem cells in red marrow that differentiate as they continuously divide, becoming red blood cells, white blood cells, or platelets. Red blood cells transport oxygen from the lungs to cells throughout the body and carry away carbon dioxide. Platelets, small pieces of blood cells, help blood to clot. The white blood cells produced in bone marrow play a key role in fighting disease and keeping the body's immune system fit. This is why bone marrow transplants can help people with diseases or treatments that attack and weaken the immune system.

Bone Structure

Bones are so light that they account for only about 14 percent of a human adult's total body weight, and yet they are about as strong as reinforced concrete. Deposits of hard minerals such as calcium and phosphorus make the bones firm. The minerals form a complex called apatite salts. The salts stick to the flexible collagen fibers, forming a rigid casing. Bones are densest and toughest near the external surface, where they bear the most stress. Inside, spongy bone forms a lattice-like scaffolding that helps to distribute loads while minimizing bone weight.

SPONGY BONE
Inside the hard outer layer of bones is spongy tissue that holds the gel-like bone marrow.

BONE MARROW
Bone marrow churns out billions of new blood cells each day.

COMPACT BONE
Osteocytes (dark spots) branch out to connect with other cells.

The end of the bone, where it meets another bone, is coated with articular cartilage both to absorb shock and reduce friction at the joints. The rest is covered by the periosteum, a double-layered, white membrane of fibrous connective tissue that contains osteoblasts supplied by lymph vessels, nerves, and blood vessels—essential for bone growth, repair, and nutrition. Immediately beneath this lining is a layer of densely packed compact bone, and, underneath that, is a layer of spongy bone (also called cancellous bone).

Other than tooth enamel, compact bone is the toughest material in the body. It contains mature bone cells (osteocytes) that live in individual lacunae, or pockets of bone. Sheets of lacunae, arranged in concentric layers, or lamellae, form tubular networks called osteons, or Haversian systems; the Haversian canal at the core of the system serves as a conduit for blood vessels and nerves from the periosteum to supply the bone tissue. Smaller canals, canaliculi, contain interstitial fluid that carries hormones, nutrients, and wastes between outlying osteocytes.

If the skeleton were made entirely of compact bone, it would be far too heavy to move.

That's where spongy bone comes into play. Despite its name, this tissue is not soft. Rather, spongy bone looks somewhat like a honeycomb; it is a web of tiny spikes, called trabeculae, that is relatively light but can endure extraordinary stress, because of the way the spikes are angled and arranged. Spongy bone is thickest closest to the joints, which absorb the brunt of bumps and jolts.

Bone Shape

Bones are typically classified by shape. Flat bones shield the brain and the organs in the chest cavity and pelvis. They are generally thin and composed of two nearly parallel plates of compact bone, providing generous space for muscles to attach. In flat bones, the spongy bone is called the diploe ("folded")

COMPACT BONE

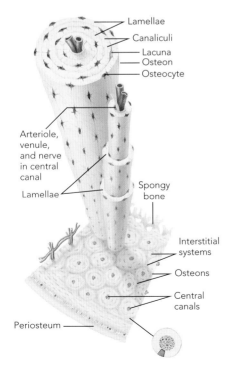

Lamellae
Canaliculi
Lacuna
Osteon
Osteocyte

Arteriole, venule, and nerve in central canal

Lamellae

Spongy bone

Interstitial systems

Osteons

Central canals

Periosteum

KEEPING HEALTHY
PREVENTING OSTEOPOROSIS

There is more to preventing osteoporosis than calcium intake. Studies are finding links between bone health and these diet and exercise habits:

+ **Tea and C:** Antioxidants called polyphenols, which are found in tea and some foods, and vitamin C may help strengthen bones.

+ **Start early:** Teens gain about 25 percent of their adult bone mass during their adolescent growth spurts. Calcium, vitamin D, and exercise during this stretch can go a long way to reduce their osteoporosis risk later in life.

+ **Start and stop:** Weight-bearing exercises that involve bursts of activity—such as soccer, dancing, or weight lifting—build bones the fastest.

+ **High salt = high risk:** Women who eat very high sodium diets—the equivalent of four teaspoons of salt a day—may run a much higher risk of osteoporosis. The recommended daily intake of salt is less than a teaspoon a day, and the average American consumes much more than that.

and is sandwiched between the layers of compact bone.

Long bones are powerful, weight-bearing bones, chiefly those of the arms and legs, the clavicles, and those of the hands and feet. They have greater length than width and consist of a tubular shaft, or diaphysis, and a variable number of epiphyses, or ends. They are usually curved somewhat to distribute the stress of the body's weight at several points, helping to prevent fractures. The diaphysis consists mostly of compact bone tissue surrounding the medullary cavity, a central opening lined by tissue called endosteum. Spongy bone fills the epiphyses and, interrupted by the cartilaginous epiphyseal plate, extends a small distance into the shaft.

Short bones are small cubelike bones found in the wrists (carpals, other than the pisiform, which is sesamoid) and ankles (tarsals, except for the calcaneus, which is irregular). Range of motion is very limited. Short bones consist of spongy bone surrounded by a thin layer of compact bone.

Irregular bones come in a variety of shapes and usually have projections that muscles, tendons, and ligaments can attach to. They vary in the amount of spongy and compact bone present. Such bones include the vertebrae of the spine and some facial bones.

Sesamoids are a type of short bone that develops in tendons where there is considerable friction and stress, such as in the palms and soles, to help reduce wear. They are usually sesame seed–shaped (hence their name). Sesamoids are not always completely ossified and typically only a few millimeters wide. Notable exceptions are the kneecaps.

Healing & Repair

When a bone breaks or fractures, blood vessels in the bone and periosteum tear and bleed. A hematoma (blood clot) forms at the fracture site. Soon bone cells deprived of nutrition die, and the tissue at the site becomes red, swollen, and painful.

Within days, several events lead to the formation of soft granulation tissue. Capillaries grow into the hematoma, and germ-fighting white blood cells called phagocytes race to the region and begin sweeping up debris to prevent infection. Meanwhile, fibroblasts and osteoblasts arrive from the nearby bones and begin reconstruction.

The fibroblasts produce collagen fibers that fill in the damaged areas. Some differentiate into cartilage-making cells called chondroblasts. The blood clot is replaced by fibrous tissue (made by osteoblasts), which forms a callus. Blood vessels grow through the callus, and osteoblasts start forming new bone tissue and collagen. The missing bone will be almost completely replaced by new bone within about two months. Osteoclasts complete the healing process by absorbing any remaining debris.

Most fractures in children and teenagers result from trauma that yanks or twists the bones (sports injuries, car crashes, and falls). In the elderly, more fractures occur as bones thin and weaken.

When the break is simple (doesn't puncture the skin), a physician will manipulate the broken edges back into place by hand and put a stabilizing cast on the area to immobilize it until the bone knits properly. Some injured bones don't need special assistance. A broken rib, though painful, will heal on its own in time.

More serious fractures may require surgery. A surgeon may insert metal plates, rods, or

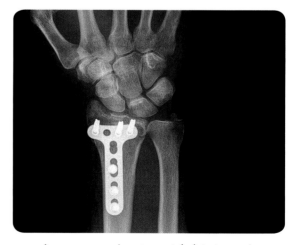

A surgeon may insert a metal plate to repair some serious bone fractures.

BROKEN BONES

FRACTURE TYPE	DESCRIPTION
Simple	Bone breaks cleanly, does not penetrate skin
Compound	Broken ends of bone protrude through skin, soft tissue
Greenstick	Bone splits and bends, does not break, like a green twig. Seen only in children
Stress	Hairline fracture caused by extensive or repeated stress
Spiral	Ragged break caused by twisting motion. Most commonly seen in sports injuries
Compressor	A bone weakened by a degenerative condition (like osteoporosis) collapses
Depressed	Broken bone presses inward. Often seen in skull fractures
Impacted	Broken bone ends are forced into each other. Typically occurs when breaking a fall
Comminuted	Bone shatters into smaller pieces. Can be caused by high-impact trauma (like a car accident) or by a degenerative condition (such as osteoporosis)

screws to piece together and secure the broken pieces. The skin is sutured and a cast or splint is applied to protect the repaired area; sometimes traction is used to maintain proper position while the bone fuses. Fractures in small and medium bones typically heal in six to eight weeks. Healing can take longer in large, weight-bearing bones and in bones of the elderly (because of poorer circulation).

New Treatments

Some medical advances may improve healing after a break. Sending minute electrical currents through the break sites can speed healing. Daily exposure to low-intensity ultrasound waves can reduce healing time in arms and shinbones by 35 to 45 percent.

Free vascular fibular bone graft technique uses pieces of the fibula to replace missing or severely battered bone. In the past, extensive bone grafts (typically, pieces of hip bone) failed because blood could not get into the substitute and the graft died. This new technique consists of a bone graft with its blood vessels being attached to vessels at the recipient site.

Scientists are also researching bone substitutes. In the past, a paste of crushed bone from cadavers or synthetic material was sculpted into the shape of the desired bone or packed into spaces in broken bones, but both had drawbacks: As with any transplant, cadaver bone carries a risk of hepatitis or HIV infection, and foreign material may be rejected by a recipient's immune system. A new material made from coral has proved more successful than previous bone fillers. The derivative is heated to kill living coral cells and convert its mineral structure to hydroxyapatite, the salt in bone. The graft is then shaped, sterilized, and coated with a natural growth substance. Other types of artificial bone, including tricalciumphosphate (TCP), a biodegradable ceramic, are being explored, as are stem cell therapies.

CONNECTIONS

Joining the Body Together

The human body can perform an incredible array of movements, from typing to threading a needle, to pitching a curveball, to tumbling, jogging, skiing, and literally jumping through hoops. These actions are possible because of movable joints, the connections between the bones of the skeleton.

There are more than 400 joints, also known as articulations, in the body. Muscles span joints, crossing from one bone to another. When these muscles contract, they tug on the bones, producing skeletal movement at the joint. Muscle contraction and movement are controlled by signals sent from the brain or ordered by reflex action.

Bones work like mechanical levers, and the joint between them is the fulcrum, or fixed point, on which they move. Bands of strong tissue called ligaments support the bones around a joint. Cords called tendons join muscles to the bones, which move when the muscles contract. Resilient tissue called cartilage cushions the bones' ends and helps joints function smoothly.

Ligaments

Ligaments are bands of tough, fibrous collagen and stretchy elastin that connect bones to one another; they also surround and bind the joints to help strengthen and stabilize them, permitting movement only in certain directions. For instance, ligaments on either side of the fingers prevent side-to-side bending, and ligaments that stretch across the palm keep fingers from hyperextending.

Ligaments are the most important stabilizers in the complex knee joint. These powerful, tight cords connect the thighbone (femur) and shinbone (tibia), keeping the knee from wobbling. Unlike the hip, for example, which is securely anchored in a deep socket, the knee has virtually no built-in bony stability and would roll around like a ball were it not for ligaments, along with the upper and lower patellar tendons and hamstring muscles, to rope it in place. Collateral (side) ligaments—the medial

DID YOU KNOW?

WATER ON THE KNEE

After a trauma to the knee, synovial fluid builds up within the joint, causing pain, inflammation, and stiffness, a condition often called water on the knee. The fluid can be drained with a needle; an injection of corticosteroids can reduce inflammation and prevent new buildup. Football great Joe Namath at times had his knees drained during a game.

SHOULDER JOINT

(lateral view, humerus removed)

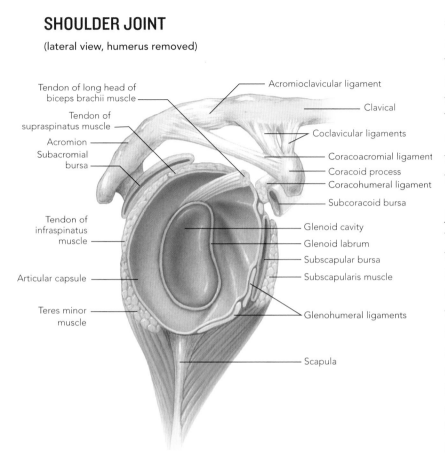

Tendon of long head of biceps brachii muscle

Tendon of supraspinatus muscle

Acromion

Subacromial bursa

Tendon of infraspinatus muscle

Articular capsule

Teres minor muscle

Acromioclavicular ligament

Clavical

Coclavicular ligaments

Coracoacromial ligament

Coracoid process

Coracohumeral ligament

Subcoracoid bursa

Glenoid cavity

Glenoid labrum

Subscapular bursa

Subscapularis muscle

Glenohumeral ligaments

Scapula

collateral (MCL) and lateral collateral (LCL)—prevent the knee from wobbling too far to either side; cruciate (crossing) ligaments—the anterior cruciate (ACL) and posterior cruciate (PCL)—prevent the knee from slipping too far backward or forward. (Ligaments also support various organs, including the liver, bladder, and uterus, as well as the female breast.)

Tendons

Tendons are rigid cords of tissue, containing mostly collagen, that secure muscles to bones and to other muscles. Usually they are round or flat bundles, but a few, such as those in the abdominal wall, are flat sheets called aponeuroses. Tendons are covered by lubricated sheaths that allow them to glide easily without disturbing surrounding tissue. When a muscle contracts, it pulls on a tendon, which then tugs on the attached bone and—presto—the bone moves at the joint. The fingers, for instance, are moved mainly by muscles in the forearm that are connected to them by long tendons.

Tendons also help stabilize joints. In the knee, for example, the upper and lower patellar tendons help keep the patella (kneecap) from slipping out of place. The Achilles tendon, which attaches to the heel bone and connects the calf muscles to the foot, is probably the best known tendon in the body. Easily damaged during overstrenuous exercise, it is named after the mythical Greek hero Achilles, who died when an arrow pierced his heel, his only vulnerable spot.

Cartilage

No joint in the body is under more stress and pressure than the knee, which is why it needs so much support. In addition to ligaments and tendons, it has thick pads of protective cartilage. Cartilage is a dense, rubbery connective tissue composed of collagen fibers, water, and a gel-like substance. It is key to preventing wear and tear on joints. A layer of cartilage covers the ends of joint bones, cushioning them and reducing friction during movement. Cartilage also supports and shapes various structures such as the ears, nose, and intervertebral disks.

A VULNERABLE JOINT

The knee is the most complex joint in the body. An engineering marvel, it supports while allowing for a wide range of movement; but it is this design that also makes it vulnerable to a range of injuries. A knee injury can happen quickly, as in a hyperextension, or it may occur over a lifetime, as with osteoarthritis. Prevention and care are crucial to keeping the knees healthy.

HYPEREXTENSION

Hyperextension is an injury that occurs when the knee is extended beyond its normal fully straightened position. It is relatively common in sports that feature a lot of twisting and jumping, such as basketball, football, or gymnastics. It often results from an off-balance landing or when the upper and lower parts of the leg pull in opposite directions.

The injury may cause swelling and pain when the knee is extended. Treatment depends on the severity of the damage. A hyperextended knee may also lead to a partial or complete ligament tear, especially in the anterior cruciate ligament (ACL). A person should seek medical attention if an injury causes severe pain, swells dramatically, impairs mobility, or causes the knee to lock or give out.

LIGAMENT TEAR

Ligaments are strong bands of tissue that connect bones and support the knees and various organs. They can be bruised if they're suddenly twisted or pulled. Injuries range from minor stretches and strains to total tears. Commonly injured ligaments are the medial collateral ligament (MCL), which runs along the inner knee, and the anterior cruciate ligament (ACL) in the front center of the knee. The MCL usually heals on its own. If the ACL rips, it generally requires surgery; if it isn't repaired, it can result in permanent instability, meaning the knee might give out at any time.

About 70 percent of traumatic knee injuries involve the ACL. They frequently occur during sports in which the feet are planted while the body sharply shifts direction. An ACL tear can be a career-ender for many professional athletes, especially football and soccer players.

When a ligament is torn, the pain is swift and severe. If the knee swells, a physician will likely aspirate it—remove fluid with a needle. Blood in the fluid may indicate a torn ligament or a fractured patella. In mild cases, a knee brace can

When cartilage in the knee is suddenly torn, the injury may appear crisper compared to cumulative damage.

keep the ligament still enough to heal, but for many people surgery is the only option.

The ACL can be reconstructed by creating a new ligament from a tendon near the knee. The operation is usually performed using an arthroscope (a tiny, fiber-optic camera). In arthroscopic knee surgery, small incisions are made for inserting the arthroscope and an irrigator, which pumps sterile fluid into the joint to expand it. Other surgical instruments may be needed to cut, shave, or repair tissue. Recovery from ligament surgery is a long process: The patient may be on crutches for several weeks and require several months of physical therapy to improve the knee's range of motion and to regain muscle strength.

MENISCUS TEAR

The meniscii, wedges of protective cartilage sandwiched between the femur (thighbone) and tibia (shinbone), one on each side, may be torn if the knee is twisted too far while bearing weight—say, when a person abruptly turns to hit a tennis ball. Symptoms of a torn meniscus include swelling, pain (especially when straightening the knee), buckling, clicking, and locking.

Magnetic resonance imaging (MRI) is used to confirm the diagnosis. Traditional x-rays cannot detect meniscus injuries, but physicians may take them to rule out degenerative damage or any arthritic change. For years, surgeons treated such injuries by removing the meniscus. They discontinued that treatment after discovering that some patients developed arthritis or deformities later on as a result.

Today, physicians repair and preserve as much of the meniscus as possible with arthroscopic surgery, removing damaged sections and leaving the healthy portions in place. Recovery from such surgery typically takes about a month.

OSTEOARTHRITIS

Osteoarthritis is the most common form of the 100-plus types of arthritis, affecting some 21 million Americans. It mostly occurs in people over 60 but can strike at any age. Osteoarthritis used to be considered a natural part of aging caused by normal wear and tear on cartilage. Osteoarthritic cartilage, however, is chemically different from cartilage with age-related changes. Many experts now believe that microscopic changes in cartilage structure and composition are often the precipitating factor. Other factors include heredity, obesity, or injury.

In healthy joints, cartilage covers the ends of bones, reducing friction and absorbing shock. Osteoarthritis causes that cartilage to erode. As a result, affected bones grind directly against one another, causing accelerated wear. As the disease advances, bone density decreases and osteophytes (bone spurs and cysts) may develop. This is because the body tries to repair the damage but manufactures bony protrusions instead of normal bone. Such growths irritate surrounding soft tissues, often causing swelling.

Osteoarthritis most often affects the joints of the knees, hips, lower spine, and neck. Men are more likely to develop osteoarthritis in their hips, knees, and spines; women suffer more in their hands and knees.

Symptoms may develop slowly over many years. The first sign may be a bit of pain and stiffness. Symptoms of arthritis in the knee include pain, swelling, stiffness, and a locking sensation, especially during attempts to straighten it. Treatment may include exercise to strengthen supportive muscles; physical therapy with heat to reduce stiffness; treatment to restore cells in cartilage with small defects; injections of a joint fluid compound into the joint; and, in severe cases, joint replacement surgery.

There are three types of cartilage. Hyaline cartilage is the most widespread type found in the body. It covers the articular bone surfaces, where one or more bones meet at a joint; extends between the ribs and sternum, cushioning them against impact; and is the flexible tissue that forms the external nose. It is also the cartilage inside bones that is the foundation for bone growth, making up most of the embryonic skeleton. Fibrocartilage is a tough cartilage located in areas needing dense cushioning, such as between the vertebrae. It is also found in the knees where it forms two disks, each called meniscus, between the femur and tibia. Participants in contact sports often injure this tissue (see "A Vulnerable Joint," page 80).

The third type, elastic cartilage, is found in the external ears and the epiglottis. As its name suggests, this type of cartilage is both flexible and resilient, because of the abundance of elastic fibers it contains. There is little blood in the tissue of tendons and ligaments and none in cartilage. Consequently, healing in those tissues can be slow when compared to the rest of the body.

ELLIPSOIDAL JOINT
The bones in the wrist permit side-to-side and up-and-down movements.

HINGE JOINT
In addition to the knees and elbows, hinge joints are found in the fingers and toes.

PIVOT JOINT
The neck has a wide range of versatile motion because of this joint's structure.

Joint Types

Each joint is constructed differently, based on its function, and each has its own range of movement. Some joints move freely, some move very little, and some don't move at all. Some joints are big and strong enough to bear all of the body's weight, while others are so small (like those between the tiny ossicles, or ear bones) that they're barely visible.

The skeletal system consists of three types of joints: fixed (also called sutures), cartilaginous (symphysis), and synovial. Fixed joints are slim bands of fibrous tissue that cement one bone to the next and do not move. Among them are the jagged sutures that seal the platelike bones of the skull; a thin layer of connective tissue joins them.

A symphysis or cartilaginous joint occurs when two bones join and are held firmly together so that they function as one bone. Such joints are made up of tough cartilage plates that flex and permit limited motion. An example is the sacroiliac joint, where the sacrum and pelvis join. The disks between the vertebral bones in the spine are also cartilaginous joints, but they are thicker and allow more movement than the sacroiliac joint. In these joints, there is no so-called synovial space between the bones, which means that the adjoining bones have little room to move. In the spine, however, while each of the joints between neighboring vertebrae has limited flexibility, working altogether they allow a lot of movement.

Synovial joints are the most common and most mobile joints. They include the shoulders, elbows, wrists, fingers, hips, knees, ankles, and toes. The ends of bones in these loose and limber joints are wrapped in slick, resilient cartilage, which reduces friction when they slide over one another. Synovial joints have a lining of tissue, a synovial membrane. The joint is filled with synovial fluid that keeps the bones lubricated so that they glide as smoothly as the parts of a well-oiled machine.

Mechanics

The way joints move depends on their shape. Some are hinged like doors, allowing limited back-and-forth movement. Others are constructed more loosely, allowing motion in many directions.

While everyone has the same number of joints, the range of motion may vary slightly from person to person. Some, for instance, are said to be double-jointed because they have unusually

BALL-AND-SOCKET JOINT
This construction is present in both the hips and the shoulders, which enjoy a wide range of motion.

PLANAR JOINT
Bone surfaces in these joints tend to be smooth, allowing for seamless gliding motion.

SADDLE JOINT
The thumb can rotate and bend forward and back because of its connection at the wrist.

flexible fingers or other joints. Others, such as gymnasts and ballet dancers, are more limber. These differences, however, are largely attributable to differences in ligament length and suppleness.

The most common and movable types of joints are synovial joints. In most of these joints, the bones either hinge onto each other or rotate. There are six types of synovial joints. A pivot (or swivel) joint is one in which a cylinder-shaped projection on one bone snaps into a ring of another bone to allow movements like a wheel spinning on an axle. There's a pivot joint between the first (atlas) and second (axis) cervical vertebrae (neck bones) at the top of the spine that allows the head to turn left and right.

In ball-and-socket joints, such as the hip and shoulder, the rounded end of one bone sits in the cup-shaped socket of the other and can move in almost any direction. This kind of joint with loose ligaments allows a wider range of movement than any other in the body.

An ellipsoidal, or condyloid, joint is one in which the dome of one bone fits into the recessed, oval-shaped socket of another, allowing limited movement.

The joints of the wrists between the radius and carpals are an example.

The bones of saddle joints interlock like a pair of curved saddles. This type of joint occurs between trapezium of the wrist and metacarpal of the thumb where it joins the hand. Its loose ligaments allow great mobility with considerable strength (as in thumb wrestling).

Planar joint bones have nearly flat surfaces where they meet. They are held together tightly, so that they have only limited side-to-side and back-and-forth gliding movement. This kind of joint can be found in the carpals of the hands, the tarsals of the feet, and between the clavicle and sternum.

Hinge joints have one bone that rotates inside a cylindrical hollow. These joints are located in the knees and elbows and between some of the bones in the fingers and toes. Hinge joints allow the moving bone to flex and extend.

The largest and most complex joint in the human body is the knee, which is designed for its own protection. It is completely surrounded by a joint capsule that is strong enough to hold the joint together but flexible enough to allow motion. The capsule is lined with synovial tissue, which secretes synovial fluid (clear in healthy joints, cloudy in diseased joints, and bloody in injured joints) that bathes the joint.

Pads of cartilage coat the ends of the thighbone (femur) and shinbone (tibia) that join to form the knee, protecting them from wear. Two menisci (cartilage pads) act as cushions between the two bones and help distribute body weight in the joint. Ligaments along the sides and back of the knee reinforce the joint capsule, adding stability. The kneecap (patella) protects the front of the joint.

Problems

Synovial joints are tough but their mobility makes them easier to damage when compared to other parts of the skeleton. Sudden wrenches or twists may result in a sprain—a painful injury involving damage to muscles and ligaments. During strenuous exercise, the knee's internal ligaments can be torn, and pieces of cartilage may break away, jamming or locking the joint (see "A Vulnerable Joint," page 80). Injuries can also result in joints being displaced: Luxation is the full dislocation of a bone from its joint, and subluxation is a partial displacement.

Synovitis is an inflammation of synovial membrane that results in swelling and pain. It can

BREAKTHROUGH ARTIFICIAL JOINTS

To relieve joint pain and loss of movement caused by osteoarthritis, other diseases, or injury, surgeons can often replace the damaged joint with an artificial one. Hips and knees are the joints most commonly replaced. A prosthesis is joined to natural bone with an adhesive, or the surgeon may use a replacement joint made of porous material into which new bone cells migrate. Eventually, enough new bone is produced to firmly anchor the shaft. In joint replacements (as in a hip replacement, right), points of contact between prosthesis and bone are coated with a special plastic to reduce friction.

Rheumatoid arthritis can make even common household tasks difficult and painful.

be caused by an injury, infection, or irritation produced by damaged cartilage.

A bunion (a condition also known as hallux valgus) is an abnormal angulation of the joint at the base of the big toe. This misalignment puts pressure on the other toes as the big toe is forced sideways. Those who wear tight-fitting, pointed, or high-heeled shoes put themselves at risk for this condition, which can often only be treated surgically.

A group of degenerative diseases that inflame or damage the joints are collectively known as arthritis. The most common form, osteoarthritis, is most common among the aged and is attributed to normal wear and tear. Articular cartilage is broken down, and new tissue does not replace it. Bones begin to grind directly against one another, eventually wearing down the cartilage and forming bony spurs where the exposed bone tissue has thickened. A crunching noise, called crepitus, may be heard when the roughened surfaces rub against each other. Today, an estimated 21 million Americans live with the disease.

Another is an inflammatory, autoimmune disease, rheumatoid arthritis. In it, the immune system attacks the body's own tissues. The synovial membrane proliferates and triggers production of enzymes that break down cartilage, damaging the joint. Symptoms may range from stiffness and pain to restriction or loss of movement.

THE SKULL

The Most Complex Bones in the Body

The most complex portion of the skeleton is the skull, a network of bones that gives shape to the head and face as it cradles and protects the brain. Given the nature of its important function, many assume the skull to be a heavy mass of solid bone. But, appearances are deceiving. The skull's structure is lighter, more complicated, and more specialized than it looks.

The skull looks at first glance as though it were made of one bone, but it's composed of 22 separate, mostly flat bones held together by five rigid joints (called sutures), which resemble jagged stitching. The only bone in the skull that is freely movable is the mandible, or lower jaw. The condyloid joints that connect it to the skull enable the multiple angular motions that allow humans to chew and talk.

There are two sets of bones in the skull: the cranial and facial. Eight cranial bones form the cranium, the upper dome, or cranial vault, that encases, supports, and shelters the brain and hearing apparatus; it also provides attachment sites for head and neck muscles. The 14 facial bones form the base for the face and jaw, provide openings for air and food, secure the teeth, and connect to facial muscles that enable the myriad expressions of which the human face is capable.

The skull has four large cavities—the cranial, which contains the brain; nasal (nose); and two orbits, or eye sockets, which house the eyeballs (cushioned by fatty tissue). There are also many smaller cavities, including those of the middle and inner ear.

Some of the bones of the skull, most notably surrounding the nasal cavity, contain sinuses. These hollow air chambers serve to lighten the load of the skull, but as allergy and cold sufferers know, they can make the head feel heavy and cumbersome when they fill with fluid. Sinuses also act somewhat like echo chambers, making the voice sound richer and more resonant.

DID YOU KNOW?

SOFT SPOT

The cranial bones of newborns are not fused but connected by fibrous membranes called sutures. At six places where the bones abut each other, the sutures are expanded. These "soft spots"—two on top, two on each side—are called fontanelles. The fibrous sutures and fontanelles allow the bones to overlap during the birth process and to expand evenly as the brain grows. The fontanelles normally are replaced by bone when the infant is six months to two years old; sutures ossify by early adulthood.

THE SKULL

Bone is abbreviated b.

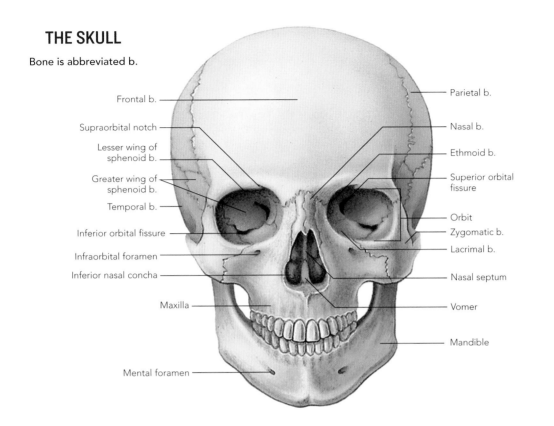

Frontal b.
Supraorbital notch
Lesser wing of sphenoid b.
Greater wing of sphenoid b.
Temporal b.
Inferior orbital fissure
Infraorbital foramen
Inferior nasal concha
Maxilla
Mental foramen

Parietal b.
Nasal b.
Ethmoid b.
Superior orbital fissure
Orbit
Zygomatic b.
Lacrimal b.
Nasal septum
Vomer
Mandible

During speech or other vocalization, such as singing, the sound waves reverberate in the sinuses. The reverberation prolongs and intensifies the sound of the voice. By the same token, when a person has a cold and the sinuses are blocked, the voice can sound muffled and dull.

There are small holes in the skull to allow blood vessels and nerves to pass through: for example, the optic nerves to the eyes and the olfactory tracts to the nose. The largest opening is the foramen magnum in the base of the occipital bone at the back of the skull, through which the spinal cord passes to join the brain.

Cranial Bones

The brain's bony protector includes the two parietal bones, which form most of the roof and upper sides of the skull. These bones make up the bulk of the cranial vault, or calvarium, where the brain sits safely enclosed within.

The shell-shaped frontal bone forms the frontal squama (forehead), the upper margin of the eye sockets, and a portion of the nose, or nasal cavity. Temporal bones form the sides (just above the ears) and part of the base of the cranium. The butterfly-shaped sphenoid bone forms part of the base of the skull and part of the bottom and sides of the eye sockets. A sunken portion of the sphenoid bone called the sella turcica ("Turk's saddle") shelters the pituitary gland.

The delicate leaf-shaped ethmoid bone forms part of the nose, orbits (it comprises most of the area between the eyes), and the floor of the cranium. It contains several sections called plates, including the cribriform plate, which shapes the top of the nasal cavity; nerves from the olfactory (smell) receptors in the nasal cavity pass through tiny holes in the cribriform (pierced) plate to the brain. The perpendicular plate forms part of the nasal septum, which divides the nasal cavity into right and left nasal passages. The occipital bone forms the back wall of the skull and, with the sphenoid bone, the base of the cranium.

Facial Bones

The remaining skull bones form the skeleton of the face; all but the mandible (lower jaw) and vomer come in pairs, giving the face its symmetrical structure. Zygomatic bones form the cheekbones and sides of the eye sockets. The nasal bones, two thin rectangles, form the bridge of the nose, one on either side; the lower, movable portion of the nose is supported by flexible cartilage. Inferior nasal conchae are the thin, scroll-like bones that form part of the interior lateral wall of the nose. Palatine bones form the floor of the nasal cavity and part of the hard palate (roof) of the mouth.

The skull contains four cavities, including the cranial cavity, which holds and protects the brain.

The slender vomer bone, along with the ethmoid bone and a section of cartilage, forms the nasal septum. The nasal septum is the structure that divides the nasal cavity into right and left airways; it's also the part of the nose that can be easily pushed off track (called deviation) by a swift blow. More commonly, uneven growth rates between bone and cartilage can cause the septum to deviate. This often occurs during the growth spurt that takes place in the teen years. The deviation can block one of the nasal airways, resulting in mild to severe difficulty in breathing. It can also produce a bump in the bridge of the nose. Surgery can usually correct the problem, unblocking the air passage and restoring an even appearance to the nose.

Maxillary bones, or maxillae, form most of the upper jaw. The mandible, or lower jawbone, and maxillae both contain sockets for the teeth. The horseshoe-shaped mandible is the largest, strongest facial bone and the only movable bone of the skull. The mandible is attached to the temporal bones, one on each side of the skull, just in front of the ears, at the two temporomandibular joints.

The temporomandibular joint, or TMJ, is one of the most complex joints in the body. Supported by muscles, tendons, and ligaments, it can open and close like a hinge but also move from side to side and slide forward and backward. A disk of cartilage, similar to the meniscus disk in the knee, provides extra padding at this much-used joint. TMJ disorders can produce a range of symptoms including muscle pain, headache, and an inability to open the jaw wide.

Shaped like fingernails, the lacrimal bones form the inside walls of the orbits (eye sockets) at the inner angle of the eyes. Ever wonder why your nose runs when you cry? It's

Whiplash, which can occur when the head jerks suddenly, may cause the brain to shift inside the skull.

because tears flow from the eyes to the nasal cavity through a passageway called the naso-lacrimal canal, located between the lacrimal bones and the nose.

Bumps & Breaks

The domed structure of the skull is able to distribute loads, making the skull very strong while allowing it to be relatively light. The skull is thick and dense at the top, where there are no muscles to pad it, as there are on the sides and back, areas where the bone is thinner. It's nevertheless advisable to pro-tect the skull in situations where it could receive a severe blow, as when playing sports such as football and ice hockey, auto racing, and bicycling.

If a person gets a bump on the head, the skull is designed to absorb the impact, working with the cerebrospinal fluid that surrounds the brain inside. But if a person takes a severe blow, injuries to the brain, such as concussions, can occur—even if there is no skull fracture. Symp-toms of a concussion can include dizziness, vomiting, headaches, ringing in the ears, and blurred vision. If a concussion is suspected after a head injury, it's wise to be examined by a doctor.

When the skull fractures, surgery is not always necessary. If the fracture is depressed, however, with bone pressing inward, brain damage may occur. The bone may also tear the tissue layers covering the brain, allowing bacteria to enter and causing infection and brain damage. In these situations, surgery is necessary to clear away debris, make what repairs are possible, and restore fragments to their proper position.

THE TEETH

Built to Break Down Food

A bright, healthy smile starts with a set of pearly white teeth, the living, calcified structures embedded in the jawbones. But the teeth are more than pretty; they also play a vital role in digestion: breaking up food through chewing.

Composition

The teeth sit in the jawbones but are not composed of bone. The exterior of the visible portion, or crown, is enamel, the hardest tissue in the body. Beneath the enamel shell is a framework of bonelike dentin that makes up the bulk of the tooth. This hard, yellow tissue extends into the long root portion of the tooth, below the gumline. A chamber within dentin contains a soft pulp, rich in blood vessels, nerves, and nerve endings that detect pain, cold, and heat. Blood vessels course through the jawbone and into the pulp, supplying it with nutrients. The pulp transmits sensations such as temperature and pain to the roots of the tooth, embedded in a socket in the jawbone.

Cementum is a thin, bony layer of tissue covering the dentin in the root portion. It serves as an area of attachment for the tooth, which is connected to the jaw by the periodontal ligament, a two-sided membrane that lines the socket. Collagen fibers from the ligament attach to the cementum, on one side, and to the bone, on the other side, both anchoring and cushioning the tooth. The teeth are surrounded by soft smooth tissue called the gums.

Tooth Type

Humans have two sets of teeth. Primary, or baby, teeth are the first to develop. These typically emerge between 6 and 24 months. There are 20 primary teeth (10 upper, 10 lower), which are then naturally shed, starting at about age 6.

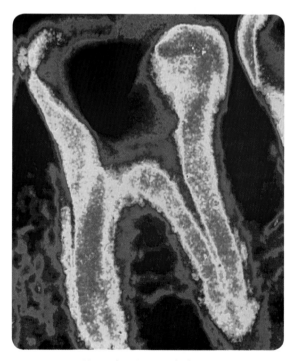

X-ray showing tooth decay

THE TEETH

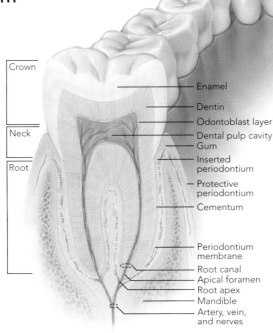

Crown

Neck

Root

- Enamel
- Dentin
- Odontoblast layer
- Dental pulp cavity
- Gum
- Inserted periodontium
- Protective periodontium
- Cementum
- Periodontium membrane
- Root canal
- Apical foramen
- Root apex
- Mandible
- Artery, vein, and nerves

A tooth consists of four layers

Enamel is the white, highly calcified outer layer. It is the hardest substance in the body and highly resistant to acids and other corrosive agents. Enamel is without feeling.

Dentin is a hard, yellow layer of tissue beneath the enamel that forms the bulk of the crown. It is softer than enamel and transmits sensations such as temperature and pain to the root.

Cementum is a thin, bony layer covering the root portion of the tooth. It is connected to the jaw bone by collagen fibers that pass through the periodontal membrane to hold the tooth in place.

Pulp is the soft tissue in the inner cavity of the tooth. It contains the nerve fibers and blood vessels and supplies nutrients to the tooth. Pulp extends into the jaw bone and is highly sensitive to pain and temperature.

Healthy baby teeth play a key role in a child's ability to chew, to develop normal jaw structure and facial characteristics, and to speak clearly.

The adult jaw holds 32 secondary teeth arranged in two arches, 16 in the upper jawbone and 16 in the lower. The incisors (four upper, four lower) are chisel-edged, center-front teeth used to cut food. Canines (cuspids) work with the incisors to rip food, support the lips, and guide jaw alignment. Next to each canine are two premolars (bicuspids) with broad, pointed surfaces and rounded cusps for crushing and grinding. Six pairs of molars (three upper, three lower) follow the premolars. The four rearmost molars, one in each back corner, are the wisdom teeth, the last to grow.

Decay

After a tooth erupts from the gum, the cells that make enamel become susceptible to decay. Tooth decay is caused mainly by plaque, a thin, sticky film of food debris, mucus, and bacteria. Plaque forms an acid that gradually erodes enamel. Once a cavity begins, erosion continues into the dentin and, if left untreated, spreads to the pulp and nerve, causing a toothache or, more seriously, gum disease.

Protective bacteria in the saliva help neutralize the acids. Brushing and flossing the teeth twice a day can also help control acid levels and remove food particles.

DID YOU KNOW?

BY THE NUMBERS

Thirty-two: a full set of adult teeth

+ **Eight** incisors for chopping

+ **Four** canines for tearing

+ **Eight** premolars for crushing

+ **Twelve** molars (including four wisdom teeth) for grinding

THE SPINE

The Twists and Turns of the Flexible Backbone

The spine is a chain of small, but tough, drum-shaped vertebrae, or bones, that extends from the base of the skull down to the hips. Also known as the backbone or vertebral column, it has multiple functions: supporting the body while also protecting the spinal cord, the nervous system's main conduit to the brain.

Along with other bones and muscles, the spine provides strong support for the head and trunk, and, at the same time, allows the back to bend so that the body can perform a wide range of twists and turns. It also provides areas of attachment for the muscles that hold the back upright.

Composition

Each vertebra is linked to the next by symphyses, which are cartilaginous joints. The individual amount of movement between a vertebra and its neighbors is actually very small, but altogether the vertebrae allow a large range of movement. The spine can bend in many directions. In general, the movements that can occur between vertebrae are flexion (bending forward) and extension (bending backward), lateral flexion (bending the upper body to the right or left), and rotation (the vertebrae rotate on one another in the longitudinal axis of the spine). Not all movements are equal: Humans can bend farther forward than

THE BACKBONES

SECTION	NUMBER	LOCATION	DESCRIPTION
Cervical	7 bones	Neck	Smallest, lightest vertebrae
Thoracic	12 bones	Chest (or thorax)	Medium-size vertebrae. Feature facets to connect to the ribs
Lumbar	5 bones	Lower back	Largest, strongest vertebrae. Support bulk of body's weight
Sacrum	1 (formed by 5 fused bones)	Upper pelvis	Fusing begins in the late teens and is complete by age 30
Coccyx	1 (formed by 4 fused bones)	Lower pelvis	Name means "cuckoo" in Latin. Its triangle shape resembles a bird's beak

THE VERTEBRAE

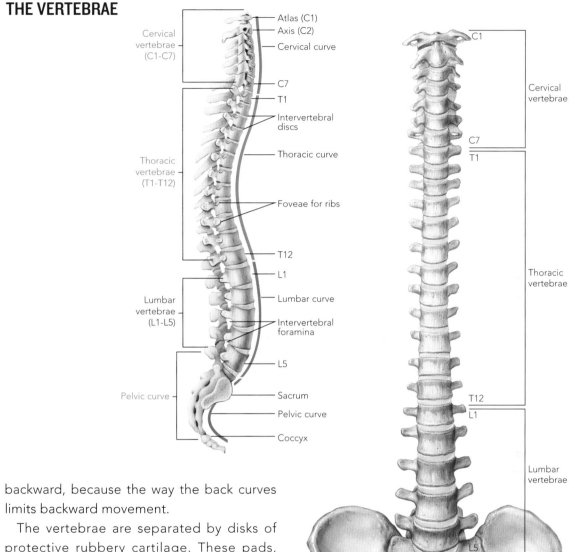

Cervical vertebrae (C1-C7)

Atlas (C1)
Axis (C2)
Cervical curve

C7
T1
Intervertebral discs

Thoracic vertebrae (T1-T12)

Thoracic curve

Foveae for ribs

T12
L1

Lumbar vertebrae (L1-L5)

Lumbar curve

Intervertebral foramina

L5

Pelvic curve

Sacrum
Pelvic curve

Coccyx

C1

Cervical vertebrae

C7
T1

Thoracic vertebrae

T12
L1

Lumbar vertebrae

L5

backward, because the way the back curves limits backward movement.

The vertebrae are separated by disks of protective rubbery cartilage. These pads, sandwiched between the bones, act as shock absorbers for the spine, cushioning the bones to keep them from colliding with and grinding against each other and wearing away. During the active day, these disks are compressed, and they expand again at night when at rest. This means that we're actually slightly taller in the morning than in the evening.

The spine, pelvic girdle, ribs, and shoulder blades form the framework of the skeletal posterior. Viewed from the side, the spine is S-shaped, a design that gives it the resilience to absorb jolts when we move. The spine curves four times from anterior to posterior as it winds down the back, much like a skier slaloming down a slope. The swerves, aided by the cartilage disks between vertebrae, give the backbone spring that it would not have were it a straight rod. A curved spine also makes it easier to balance by positioning the

KEEPING HEALTHY
PREVENTING "TEXT NECK"

Your head weighs about ten pounds. When you lean over to check your text messages or use your tablet computer, it's like trying to keep a bowling ball from rolling down a steep hill. That puts a lot of pressure on the muscles and other structures in the spine and leads to neck and back pain.

Here are some tips to prevent these injuries:

+ **Strengthen back and neck muscles:** Take breaks to do range-of-motion stretches including looking over each shoulder, extending your chin to your chest, or reaching your ear toward your shoulder.

+ **Check your posture:** Make sure you are sitting or standing straight, with your shoulders relaxed and your belly button pulled in. Slouching increases the risk of neck and shoulder injuries.

+ **Build your core:** Strong stomach muscles improve posture and support the spine. Doing yoga, Pilates, and other exercises can help your core.

upper body directly over the legs and better distributing the weight of the skull over the pelvic bones, which is key to walking upright. Without this stabilizing feature, humans would be top-heavy.

Cord Protection

A bony ring attached to the back of each spool-like vertebral body contains ridges called spinal processes. These projections link to form a tube, or canal, that shelters the delicate spinal cord, the route for all nerve signals traveling between the body and the brain. The spinal cord, on average, is 17 inches (43 cm) long and 0.4 inches (1 cm) thick. It stops growing when a child is about five years old.

Damage to the spinal column can lead to damage to the spinal cord. If the spinal cord is injured, the nerves' ability to carry messages from the brain to the muscles can be impaired or cut off completely, depending on the severity of the injury. Spinal injuries below the neck can cause paralysis below the waist, or paraplegia; neck injuries may cause quadriplegia, paralysis from the neck down.

Spinal Sections

The spine is divided into five sections: cervical, thoracic, lumbar, sacrum, and coccyx. The cervical vertebrae are the first set of seven bones that form the neck. They are also known as C1 through C7. The thoracic vertebrae make up the second set of 12 vertebrae. They form the outward curve of the spine and are known as T1 through T12 (from top to bottom). They are bigger than cervical vertebrae but smaller than the lumbar. Lumbar vertebrae compose the third set of vertebrae. There are five, which are known as L1 through L5. The lumbar bones—forming the inward curve of the spine—are

The cervical vertebrae are the smallest bones in the spine and are vulnerable to injury.

the largest and strongest of the vertebrae, as they have to support the most weight. The sacrum is a slightly curved triangular bone near the base of the spine, and is followed by the coccyx.

The sacrum, below the lumbar region, joins the spine to the pelvic girdle. At birth, the sacrum is composed of five separate sacral bones, but these will fuse together in early childhood to form a single bone.

The coccyx (or tailbone) is made up of four small, fused vertebrae and forms the very bottom tip of the spine. The coccyx is popularly called the tailbone, because it is believed to be the vestigial remnant of a tail that disappeared as *Homo sapiens* evolved.

Overall the spine appears flexible, allowing a wide range of movements. However, each separate region of vertebrae is slightly different and specialized according to its location. The cervical vertebrae are the smallest and lightest of the vertebrae and allow the greatest flexibility of the spinal bones, a feature that can be seen in movements of the human neck. The top two cervical bones, the atlas and the axis, allow the head both to nod up and down and to shake side to side. The atlas (C1) allows the head to move forward and back (in Greek mythology, the Titan Atlas held the world on his shoulders); the axis (C2) projects into the hollow atlas, allowing the head to rotate from side to side.

The 12 heart-shaped thoracic vertebrae—spanning the back between the lumbar and cervical regions—increase in size from top to bottom, and each bone is attached to a pair

DID YOU KNOW?

SPINAL BONES

The spine contains 26 bones divided into three vertebrae regions (from the neck down): cervical, thoracic, lumbar, followed by the sacrum and coccyx (or tailbone). The upper vertebrae provide flexibility for the body to move, especially the seven cervical (neck) and five lumbar (lower back) vertebrae. The five vertebrae of the sacrum and the four of the coccyx are fused together.

of ribs. These permit a fair range of motion with some limitations caused by the presence of the ribs.

The lumbar region of the vertebral column, commonly referred to as the small of the back, receives the most stress of all the vertebrae—and the bones there are built to take it: They are very strong, thick, kidney-shaped structures. The lumbar region also permits limited movement. Bending forward and backward is possible, but rotation is not.

Spinal Problems

Some spines have abnormal curvatures that can cause problems. Some are congenital (present at birth); others result from disease, poor posture habits, or unequal muscle pull on the spine.

Scoliosis, which literally means "twisted disease," is an abnormal lateral curvature that occurs most often in the thoracic (midback) region. It is most common during adolescence;

The spine is flexible, allowing a wide range of movement, including this back bend.

for some unknown reason, girls are particularly prone to developing the condition. More severe cases result from abnormal vertebral structure such as uneven lower limbs or muscle paralysis. If muscles on one side of the body are not functioning properly, those on the opposite side tug on the spine and force it out of alignment. Scoliosis is treated—with body braces or surgically—before the growth period ends to prevent permanent deformity and breathing difficulties due to limited rib motion and compressed lungs.

Kyphosis, or hunchback, is an exaggerated backward curvature of the spine in the thoracic region. It is most common in older individuals because of osteoporosis (see "Keeping Healthy," page 75) but may also be caused by tuberculosis, rickets (typically, a childhood disorder resulting from calcium and vitamin D deficiency), or osteomalacia (abnormal softening of bones caused by disease).

Lordosis, or swayback, is an accentuated lumbar (lower back) curvature. It, too, can result from spinal tuberculosis or osteomalacia. Temporary lordosis is common in those carrying a "large load up front" such as pregnant women and men with sizable potbellies, who throw back their shoulders in an attempt to preserve their off-kilter center of gravity.

Rib Cage

Attached to the backbone near the chest is the rib cage. This structure (also called the thoracic cage because of its location in the thoracic, or chest, cavity) consists of the thoracic vertebrae, the ribs, and the sternum (or breastbone). The ribs are the thin, flattish bones that curve around the chest to protect the heart, lungs, and other vital organs. The spongy lungs fill up most of the thorax, or chest. Part of air

In scoliosis, the spine curves sideways.

exchange depends upon the rib cage's ability to enlarge and reduce the size of the thoracic cavity as the lungs expand and contract.

There are a dozen pairs of ribs. Seven pairs are called true ribs: Each rib is attached to the sternum by cartilage in front and curves around to join one of the vertebrae that make up the backbone. There are three pairs of false ribs, which are not directly attached to the breastbone. In the case of ribs eight through ten, each rib is attached to the rib above it by cords of costal (rib) cartilage, which connects to the cartilage of the seventh rib. This forms a lower edge called the costal margin.

Ribs 11 and 12 are called floating ribs. These are connected only to the vertebrae of the backbone. The gaps between the ribs are called intercostal spaces, and they contain thick sheets of muscle that expand and relax the chest during breathing. The floating ribs protect organs including the liver, kidneys, stomach, and spleen. In addition to their other functions, the bones of the rib cage contain red marrow and are one of the body's major blood-cell suppliers.

HANDS & FEET

Intricate and Practical

From the cradle on, the hands and feet shape how a person interacts with the world. With the hands, we can manipulate objects—touching, feeling, grabbing the things around us. The feet bear the body's weight, taking it where it wants to go. Intricate connections between the bones and the connective tissues make these appendages highly effective tools, specialized in helping the body navigate its environment. Packed with sensitive nerve endings, they are also the place where the body takes in sensory information and relays it to the brain.

The Hand

The hands are among the busiest, most versatile anatomic avenues in the body, together containing more than a quarter of the body's bones and a vast network of muscles. The mix of 27 bones, 37 muscles, and a highly mobile thumb in each hand allows a wide range of motion. The bones of the hand are divided into three groups: carpals, metacarpals, and phalanges. The carpals are the group closest to the arm; next, the metacarpals lie under the palm, and then the phalanges make up the fingers and thumb.

Carpals are the eight pebble-like short bones located at the end of each arm formed by the tips of the radius and ulna. The

BONES OF THE HAND

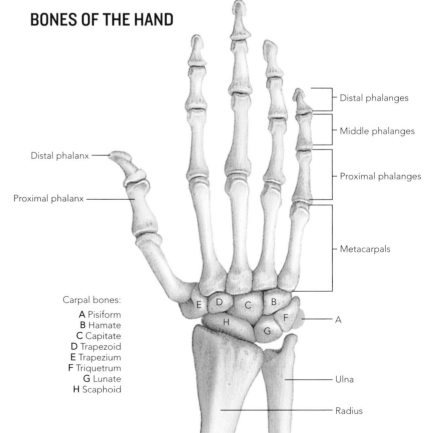

Distal phalanx

Proximal phalanx

Distal phalanges

Middle phalanges

Proximal phalanges

Metacarpals

Carpal bones:
  A Pisiform
  B Hamate
  C Capitate
  D Trapezoid
  E Trapezium
  F Triquetrum
  G Lunate
  H Scaphoid

A

Ulna

Radius

carpals make up the wrist, or carpus, and are arranged in two rows of four. Closest to the arm, proceeding from ulna to radius, are the pisiform, triquetrum, lunate, and scaphoid. The front-row carpals, which connect with the five metacarpals that form the body of the hand, are the hamate, capitate, trapezoid, and trapezium. Only the scaphoid and lunate actually articulate with the radius to form the wrist joint.

The carpals are tightly bound by ligaments. So-called gliding joints between them, however, enable the wrist bones to slide smoothly, making the wrist quite flexible, although movement at each carpal is limited.

Palms & Fingers

The five metacarpals extend from the wrist like spokes on a wheel to form the broad metacarpus, or palm, of the hand. These miniature long bones are identified not by names but by numbers, 1 to 5, counting from the thumb to the little finger. The metacarpals attach to the carpals and the phalanges, the bones of the fingers and thumb. (The bulbous heads of the metacarpals become knuckles when the hands are clenched.) Metacarpal 1, associated with the thumb, is the shortest and most mobile of the palm bones. There's a special link between it and the trapezium carpal called a double saddle joint, which allows the thumb to flex enough to touch the tips of the other digits, giving the hand its ability to grasp and manipulate even the tiniest of objects.

Fingers, or digits, of the hand, are also numbered 1 to 5, beginning with the thumb, or pollex. The 14 phalanges are miniature long bones similar to the metacarpals. Four of the fingers feature three bones: (from the top down) distal, middle, and proximal; the thumb lacks the middle. Each hand also has an assortment of sesamoid bones, small ossified nodes embedded in the tendons to provide extra leverage and reduce pressure on underlying tissue. Many exist around the palm at the bases of the digits, but the exact number varies among people.

Movement

The hands' wide range of fine movements is made possible by numerous small joints, long and short muscles, and extensor and flexor tendons that allow the fingers and thumb to straighten, bend, and flex. Muscles pull with precision on multiple movable joints—14 of them, the knuckles between the finger bones and metacarpals. The knuckles at the base of each digit (between the phalanges and metacarpals) are condyloid joints, like the wrist. These joints provide greater mobility than the other finger knuckles, which are simple hinge joints that allow the fingers to bend toward the palm but not toward each other.

DID YOU KNOW?

HAND FACTS

+ In all, each hand has 27 bones.

+ There are three kinds of bone in the hands: carpals, metacarpals, and phalanges.

+ Carpals: These eight marble-size short bones are located in the wrist.

+ Metacarpals: There are five of these bones supporting the palm.

+ The fingers and thumb together contain 14 bones, called phalanges.

+ Each finger has three phalanges, except for the thumb, which has only two.

+ On average, the longest digit on the hand is the third one.

1 Distal phalanges
2 Middle phalanges
3 Proximal phalanges
4 Metatarsal bones I to V
5 Sesamoid bones*
6 Lateral cuneiform b.
7 Intermediate cuneiform b.
8 Medial cuneiform b.
9 Cuboid b.
10 Navicular b.
11 Talus
12 Calcaneus
13 Articular capsules
14 Deep transverse metatarsal II.
15 Plantar metatarsal II.
16 Fibularis longus t.
17 Plantar calcaneocuboid (short plantar) I.
18 Long plantar I.
19 Plantar tarsometatarsal II.
20 Plantar calcaneonavicular (spring) I.
21 Tibialis posterior t.

*A **sesamoid bone** is a bone that forms in a tendon over a joint. The patellae (kneecaps) are large sesamoid bones that all people have. Most people have extra, much smaller sesamoid bones near other joints, often in the hands and feet.

BONES AND LIGAMENTS OF THE LEFT FOOT

In a firm grip, most of the power comes from muscles in the forearm that are linked to bones by very long tendons, which can be seen in action on the back of the hand as it opens and closes. The fingers have two long flexors, fastened to the underside of the forearm. The thumb has one long flexor and a short flexor; it also has other muscles moving the thumb in opposition, making grasping possible. A fibrous band, called the retinaculum, encircles the wrist and keeps the tendons in place.

Hurt Hands

Many bones in the hands, including the wrist bones and the radius and ulna (where they connect to the wrist), are vulnerable to fractures. The two most common breaks here are Colles and scaphoid fractures. A Colles fracture is a complete crack in the end of the radius, which often occurs when the hand is flexed to brace the body during a fall. Falls can also cause a break in the bones of the palm of the hand, where the scaphoid carpal connects with the radius. This injury may initially be confused with a bad sprain and tends to heal slowly—and sometimes incompletely—due to limited blood supply.

Broken fingers are also common, resulting from accidents (such as slamming one in a car door) to sports injuries. Luckily, taping the broken finger to a healthy neighboring finger or splinting it is usually all that's necessary for a fracture of the finger bones to heal. Surgery may be required in cases of serious damage.

The Feet

The foot is an engineering wonder that supports body weight and acts as a lever to propel the body forward. It is both strong enough to push off with each stride, supple enough to absorb the impact of every step, and facile enough to negotiate uneven surfaces. The foot is designed in much the same way as the hand. There are three regions: the tarsals (closest to the leg), metatarsals (underneath the sole), and phalanges (the toes).

The ankle is formed where the talus articulates with the tibia and fibula (the lower leg bones). This hinge joint allows the foot to flex and to rotate when tugged on by muscles. The ankle, or tarsus—akin to the wrist (carpus)—consists of seven tarsal bones. Only one, the talus, actually helps form the hinge-type joint. The large, topmost tarsal, it transfers weight and pressure from the leg to the foot and plays a key role in all movements involving the feet and ankles. It sits above another tarsal, the calcaneus (or heel bone)—the largest and strongest bone in the foot—which also supports body weight, and it is the site of attachment for the Achilles tendon. The Achilles connects calf muscles to the heel bone. (Calf muscles pull up on the calcaneus to bend the foot downward when moving or standing on tiptoes.)

Together, the talus and calcaneus compose the hindfoot. The midfoot contains five interlocking smaller tarsal bones. The midfoot is connected to the forefoot and hindfoot by muscles and the plantar fascia ligament, which stabilizes the foot and helps maintain its arch. The forefoot contains the metatarsal bones and phalanges, which are connected at the ball of the foot by five metatarsophalangeal joints.

Moving down the foot, there are strong bands of ligaments and tendons that pull the metatarsal and tarsal bones into three arches. These form a dome that distributes weight to different parts of the feet. About half of standing and walking weight goes to the heel bones and half to the heads of the metatarsals in the ball of the foot. The instep, or metatarsus, consists of five metatarsal bones. The metatarsals are linked to the 14 phalanges, or toe bones.

The phalanges of the toes are similar to those of the fingers but are much smaller and, thus, less nimble. The big toe, or hallux—like the thumb—consists of only two phalanges, while the other four digits each boast three bones: the proximal, middle, and distal. (The big toe lacks the middle one.)

Flat Feet

In flat feet, or fallen arches, there is little or no arch between the heel and toes, resulting in a sole that rests almost flat on the ground. Nearly everyone has flat feet at birth; the arches do not develop fully until the ligaments and muscles in the soles are completely formed. Some people never develop arches; others get flatter feet over time as the ligaments loosen or stretch, often due to weight gain. A person with flat feet is more prone to damage the bones of the feet, because of increased pressure on them.

Flat feet also can contribute to knee, hip, and lower-back pain, as well as a painful condition called plantar fasciitis. This inflammation of the plantar fascia is most often experienced as heel pain and stiffness when standing or walking. Plantar fasciitis can also occur when the ligament is strained from overuse or from ill-fitting shoes.

DID YOU KNOW?

FOOT FACTS

+ There are 26 bones, 100 ligaments, and 35 muscles in each foot.

+ Among the bones are 14 phalanges (toe bones), 5 metatarsals (sole bones), and 7 tarsals (ankle bones).

+ The heel bone (or calcaneus) is the largest bone in the foot.

+ The knobby protrusions on each side of the ankle are actually the ends of leg bones. Each is called a malleolus: The inner one is part of the tibia; the outer one belongs to the fibula.

Chapter 4

BODY MOVERS

THE MUSCLES

Giving the Body Strength

The body employs muscles in almost every possible activity: sitting and standing, walking and running, smiling and frowning—and on and on. Muscles create the shape of every human body. They turn intentions—and nerve impulses—into actions. They form a strong, protective wrapping for the underlying skeletal and organ systems. Muscles hold the body's joints firmly in place so they can operate.

Muscles power the body's essential internal operating systems as well. Without muscles, the body could not breathe air or digest food. Without muscles, blood could not circulate from the heart in the core of the body, out through the limbs to the extremities, and back again. Every movement, every forceful exertion, whether voluntary or not, occurs in a body through the work of its muscles. As a result of this energetic motion of muscles, the body also stays warm.

Every muscle, no matter where it is in the body, operates through the basic principles of contraction and relaxation. Sometimes we choose to make muscles move, and this involves a deliberate combination of contracting and relaxing them. The simple action of reaching out, grasping a glass of water, and lifting it to the lips, for example, requires the voluntary and coordinated contraction and relaxation of dozens of muscles.

Night and day, whether we are asleep or awake, muscles in our bodies are contracting and relaxing on their own as well, without any awareness or deliberate thought process on our part. These involuntary muscles make the respiratory, digestive, and reproductive systems function.

Muscle Types

The human body contains three kinds of muscles: skeletal, smooth, and cardiac. Each muscle type performs a different job. Skeletal muscles (also known as striated muscles, because of their striped appearance) make up the bulk of the body. They move voluntarily—a person can control their operation—and come in various sizes and shapes to perform many kinds of actions. Smooth muscles occur in organs and blood vessels. They contract and relax involuntarily, and they motivate the essential life systems. Cardiac muscle—the muscle in the heart—is also involuntary muscle. It works differently from either skeletal or smooth muscle. From before birth until the moment of death, cardiac muscle cells contract and relax with a rhythm embedded in their cells, creating the pulsing force that circulates blood through the heart.

Of all the functions performed by muscles in the body, physical movement is the easiest to observe and understand. Every movement, from a tiny blink of the eyelid to a sprinter's initial burst of speed, happens because of the coordinated contraction and relaxation of skeletal muscle cells operating together. Lack of movement involves many skeletal muscles as well. Standing still, sitting up straight, holding a pose—all these activities require the voluntary exertion of

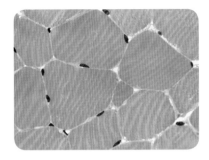

SKELETAL MUSCLE
Contractile proteins form an alternating pattern of light and dark bands called striations.

CARDIAC MUSCLE
Rich with capillaries, connective tissue (white) supplies blood to the elongated muscle cells (purple).

SMOOTH MUSCLE
These spindle-shaped cells, each containing a dark nucleus, are grouped in irregular bundles.

many body muscles. These muscular exertions also generate heat, which the body needs for growth, health, and comfort.

Harder to perceive but essential to keeping a body alive and well, smooth muscles are performing tasks all the time inside the body. Smooth muscles at work in the esophagus, stomach, and intestines move food through the digestive system. Smooth muscles in the bronchi, the air tubes in the lungs, convey oxygen and other gases through the respiratory system. Smooth muscles in the lining of the blood vessels (like the veins and arteries) help maintain blood pressure in the circulatory system.

Structure
Muscle tissue, whether skeletal, smooth, or cardiac, is made up of many thousands of long, thin cells, also called fibers, working together. When a muscle contracts, some components within each fiber slide in on top of others, shortening and tightening the fiber. When a muscle relaxes, those components release and separate, lengthening and loosening the fiber. The elasticity of a muscle fiber—the ease with which it is able to contract and relax—changes

according to muscle groups. It also varies among individuals and with age. Regular exercise can increase muscle elasticity.

The nervous system makes skeletal muscles contract and extend. Electrical signals travel from the brain, through the spinal cord, and out to the nerves branching through the body. The signals ultimately arrive at individual muscle cells. When a neurotransmitter pings a muscle cell, that cell responds accordingly. Signals sent to a group of muscle cells elicit a coordinated set of interactions, and the muscles contract or relax, resulting in movement.

Simultaneously, neurotransmitters ping back, telling the brain what has occurred. Deliberate movements of the body involve thousands of such cross-communications between the nervous system and skeletal muscles. It is a complex cycle of feedback and adjustment that depends on nearly instantaneous electrochemical impulses between the nervous system and the muscle tissues.

All the while, the autonomic, or involuntary, nervous system communicates in the same way, electrochemically, with the body's smooth, or involuntary, muscle cells. Without a person's knowing it or deciding to make it happen, the brain sends and receives controlling signals tending to the operation of the blood vessels, lungs, stomach, intestines, bladder, uterus, and other internal organs.

Smooth Muscle

Smooth muscle gets its name from the sleek surface of its muscle fibers, compared with the striped appearance of skeletal muscle fibers. Smooth muscle fibers are also usually organized in thin sheets, unlike skeletal and cardiac muscle cells, which are arranged in bundles. Smooth muscle is at work, for example, in the walls of

The stomach's three muscular layers are composed of smooth muscle cells.

blood vessels, in the walls of hollow organs such as the stomach, intestines, and bladder, in the airways leading into the lungs, and inside the iris of the eye. All these body parts operate by means of the forces created as smooth muscle cells involuntarily contract and relax.

Groups of muscle cells often work together. In many of the body's vessels and organs, for instance, their orientation and coordinated operation make material pass through efficiently. The esophagus—the tube that sends food down into the stomach after a person swallows—is a good example because it is composed of layers of smooth muscle. One layer has muscle cells oriented lengthwise, and the next has muscle cells perpendicular to that, oriented like rings around the inside of the tube. The interaction of these longitudinal and circular muscle groups is essential to the way that food travels down the esophagus and toward the stomach to be digested.

After a person chews and swallows, an involuntary digestive action begins in the esophagus, performed by groups of smooth muscle cells. Operating in slow waves, they contract and relax in rhythmic coordination. Cells in front of the wad of food relax, opening a pathway, while cells behind the food contract, pushing it farther down the tract. The mechanical force of these muscle groups is so strong that food would still continue toward the stomach even if a person were standing on his head.

This process is called peristalsis. It occurs not only during the digestive process in the esophagus and intestines, but in other body processes involving the ureters, the bile ducts, and the fallopian tubes.

Cardiac Muscle

Cardiac muscle—the muscle that makes up the heart—is composed of a type of muscle cell different from those found in smooth and skeletal muscle. Like smooth muscle, cardiac muscle operates involuntarily. Like skeletal muscle, however, it is striated, with dark bands of color visible in the muscle cells when viewed under the microscope. Although it consists primarily of cardiac muscle, the heart is not part of the musculoskeletal system. Along with the blood vessels, it forms the circulatory, or cardiovascular, system.

Cardiac muscle cells are short and squat when compared with other muscle cells. They connect to one another with a special bonding design that makes the heart a dense, strong organ and allows tiny electrochemical impulses to pass rapidly and easily from cell to cell.

The rhythmic throb of the heartbeat happens when, in response to those impulses, all the cardiac muscle cells work together, contracting and relaxing in harmony. Muscle relaxation allows the blood into the heart organ. Muscle contraction forces the blood out into the rest of the body.

Some special cardiac muscle cells serve as the heart's internal pacemakers. These cells keep up a constant rhythm of contraction and relaxation, and they generate the electrochemical impulses that, almost instantaneously, transfer that rhythm to the other cardiac cells, coordinating and regulating their never-ceasing contractions.

The influence of that rhythm is felt through the entire body, moving blood out to the extremities and then returning it to the heart. You are feeling that rhythm when you take your pulse at your neck, wrist, or ankle. Although the brain and nervous system signal the pacemaker cells to speed up or slow down as the body demands, the basic motor movement of the heartbeat originates in the cardiac muscle itself.

THREE TYPES OF MUSCLE

	SKELETAL	SMOOTH	CARDIAC
LOCATION	Throughout body, primarily attached to bones by tendons	Walls of hollow organs and tubes such as blood vessels	Heart only
CONTROL	Voluntary	Involuntary	Involuntary
CELL SHAPE	Long cylindrical fiber, striated	Long, thin fiber, tapered at ends, smooth	Branched cylindrical fiber, striated

SKELETAL MUSCLE

Strong and Flexible

When thinking of muscles, skeletal muscles are what usually come to mind. Striated tissue forms distinct masses, each one a muscle. They can be big or little, thin or thick, round or flat—all shaped to perform the many diverse tasks a person undertakes in the course of a day. As the name suggests, skeletal muscles usually attach to the bones of the skeleton, where they facilitate movement. This type of muscle works voluntarily, and its actions—coordinated sequences of contracting and relaxing—move the bones at their joints.

The body's muscular system interacts intimately with the nervous and circulatory systems, as can be seen at the cellular level in skeletal muscle fiber. Skeletal muscle cells are long and cylindrical. Generally, one nerve, one artery, and at least one vein connect to each muscle cell. A network of capillaries weaves along the skeletal muscle cells. Blood flow to muscle cells furnishes oxygen and nutrients; blood flow from muscle cells removes waste products created during muscle activity.

Nerve fiber runs up and down the length of skeletal muscles. The two systems meet at neuromuscular junctions. A neuromuscular junction is a receptive spot where the axon, or nerve ending, connects into the muscle cell. There, whenever the nervous system sends a signal for a muscle to move, the nerve delivers minute amounts of biochemical messengers into the muscle to incite it to action.

Ties That Bind

A muscle generally is attached at one end to a fixed bone, then crosses over a joint, and is attached at the other end to the bone it moves, with the joint connecting the two bones. A muscle attaches to bone by means of a tendon, a tough, ropelike connective tissue rich in the protein collagen. (Sometimes, if the muscle is wide and thin, the connector is an aponeurosis,

DID YOU KNOW?

WALKING

How many leg muscles does it take to walk?
The answer is 12. Motor neurons stimulate the quadriceps femoris, four thigh muscles; the hamstrings, three muscles at the back of the leg; the buttocks, three muscles; and two calf muscles. (Walking also involves the muscles in your feet, the muscles to keep you upright, and those to swing your arms.)

SKELETAL MUSCLE

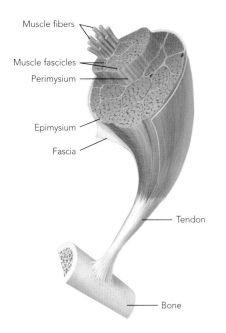

Muscle fibers
Muscle fascicles
Perimysium
Epimysium
Fascia
Tendon
Bone

or flat tendon, which is sheetlike. Muscles can also attach directly to bone with their own collagen fibers.) While muscles are well supplied with blood, tendons receive comparatively little, taking their gray or white color from collagen, and contain little or no nerves.

When the muscle contracts, the tendon transfers the force to the bone, causing it to move around the joint. Whereas muscles are elastic, tendons are fairly stiff; if it were otherwise the muscle's force would be dissipated in the tendon's own movement. Collagen fibers originating in the muscle continue into the tendon. At the bone, the tendon grows into the bone tissue, completing the connection.

Damage Control

Human body tissue has the inborn capacity to regenerate when it has been wounded, but there is a limit to how well muscle mass can heal. Smooth muscle and tendons replenish themselves better than skeletal muscle. Cardiac muscle cells rarely regenerate, so physical damage to the heart usually results in scar tissue, not new heart muscle. Skeletal muscle tissue also has severe limits to how well it can build new cells.

More often, the healing of damaged muscles involves fibrosis, or the development of fibrous scar tissue. An area of tender new tissue develops at the site of the wound. New capillaries bring blood to the site. A matrix of collagen fibers and other materials knits together, filling in the gap made at the wound.

This matrix, the developing scar tissue, does not operate like the lost muscle it is replacing. Not fully infused with blood, it tends to be lighter in color than the surrounding area. Scar tissue, however, is tough and resilient, being composed almost entirely of collagen. It serves to knit together the edges of the original wound and fill them in with sturdy tissue.

Making a Move

It is the business of normal, healthy skeletal muscles to contract and relax in order to move all the parts of the body. These movements come as responses to signals sent to the cells by the nervous system.

Here, at the microscopic level, is how contraction works in a group of skeletal muscle cells. The decision to move sends a nerve impulse through the motor neuron and into the muscle cell at the neuromuscular junction. The impulse travels through the membranes wrapping each muscle fiber and into the organelles at the core of that fiber, called

These athletes are using primarily aerobic respiration to replenish the ATP needed to complete a vigorous spinning class.

myofibrils. Inside every myofibril are two types of filaments, thick and thin. On the signal of the nerve impulse, the thin filaments slide over the thick filaments. The net effect is that the muscle cell shortens and bulks up. When this event occurs simultaneously in millions of muscle cells, a muscle group contracts.

At the same time, nerve signals tell a related muscle group to pull the thin filaments back away from the thick filaments—and thus relax. For every muscle group that contracts, an opposite muscle group relaxes. When a weight lifter flexes his biceps, for example,

he is at the same time relaxing his triceps, the muscle group on the underside of his lifted arm. Muscles are only working when they contract; relaxation is merely the effect of the opposite muscle contracting. Thus, muscles only pull, never push.

Fueling Up

Like other living cells, skeletal muscle cells require adenosine triphosphate (ATP) to transform potential chemical energy into actual mechanical energy, the essential process of muscle metabolism. Muscle cells store only small amounts of ATP, so cells are constantly replenishing that supply, either by aerobic ("with oxygen") respiration or anaerobic ("without oxygen") respiration.

Through aerobic respiration, cells use oxygen to break down glucose molecules, resulting in energy and new ATP molecules. In anaerobic respiration, cells perform those same tasks but much less efficiently. Blood brings the necessary glucose, a sugar, to the muscle cells. When demand is high, cells may also use glycogen, a sugar stored in muscle itself.

These two metabolic processes, aerobic and anaerobic respiration, occur in living cells throughout all body systems. Aerobic respiration is the usual method of ATP replenishment in muscle cells. It takes place when a person's body is at rest or undergoing light exertion.

When a person begins to exercise more strenuously, anaerobic respiration kicks in, too. Even during moderate exercise, though, aerobic respiration provides about 95 percent of the ATP needed by muscle cells. At the highest levels of exertion, muscle cells use creatine phosphate, a molecule found only in muscle, to generate even more ATP and keep working.

With all these methods of producing energy, muscle cells actually create more energy than they need. About 40 percent of the energy is used for the work they are called on to do, and the rest is expended as heat. That is the reason why a person's body temperature rises when she exercises vigorously. Excess energy generated in the muscle cells is dumped into the bloodstream, which brings that heat to the skin's surface as the body attempts to return to its normal temperature.

Sometimes physical effort overtaxes the muscle cells. The cells need more sugar than either the blood or their own glycogen stores can supply, and so they cannot keep up ATP production to meet the continued demand. When this happens, the body undergoes muscle fatigue. Nerve stimuli still reach the cells, but the muscle fibers are unable to contract in response. The muscle cannot, and does not, do what the brain is asking it to do.

When pushed to an even greater extreme, so that they run out of ATP altogether, muscle cells undergo contracture—a state of continuous contraction. That's what happens when a person's hand seizes up in the phenomenon known as writer's cramp. The signs of rigor mortis—the stiffening of muscles after death—arise from the same kind of action.

Which Twitch?

Just about every skeletal muscle in the body is made up of two sorts of muscle fibers: fast-twitch and slow-twitch. The type of fiber determines what sort of work the muscle can do.

Fast-twitch fibers contract rapidly but tire quickly. For energy, they use ATP and glucose but need little oxygen, protein, or fat. The fast-twitch muscles come into play for sudden, rapid movements, like jumping, sprinting, or blinking an eye. It's also the fast-twitch muscles that are exerted in anaerobic exercise such as weight lifting or isometrics.

DID YOU KNOW?

RIGOR MORTIS

Three to four hours after death, the body lapses into rigor mortis (Latin for "stiffness of death"). The muscle cells no longer receive chemicals they need to keep up the cycle of contraction and relaxation. Without new ATP, they get stuck in a contracted position. Rigor mortis usually lasts 36 to 60 hours, then reverses as the effect subsides.

When running, changing speeds and varying the pace can work both slow- and fast-twitch muscle fibers.

Slow-twitch fibers contract slowly, and they can keep at it for a long time. They need plenty of oxygen, and they use up protein and fat, turning their chemical energy into mechanical energy and heat. In an endurance exercise, such as running, rowing, or swimming, the body calls on the slow-twitch muscle fibers to do most of the work.

A vivid example of the two sorts of muscle fibers can be seen in the musculature of a chicken. What we call white meat is primarily fast-twitch muscle fiber, found in the breasts and wings, which exert quick bursts of movement. What we call dark meat is primarily slow-twitch muscle fiber. Found in the legs, which perform slow, steady exertion, these muscles look darker because they are infused with more blood.

Everyone has a combination of slow-twitch and fast-twitch muscle fibers, but some bodies are genetically inclined toward one or the other type. When coaches advise athletes to cross-train, they are in essence encouraging them to exercise the type of muscle fiber that is less predominant in their body. A runner might be advised to lift weights, or a weight lifter might be advised to use a rowing machine, to build up an optimal combination of strong fast-twitch and slow-twitch muscle fibers.

Toning up

Exercise does not increase the number of muscles a person has, but it does increase the size and capability of the muscles they do have. The bigger circumference a muscle has, the more tension it can develop, the more work it can perform, and the longer it can sustain that work at a certain performance level. Bodybuilders are walking examples of the physical phenomenon called hypertrophy: muscle mass increased in size through concerted exercise. The opposite, atrophy, can occur as well. Muscles that are not used will shrink in size over time.

Exercise also increases the tone of muscles. We are all aware of the concept of muscle tone, which plays a large part in the modern emphasis on fitness and appearance. A body with good muscle tone holds firm, even when at ease. But how does muscle tone look at the cellular level?

DID YOU KNOW?

CRAMPS

When a single muscle within a group contracts suddenly and involuntarily, it's called a spasm. Tics of the eye, for example, are spasms. When a muscle spasm causes pain, it's called a cramp. Cramps can occur after exercise or while a person is sleeping. Their cause is not clearly known. The best response is to rehydrate, relax, and massage the tightened muscles.

Even when a muscle is relaxed, swift, tiny, involuntary contractions take place all the time. Called tonic contraction, this is a state of being we often take for granted. The neck holds the head up straight, for example, thanks to the tone of its muscles. Good posture, standing straight and tall, is accomplished thanks to muscle tone. Smooth muscles as well as skeletal muscles experience variations in tone. When digestive organs and blood vessel walls have good muscle tone, they are able to work more efficiently.

Regular exercise also increases the ability of cells to receive the oxygen that they need to accomplish work. During aerobic exercise, the constant call for oxygen by muscle cells can exceed the amount available. When a runner gets winded and gasps for air, those gasps are the body's response to its oxygen debt and its effort to bring in more oxygen to send to depleted muscle cells.

Most people who are beginning a new exercise routine have at one time or another experienced sore muscles the next day, a sure sign that the exercise-engaged muscles had not been used so much before. For a long time, muscle soreness after exercise was blamed on a buildup of lactic acid, a chemical that is released as a by-product of fast-twitch muscle activity. Excess lactic acid dissipates in less than 24 hours, however, so it cannot explain muscle soreness that continues for a day or two. Researchers now believe that delayed-onset muscle soreness, or DOMS, is caused by small-scale tearing or swelling that happens to muscles that have been over-stressed. Gentle warm-up stretches prepare muscle fibers for the work ahead and can help avoid later soreness.

KEEPING HEALTHY
GET MOVING

Utilizing a variety of exercises can improve the muscles' performance. Aerobic, or endurance, exercise—slow, steady exertion such as walking, running, swimming, or dancing—boosts the capillaries bringing oxygen to muscle cells and the organelles that turn sugars into ATP and energy. The result is that muscles can work hard longer without fatigue. Muscle tone improves, too. Anaerobic, or resistance, exercise—such as weight lifting—bulks up muscle fiber and increases the connective tissue between muscle cells. It increases metabolic organelles, creates more myofibrils, and boosts the capacity of muscle fibers to store sugar. The result is bigger, more sharply defined muscles and increased strength. Most trainers suggest combining both types of exercise in a fitness plan.

MOVEMENT

Cooperating to Move the Body

Muscle cells are organized into bundles, or fascicles, which are organized into muscle groups. All work together to generate the complex combinations of movements that our bodies can perform. Many movements require a coordinated pairing of two nearby muscle groups, with one taking the lead and the other countering with an opposing stretch or tension.

Partners

Hold out your right arm and bend up your forearm like a circus strong man, showing off his muscles. On top of your upper arm, the biceps bulges as its muscle fibers contract and shorten. Underneath your upper arm, simultaneously, the triceps muscle lengthens. In this example, the biceps muscle is called the agonist ("leader") or prime mover, while the triceps is the antagonist.

Now start again with the flexed, strong-man pose, but now straighten out your arm at the elbow. In this movement, the two muscle groups reverse roles. The triceps, acting as agonist, contracts and shortens; the biceps, the antagonist, relaxes and lengthens. Nearly every skeletal muscle movement involves a pairing of muscles, agonist and antagonist.

There are other ways that muscles work together in groups to create the particular movements that the nervous system requests

DID YOU KNOW?

GENDER

The average male body is 40 to 50 percent skeletal muscle mass, much of it concentrated in the upper body. The female body is 30 to 40 percent skeletal muscle mass, concentrated more in the hips and legs. Males lose more muscle as they age, but females have more health problems from sarcopenia (age-related muscle loss), in part because they live longer.

THE MUSCULAR SYSTEM

Muscle is abbreviated m. Tendon is abbreviated t. Tendons is abbreviated tt.

DID YOU KNOW?

GREATEST STRENGTH

What is the body's strongest muscle? It depends on the definition of "strong." The gluteus maximus and quadriceps femoris—the buttocks and thigh muscles—exert the most force. The masseter, or jaw muscle, exerts the most force on an external object. During childbirth, the uterus exerts the most force pound for pound. For slow and steady effort, however, nothing tops the heart: Beating constantly for decades, it trounces all other muscles for staying power.

of them. When a muscle that crosses to several joints is called on to move only one of those joints, the muscles that hold the other joints immobile are called synergists, from the Greek for "working together." For example, a single flexor muscle group moves the wrists and the fingers. To flex only one finger, and not the wrist, synergist muscles contract and keep the wrist joint still.

Other muscles, called fixators, contract tightly and provide a foundation against which neighboring muscles can tense and stretch. Muscles attached to the scapula, or shoulder blade, for example, hold it in close to the chest wall, creating a stable foundation against which other muscles work when they lift the arm at the shoulder joint. Good posture depends on the operation of multiple fixator muscles.

What's in a Name?

Muscle names may seem like a foreign language, but they have logical meanings embedded in them. Once those meanings are understood, the names convey important information about where individual muscles are located in the body, what they look like, and how they do their work.

To identify the body's 650 muscles, anatomists established a naming system using root words that refer to the size, shape, location, action, or orientation of a given muscle. Terms such as *maximus* and *minimus* (largest and smallest), *major* and *minor* (larger and smaller), and *brevis* and *longus* (short and long) clearly refer to muscle size. *Deltoid* means triangular; *rhomboid* means diamond-shaped; *quadratus* means square or four-sided—these are just a few examples of names describing muscle shape.

Some names associate muscles with a particular part of the body. *Oris* (mouth), *brachii* (arm), *oculi* (eye), *femoris* (thigh), *pollicis* (thumb), and *spinalis* (spine) are examples of such names. Other names refer to muscle action. A *levator* muscle lifts, so the "levator scapulae" is the muscle that lifts the scapula. A *flexor* muscle creates flexion (from the Latin meaning "to bend"), decreasing a joint angle and bringing an appendage inward. So the "flexor digitorum profundus" in the forearm is the deep (*profundus*) muscle that flexes fingers. (*Digitorum* can also refer to the body's other digits, the toes.) An *extensor* is a muscle that brings about extension (from the Latin meaning "to stretch out"), increasing the angle at a joint and moving the appendage outward.

The biceps and triceps are each just one muscle, though their names sound as if they refer to more than one. *Biceps* means "two heads," *triceps* means "three heads." Most skeletal muscles have one head, or origin—the part of the body where the muscle attaches, usually a bone, and that does not move when the muscle contracts. The other end of the muscle attaches to the insertion, the part of the body that moves when the muscle contracts, usually another bone. Some muscles, including the biceps and triceps, have multiple origins, or heads.

Lever Systems

Muscles and bones operate together according to principles that can be described in terms of a simple machine: the lever. A lever is a rigid bar—like a bone. It moves around a fixed point, or fulcrum. As force is applied, the lever acts at the fulcrum and moves a load, whether the load is something the body is carrying or just the body's own weight.

Levers are classified according to their position relative to the fulcrum and the load. The same classification system applies to muscles and bones. In a first-class lever, the fulcrum is in the middle, with the load at one end and force exerted at the other. A pair of scissors is a good example of a first-class lever. The hand of the person using the scissors exerts force against a fulcrum, which is the joint where the two scissor blades meet. The force from the

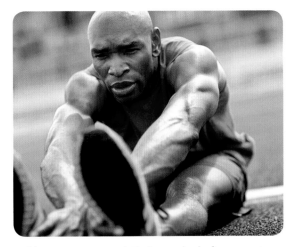

Many runners stretch their muscles before a run to prevent stiffness.

hands translates into a cutting motion against the load, in this case, the resistance of the object that is being cut.

In the human body, the neck is a first-class lever. When posterior neck muscles exert force against the atlanto-occipital joint, where the skull meets the top of the spine, the load—the head—lifts up from the chest.

In a second-class lever, the load to be lifted is in the middle between the fulcrum at one end and the point of force at the other end. A wheelbarrow is an example: The front wheel is the fulcrum, and the workman lifts the load in it by applying force on the handles.

In the human body, rising up on the balls of the feet represents an action requiring a second-class lever. The toe joints work as the fulcrum; the calf muscles exert the upward force, pulling up the heel; the load, the weight of the body, is lifted.

In a third-class lever, the fulcrum and the load are at opposite ends of the lever, with force applied in the middle. Picking up something with a pair of tweezers is an example of using a third-class lever. The tweezers' hinge is the

THE ART OF THE MUSCLE

It was the artists of the Italian Renaissance, even more than the physicians of their day, who created an entirely new vision of the human body. They did so through the art of anatomical illustration.

THE FIRST MASTERS

Fifteenth-century Florentine artist Antonio Pollaiuolo (ca 1431–1498) is credited with having been the first artist of his generation to study the anatomy of the human body in the flesh, by actually dissecting cadavers and drawing the configurations of muscles and bones as he saw them revealed beneath his hands. Pollaiuolo was, wrote 16th-century artist and art historian Giorgio Vasari, the "first master to skin many human bodies in order to investigate the muscles and understand the nude in a more modern way."

By the next generation, Italian artists had come to consider the study of real human bodies essential to their work. Leonardo da Vinci (1452–1519) and Michelangelo (1475–1564) both performed dissections and created anatomical sketches that are works of art in themselves. Michelangelo's anatomical studies also clearly inform his sculptures, perhaps most magnificently in his famous statue of the biblical King David.

Leonardo, arguably as much a scientist, engineer, and inventor as he was an artist, spent years studying the human body, dissecting as many as 30 cadavers through the course of his career and producing hundreds of anatomical drawings. He began with detailed drawings of the human skull, drafting them with an architect's eye for elevation and perspective. From those, he went on to create detailed renderings of many different bones, muscles, and organs. His drawing of a fetus inside the womb is famous not only for its grace and delicacy of line but also for its accuracy. Although none of Leonardo's anatomical drawings were published in his lifetime, they were probably seen by some contemporaries.

BOOKS ON THE BODY

The first comprehensive, accurately illustrated volume on anatomy, *De Humani Corporis Fabrica (On the Structure of the Human Body)*, was published in Basel, Switzerland, in 1543 by Andreas Vesalius (1514–1564). While the art of his volume is magnificent—Calcar, a student of the great artist Titian, may have created the images—more revolutionary for the time was Vesalius's reliance on observations of the human body instead of reference to respected ancient texts.

A lecturer at the renowned University of Padua, Vesalius performed dissections for his classes. Often, he found, the details of a dissected body contradicted statements made by the great Greek physician Galen (ca 129–216), on whose teachings—based on animal dissection—Renaissance medicine depended. Consequently, Vesalius's book outraged the Galenists and fascinated others.

Another illustrated anatomy was published in 1543: Giovanni Battista Canano's *Musculorum Humani Corporis Picturata Dissectio (Illustrated Dissection of the Muscles of the Human Body)*. Like Vesalius, Canano (1515–1579) was a lecturer at the medical school in Padua and based the illustrations in his book

Leonardo da Vinci produced this drawing of the well-muscled upper back and arms of the human male.

on observations of actual bodies. His book was not as broadly distributed or as influential as Vesalius's, but it featured some elements of anatomy that his contemporary did not study. Canano represented the complicated bones and muscles of the hand, for example, and explored how valves within the circulatory system regulate the direction of blood flow through the body. His illustrations were also the first to show the action of specific muscles.

ART OR SCIENCE?

Into the early 18th century, it was hard to distinguish between anatomy as art and anatomy as science. German anatomist Bernhard Siegfried Albinus (1697–1770), for example, mingled art and science in his volume of plates depicting the bones and muscles of the human body, published in 1747. Albinus drew full anatomical figures in exquisitely landscaped scenes. He considered that such scenery helped fill in the empty spaces of the pages and made the figures appear more three-dimensional. For him, the art was as important as the science,

but with the coming of the Enlightenment, later 18th- and early 19th-century anatomists vehemently disagreed. Scottish anatomist John Bell (1763–1820), for example, bemoaned the "vitious practice of drawing from imagination" and "the subjection of true anatomical drawing to the capricious interference of the artist." His unflinchingly clinical drawings contrasted sharply with the ornamental illustrations of Albinus.

BELLS' INFLUENCE

John Bell often collaborated with his younger brother, Charles, also a physician and anatomist. Between 1797 and 1804, the two brothers published the extensively illustrated *Anatomy of the Human Body*, the illustrated *Anatomy of the Brain*, and *A Series of Engravings Explaining the Course of the Nerves*.

Charles Bell (1774–1842) outlived his older brother and continued to practice medicine into the 1830s. In 1815 he served as a surgeon on the battlefield at Waterloo, the scene of the great British victory over Napoleon Bonaparte, and then codified the lessons learned there in paintings and lectures on the effects of gunshot wounds. His tireless studies of the human face are commemorated in medical terminology today. He identified the long thoracic nerve (called Bell's nerve), which assists in breathing and stabilizes the scapula against the chest wall. Bell's palsy, a facial nerve disorder, is also named after him.

Fascinated by the way that bones, nerves, and muscles combine to create facial expressions, Charles Bell took his study full circle. In 1806 he published *Essays on the Anatomy of Expression in Painting*, a classic in the study of art history. Just as artists created the earliest anatomies, Bell shared his knowledge as a surgeon and anatomist on great works of art history.

KEEPING HEALTHY

DONE TOO MUCH

Sore muscles after a workout are no fun, but they often mean you're getting stronger. Here are some tips for faster recovery so you can get back in action as quickly as possible:

+ Know what's normal: Muscle discomfort usually peaks about 24 to 48 hours after exercise. Pain right after exercise may signal an injury.

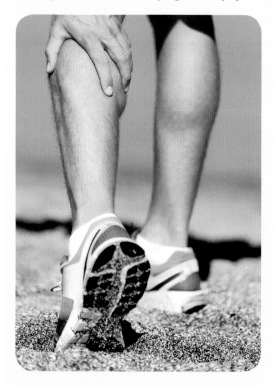

+ Try heat and ice: Get a jump on aches and pains by applying ice right after strenuous exercise. Once soreness sets in, applying heat can increase the blood flow to the muscles and help aid healing.

+ Keep moving: Don't stop exercising, but take it easier for the next few days. Try working other muscle areas while the sore ones recover. Then get ready to feel the burn all over again!

(Also see "Keeping Healthy" on page 331 for tips on how to treat sprains and strains.)

fulcrum, the force is applied in the middle, and the load is between the two gripping ends.

In the human body, lifting a weight with the forearm pressed against the side of the body represents a third-class lever system. The fulcrum (the elbow) sits opposite the load (the weight). In between, the biceps muscle in the upper arm exerts force against the fixed point. The load cantilevers out, and the muscle lifts it against the fixed point of the elbow.

Facial Muscles

Something as simple as a facial expression is also the product of muscular movements. A broad smile, a furrowed brow, a curled lip, a wrinkled nose—we tend to think of such facial expressions as messages of emotion. They can also be described, strictly speaking, as the combination of movements of the up to 50 bilaterally symmetrical muscles that operate just under the skin of the face.

Facial muscles are skeletal muscles, like those of the trunk and limbs. Unlike the other skeletal muscles of the body, however, they do not work in opposing pairs. They stretch in a pattern of layered sheets around the eyes, nose, and mouth, and across, between, and around the bones, fat, and connective tissue of the head. Facial muscles insert into the subcutaneous tissue beneath the skin or into other muscles. Contraction and relaxation of facial muscles, which are innervated by the facial nerve, produce the myriad expressions that enliven the faces of human beings around the world.

The expressive facial muscles can be divided into groups, depending on what part of the face they serve. Over and around the scalp stretches the big occipitofrontalis muscle, divided into the frontal and occipital

DID YOU KNOW?

SMILE OR FROWN

There's an old contention that it takes more muscles to frown than to smile. It's often used to encourage optimism—but is it true? A recent calculation determined that it takes 12 muscles to smile and 11 to frown. Most people smile more than they frown, however, hence the muscles for smiling are probably more fit—and operate with less effort.

bellies, the scalp muscles in the front and rear of the skull.

All around and at the edges and corners of the mouth are muscles that push and pull on the lips. They not only create expressions, but also play a critical role in speech by forming the different shapes the mouth must make when forming different sounds of speech.

One large circular muscle, the orbicularis oris (from "circling the mouth"), surrounds the mouth opening. A pair of muscles, the zygomaticus major and minor, draws the corners and the upper lip up when a person smiles or bares the front teeth. The risorius muscle—its name comes from the Latin for "laugh"—attaches to the outer edges of the mouth, pulling at those corners when someone is literally feeling down at the mouth. This muscle also is used to produce an insincere smile whose expression does not involve the eyes.

An equally complex grouping of muscles works together around the eyes. Here it's the orbicularis oculi ("circling the eye") muscles that surround the orbit of the eye. Contractions of these muscles close the eyes. Above the eye, connecting the frontal bone of the forehead to the eyebrow skin, are the levator palpebrae ("eyebrow lifter") muscles. When they contract, the eyes open. Eyebrow and forehead gestures

originate from the frontalis and the corrugator supercilii ("wrinkle of the eyebrow") muscles, which angle from above the eye toward the bridge of the nose.

While the jaw muscles participate to some extent in facial expressions, their primary job is to move the lower jaw (or mandible) while chewing. They are innervated by the trigeminal nerve. Three of the four pairs of muscles—the masseter, temporalis, and medial pterygoid—operate on a vertical plane, opening and closing, hinged on the temporomandibular joint, also called TMJ. A fourth pair, the lateral pterygoid muscles, contracts with slight horizontal movements and works with other muscles to move the jawbone from side to side during mastications. Grouped together, these four pairs are called the masticatory muscles.

Chapter 5
CIRCULATION

THE SYSTEM

Moving the Blood Throughout the Body

The circulatory system has three basic functions: transportation, protection, and heat dispersion. Blood carries oxygen, nutrients, and hormones to all the body's tissues and totes away carbon dioxide and other waste products. It also contains a wide variety of immune-system cells that fight off viruses, bacteria, and other invaders and foreign substances. Platelets and clotting agents in the bloodstream speed to the site of injuries and seal them off in minutes. And blood disperses heat around the body, a lesser known but still important job, helping to cool the brain and hardworking organs in the chest and abdomen and keeping the body from overheating during exercise.

Circulating Parts

The three main components of the cardiovascular system are the heart, blood, and blood vessels. The heart is the central engine of circulation. A fist-size, roughly cone-shaped lump of muscle, it nestles between the lungs in the middle to left part of the chest. Guided by electrical signals generated within its own muscle cells, it begins beating at four weeks after conception and does not stop until death. Its four chambers consist of two small atria on the top and two larger ventricles on the bottom. The left atrium collects oxygenated blood from the lungs and pumps it into the left ventricle, which propels it through the body. The right atrium collects oxygen-poor blood returning from the body and sends it into the right ventricle, which pumps it over to the lungs to exchange carbon dioxide gas

DID YOU KNOW?

BLOOD BY THE NUMBERS

+ A newborn baby has about 1 cup (0.25 L) of blood in circulation.

+ All the blood vessels in an adult body, strung together, could circle the Earth two and a half times.

+ In one day, blood travels 12,000 miles (19,000 km).

+ After it leaves the bone marrow, a red blood cell makes approximately 250,000 trips around the circulatory system before dying.

for oxygen. The blood then returns to the left side of the heart to start the process all over again.

Connected to the heart is a closed network of blood vessels—actually, two networks. The

ARTERIES AND VEINS

Artery is abbreviated a.
Vein is abbreviated v.

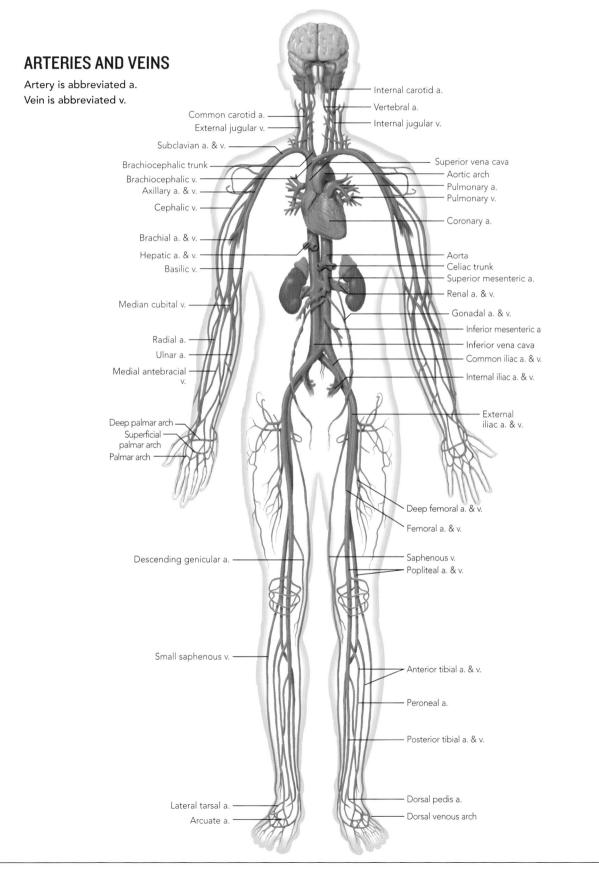

Common carotid a.
External jugular v.
Subclavian a. & v.
Brachiocephalic trunk
Brachiocephalic v.
Axillary a. & v.
Cephalic v.
Brachial a. & v.
Hepatic a. & v.
Basilic v.
Median cubital v.
Radial a.
Ulnar a.
Medial antebracial v.
Deep palmar arch
Superficial palmar arch
Palmar arch
Descending genicular a.
Small saphenous v.
Lateral tarsal a.
Arcuate a.

Internal carotid a.
Vertebral a.
Internal jugular v.
Superior vena cava
Aortic arch
Pulmonary a.
Pulmonary v.
Coronary a.
Aorta
Celiac trunk
Superior mesenteric a.
Renal a. & v.
Gonadal a. & v.
Inferior mesenteric a
Inferior vena cava
Common iliac a. & v.
Internal iliac a. & v.
External iliac a. & v.
Deep femoral a. & v.
Femoral a. & v.
Saphenous v.
Popliteal a. & v.
Anterior tibial a. & v.
Peroneal a.
Posterior tibial a. & v.
Dorsal pedis a.
Dorsal venous arch

The heart is the center of the body's closed network of blood vessels.

left side of the heart pumps blood to all parts of the body except the lungs, while the right side supplies the lungs only. The largest blood vessels are the arteries—big, elastic tubes that carry blood away from the heart. Arteries branch out into smaller channels, called arterioles, which lead into the fine webbing of tiny blood vessels called capillaries that are the site of oxygen and nutrient exchange with the tissues. Capillaries, their job accomplished, merge into tiny veins, called venules, which connect to larger veins to carry blood back to the heart.

The blood that makes its regular, pulsing way through the body consists of a fluid holding a variety of cells. The fluid is called plasma; it's a nonliving, straw-colored liquid made mostly of water but with some nutrients, hormones, gases, and other substances dissolved in it. Plasma makes up almost 55 percent of the bloodstream. Suspended in it are the "formed elements" of the blood: red blood cells, making up almost 45 percent of the blood, and a potpourri of white blood cells and platelets, constituting less than 1 percent (but essential for life nonetheless).

Moving the Blood

Every living cell in every organ in the body must be fed by the bloodstream or die. Not all organs need the same amount of blood, however, and their requirements also vary depending on what the body is doing: exercising, sleeping, digesting, and so forth. The structure and feedback mechanisms of the circulatory system neatly manage these differing needs.

The total volume of blood in the body makes up about 7 percent of the body's mass, which comes to about 4 to 5 quarts (4 to 5 L) in most adults. It is in constant motion; the heart pumps all the blood through all the tissues and to and from the lungs in the course of one minute, beating, on average, 75 times.

The amount of blood expelled by each ventricle of the heart in a minute is called the cardiac output, and it is the product of the heart rate (how often the heart beats in one minute) and stroke volume (how much blood a ventricle pushes out with each beat). Typically, the ventricles pump out about half the blood they contain, or roughly 2 ounces (70 mL), with each beat.

DID YOU KNOW?

BLOOD SPEED

+ The heart pumps oxygenated blood through the aorta (the largest artery) at a speed of about 1 mile (1.6 km) per hour.

+ By the time it gets to the tiny capillaries, blood creeps along at about 43 inches (109 cm) per hour.

Athletes, with bigger, more muscular hearts, may have lower heart rates but a larger stroke volume, so that their cardiac output is the same as or greater than the average person's. Cardiac output changes on demand. During exercise, the heart speeds up and delivers much more blood to the needy muscles: as much as seven times the resting rate, or 32 quarts (35 L) a minute.

Blood flowing through arteries and veins meets resistance, just like any fluid flowing through a tube. The amount of force needed to overcome the resistance is called the blood pressure. Blood pressure is highest in the aorta,

the big artery leading out of the heart—blood from the aorta would shoot six feet into the air if the vessel were opened up when the ventricle contracts. Pressure is lowest where the blood returns to the right atrium of the heart after traveling through the body.

The highest pressure in an artery, just after the heart beats, is called the systolic pressure, which measures about 120 mmHg in healthy adults. (The unit of measure stands for millimeters, mm, of mercury, Hg being the symbol for that element, and is derived from the earliest instruments used to quantify blood pressure.) By the time blood reaches the smallest

The heart pumps blood from the lungs to all the tissues in the body and back again to the lungs in the course of about one minute.

capillaries, its pressure has dropped to less than 20 mmHg.

On the return trip through the veins to the heart, the blood maintains a fairly low and unvarying pressure. Because veins don't have to withstand high pressures—in fact, they are usually only partially filled with blood—they have thinner, less muscular walls than arteries.

Blood flowing back to the heart meets little resistance, but it has to fight gravity as it returns from the arms and legs in an upright body. That's why veins in the limbs have valves: to prevent the blood from flowing back down the limb after each pulse forward. When muscles in the legs contract, they act as additional outside pumps, pressing on the veins and propelling the blood upward.

Blood Demands

Different parts of the body have different demands for blood. At rest, the abdominal organs receive about 24 percent of the blood supply; the kidneys get 20 percent, the skeletal muscles about 20 percent, and the brain 13 percent.

As well as pumping blood to the rest of the body, the heart tissue itself needs blood—quite a lot of it—to support its ceaseless activity. Two coronary arteries arise from the base of the aorta and encircle the heart. Blood from the aorta shoots immediately into them, flowing into smaller arteries and quickly reaching capillary beds that supply the heart muscle. The depleted blood is collected by coronary veins to empty into the heart's right atrium. This blood supply pulses through the coronary arteries only when the heart is relaxed; when the heart contracts, the arteries are too compressed to carry blood. The heart makes up only 0.5 percent of the body's weight but needs 5 percent of the body's blood.

Two organs, the lungs and the liver, have their own special circulation design. The lungs receive deoxygenated blood from the right side of the heart via the pulmonary artery. After the blood flows through the lungs' capillary beds, giving up carbon dioxide and picking up fresh oxygen, it returns to the left side of the heart via the pulmonary veins.

The liver receives about 25 percent of its blood from the hepatic artery, part of the regular systemic circulation flowing from the aorta. The other 75 percent arrives via the portal vein, which carries blood from the intestinal tract and spleen. This nutrient-rich blood is modified and

Curling like the tendrils of a tropical plant, abdominal arteries branch off the central abdominal aorta in this colored arteriograph.

detoxified by the liver before being passed on to the rest of the body.

Short-term stresses can redirect the blood supply to the tissues that need it most; during exercise, for instance, more blood flows to the skeletal muscles and skin and less to the digestive organs. The body manages this through a complex feedback system among the blood vessels and nervous system. Little receptors in the blood vessels monitor how much the vessels are stretching and the chemical composition of the blood; if a tissue needs more blood (and the oxygen it contains), these receptors send impulses through the nervous system that trigger increased respiration and dilate the blood vessels, bringing more blood and oxygen into that area.

The normal amount of blood in the body is more than enough to keep all the organs perking along. When the heart can't deliver enough blood, however, a person can quickly go into shock, a condition characterized by falling blood pressure, pale and clammy skin, and a rapid and thready pulse.

A number of crises can reduce blood flow: the blood loss of a traumatic accident, a heart attack, an excessive loss of fluid from diabetes or vomiting, or even an allergic reaction that causes the blood vessels to dilate, causing the blood pressure to fall.

Without enough blood to deliver oxygen to organs throughout the body, cells and tissues can become damaged and begin to die. In these situations, the body reacts, when it can, by increasing heart rate, dilating or constricting various blood vessels, and retaining water—but these responses can't compensate completely for a blood loss of more than 10 to 20 percent. A person in shock needs emergency medical care.

KEEPING HEALTHY
DEEP-VEIN THROMBOSIS

Bedridden patients and airplane passengers on long trips may share an unfortunate side-effect of immobility: deep-vein thrombosis. The condition is caused by blood flowing slowly or pooling in the deep veins in the lower legs or thighs of an unmoving person. The blood may not be able to wash away clotting factors, and a blood clot (thrombus) can result, blocking circulation. The limb becomes tender, swollen, hot, and red. The clot may break away and travel to the brain—causing a stroke—or to the lungs, heart, or other areas. Doctors may prescribe anticoagulants to the bedridden; people prone to the condition can wear compression stockings to prevent pooling. Plane or car passengers on a long trip should take periodic breaks to stand up and walk about, lest they suffer "economy-class syndrome." Another option is to do leg stretches to keep blood moving.

THE BLOOD

Many Components, Many Tasks

Although it is a fluid, blood is in fact a tissue—a connective tissue whose individual cells float suspended in liquid plasma. The cells perform a variety of critically important jobs in the body, and blood's mix of cells and chemicals is a key indicator of the body's overall health.

Blood Components

Blood plasma, a pale yellow liquid, makes up the largest part of blood by volume, about 55 percent. Plasma itself is more than 91 percent water. The rest of it consists mainly of proteins produced by the liver that help to maintain the proper levels of water in the bloodstream and help blood to clot when necessary. Plasma also carries small amounts of nutrients picked up from the digestive tract, electrolytes such as salt and potassium, respiratory gases such as oxygen and carbon dioxide, and waste products including urea and ammonia. It transports heat as well, keeping the body at a steady overall temperature.

Floating in the plasma are the formed elements of the blood, the blood cells, making up 45 percent of the bloodstream. These come in three types: red blood cells, or erythrocytes; white blood cells, or leukocytes; and platelets, which are cell fragments.

Red blood cells dominate the bloodstream, making up more than 99 percent of its formed elements. These cells have one job, and one job only: to carry respiratory gases to and from the body's tissues. Hemoglobin molecules within the cells contain a red pigment that gives blood its color. Platelets, the next most common formed element, are merely cell fragments, but they have the important task of repairing rips in blood vessels and promoting blood clotting. The five varieties of white blood cells account for only a tiny percentage of the blood's cells, but they play a protective role in fighting off viruses, bacteria, toxins, and other foreign invaders.

DID YOU KNOW?

BLOOD COUNT

Complete blood count, or CBC, analyzes three blood components:

+ **White blood cells:** A high count can indicate an infection or allergy.

+ **Red blood cells:** Too few red blood cells may indicate anemia; too many may impede blood flow. The test measures number, volume (hematocrit), and total hemoglobin.

+ **Platelets:** Measures clotting ability.

Blood cells don't live long, so they must constantly be replaced. Most are produced in the bone marrow, a spongy connective tissue within the bones. In a fetus, all the bones are involved in making blood, but in an adult, only certain larger bones—among them the ends of the arm and thighbones, the skull, breastbone, pelvis, and ribs—produce the cells.

Bone marrow contains stem cells, undifferentiated cells that are able to develop into many different types of blood and immune system cells. These stem cells divide, proliferate, and specialize into red blood cells and different forms of white blood cells within the marrow; the cells then enter the bloodstream via blood vessels in the bone. Lymphocytes travel to lymph glands to finish developing before entering the blood, but the other cells enter the bloodstream fully formed.

Blood Work

Blood has three main jobs in the body, the first being transportation. Blood is the river that carries oxygen from the lungs and nutrients from the digestive system to all the body's tissues so that cells can perform their metabolic tasks. Blood also takes away carbon dioxide and waste products; the lungs exhale

RED BLOOD CELLS
Lacking a nucleus, red blood cells have a hollowed-out shape that maximizes surface area for exchange of gases.

WHITE BLOOD CELLS
Mainstays of the immune system, two lymphocytes (left) cozy up to a neutrophil (right).

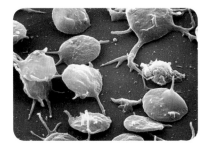

PLATELETS
Activated platelets develop fingerlike pseudopods as they prepare to form a platelet plug.

the carbon dioxide and the kidneys dispose of other wastes in urine. The bloodstream also carries hormones from the glands and organs that produce them to their target tissues.

Both the plasma and the cells of the bloodstream perform regulatory jobs as well. Blood absorbs heat from warmer areas of the body and releases it in cooler tissues, keeping body temperatures even. Blood itself is a little hotter—about 100.4°F (38°C)—than the average body temperature. It is also slightly alkaline and contains proteins that help to regulate the pH level in cells.

The blood is also the body's guardian, one of its main avenues of defense against a host of invaders. White blood cells and antibodies float throughout the bloodstream, destroying viruses, bacteria, parasites, and other pathogens that attack by the millions. Platelets and proteins in the plasma repair damaged tissues and in a remarkably short amount of time build clots that seal up cuts and tears.

You replenish your blood supply every time you drink a glass of water. Plasma, which makes up more than half the bloodstream, is more than 90 percent water taken from the digestive system. Dissolved within it, and making up the other 9 percent of its volume, are more than 100 other substances that keep the body working. Most are proteins that exert an osmotic pressure to keep the proper balance of water within the bloodstream versus water within the tissues. Other proteins in plasma include fibrinogen, made by the liver, which helps blood to clot, and antibodies that fight pathogens. Plasma also transports nutrients—such as glucose from carbohydrates, vitamins, minerals, and amino acids—from the digestive system to feed all the body's cells. It carries away wastes for disposal and distributes enzymes involved in cellular reactions and hormones that help to control metabolism and growth.

Even though plasma by itself is a pale, yellowish liquid, blood is red because it is thronged with red blood cells (erythrocytes), which make up between 40 and 50 percent of blood by volume. Between 4 and 5 million red blood cells crowd into every cubic millimeter of blood. Just 7.5 micrometers in diameter, these flattened disks are the only cells in the body to lack a nucleus; that organelle is ejected from the cell as the cell matures. As a result, red blood cells are depressed in the center, looking rather doughnut-like. They are also extremely flexible, which allows them to squeeze through the narrowest of capillaries.

Each red blood cell holds roughly 250 million molecules of the protein hemoglobin.

Lab technicians sometimes spin blood samples to separate out and analyze the different components.

DID YOU KNOW?

AS RED AS BLOOD

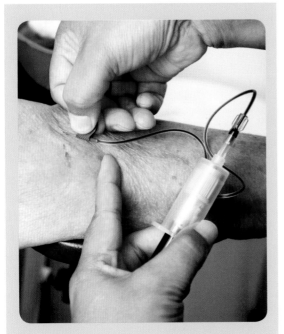

Human blood changes color as it circulates, but it is always red. After hemoglobin picks up oxygen in the lungs, the blood is bright red; when the oxygen has been unloaded in the tissues, the blood turns a dark maroon. Veins look blue because the overlying tissues filter out red light at that depth.

Hemoglobin molecules contain iron atoms that bind loosely to molecules of oxygen as blood passes through the oxygen-rich tissues of the lungs. When the blood reaches the capillaries, the concentration of oxygen in the blood is higher than it is in the surrounding tissues; therefore, the hemoglobin releases the oxygen, which diffuses through the thin capillary walls (see "Blood Vessels," page 148), moving naturally from an area of higher concentration to an area of lower concentration. The hemoglobin also picks up between 10 and 20 percent of the carbon dioxide that is a waste product of cellular metabolism and trucks it back to the lungs for disposal.

Lacking a nucleus or most other organelles, red blood cells last only about 120 days before dying at a rate of about 2 million per second. The spleen and liver remove the dead, ruptured cells from circulation; the body then thriftily recycles many of their components, such as the iron and the amino acids in hemoglobin.

Compared with red blood cells, white blood cells, or leukocytes, are few and far between in the blood—about 5,000 to 10,000 per cubic millimeter—but their importance outweighs their scarcity. Soldiers of the immune system army, white blood cells float through the bloodstream until they detect chemical signals from tissues that need protection. Then they leave the blood through the capillary walls and travel through the cells to mount an immune or inflammatory defense. Most live for only a few days.

There are five types of white cells that travel through the blood; they are classified either as granular or agranular according to the presence of little grainlike vesicles in their cytoplasm. Granulocytes include neutrophils, basophils, and eosinophils. Neutrophils, the most commonly found kind of white blood cells, attack and kill bacteria and some fungi. Eosinophils kill parasites (such as flatworms), while basophils, the rarest sort of white blood cell, release histamines during allergic reactions. They also attract other white cells to areas of infection in the body.

Lymphocytes and monocytes are the two types of agranulocytes. Lymphocytes, the second most common form of white blood cell, come in several varieties that are powerful members of the immune system, attacking everything from viruses to cancer cells. Upon entering the body's tissues, monocytes change into macrophages, which can then eat bacteria and dead or damaged cells.

More numerous than white blood cells, about 150,000 to 400,000 per cubic millimeter, platelets are the third type of "formed element" in the blood. They can't be called cells, because they are merely fragments of huge cells called megakaryocytes that form in the blood marrow. The mature megakaryocytes splinter into thousands of pieces that then enter the bloodstream as platelets. Roughly disk-shaped, they have no nucleus but contain an array of chemicals that allow them to find damaged blood vessels, form clots, and repair tissues.

The blood vessels are among the most vulnerable parts of the body. Were it not for blood's amazing rapid-healing properties, even a tiny cut in a vein or artery could quickly drain the body's entire blood supply.

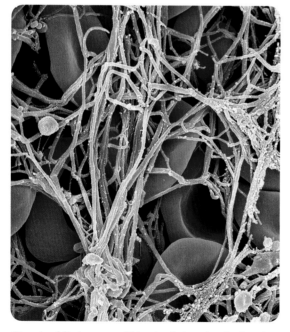

Strands of fibrin trap red blood cells in a developing clot.

DID YOU KNOW?

BLUE BLOOD?

Lobsters have blue blood (when it is exposed to oxygen), but people never do. The expression "blue-blooded," meaning aristocratic, comes from Spain, where in past centuries the oldest families of Castile claimed to be distinguished from their mixed-blood compatriots by their pure *sangre azul*, or blue blood. In actuality, their veins just showed up more conspicuously blue against their fair skin.

Healing Power

When a blood vessel is damaged or severed, a multistage, fast-acting defense mechanism springs into action. Chemicals are released by the injured tissue and by the platelets. These substances, as well as signals from pain receptors, stimulate a vascular spasm: The muscles of the blood vessel wall constrict, reducing blood flow. Meanwhile, the exposed collagen fibers in the damaged tissue prompt platelets, which normally slide smoothly along, to adhere to the wound. As they do this, the platelets change shape, becoming spiky, sticking to one another, and releasing a host of chemicals. Some of the chemicals summon more platelets to the injury; others bolster the vascular spasm reaction, while still other chemicals cause the newly arriving platelets to become sticky themselves. Soon, the accumulating platelets have formed a plug that temporarily seals the break in the blood vessel.

As this is happening, the injured tissues and the platelets are also stimulating a complex cascade of reactions that in a matter of minutes will lead to a blood clot. Substances on the surface of the accumulating platelets and others released by the injured tissue trigger a chemical reaction in the blood that forms prothrombin activator. This substance transforms a plasma protein called prothrombin into the enzyme thrombin. Thrombin, in turn, changes another plasma protein, fibrinogen, into loose, insoluble threads of fibrin. These sticky strands gather at the site of the platelet plug and trap platelets and other blood cells in a mesh of fibers—a clot, or thrombus, that stops blood from leaking through the damaged vessel.

Over the course of an hour or so, fibrin threads attached to the damaged area tighten, squeezing serum (plasma without clotting proteins) out of the clot and pulling the torn edges of the blood vessel together. The platelets release yet another chemical, a growth factor that helps the tissues of the damaged blood vessel repair themselves.

Once the tissue is on the way to being healed, the clot is no longer needed. Yet another plasma protein, plasminogen, takes care of the problem. Within two days after the clot has formed and healing begins, the plasminogen starts to release an enzyme called plasmin that eventually digests the fibrin strands and dissolves the clot.

Blood clots are lifesavers, but they can also be killers if they accumulate or if they break loose to travel through the blood system. Ordinarily, several natural anticoagulants in the blood—among them heparin (produced

Heir to the Russian throne Alexis Romanov (center) suffered from hemophilia, a condition in which the blood lacks the ability to clot.

by immune system cells), antithrombin, and activated protein C (APC)—prevent this by dissolving clots. Sometimes these factors are not enough to prevent an unwanted clot. If blood is flowing too slowly, or if the walls of the blood vessels develop rough spots due to arteriosclerosis or inflammation, then large, fixed clots that block blood flow partially or completely can develop. If this happens in the blood vessels that feed the heart tissue, it leads to a heart attack (see "When Hearts Attack," page 144).

If a clot breaks loose and floats through the bloodstream—at which point it's called an embolus—it can get stuck in a smaller blood vessel and cause an embolism, or obstruction. An embolus that becomes trapped in the lungs is called a pulmonary embolism, a life-threatening condition; one that blocks a blood vessel in the brain can cause a stroke.

Certain drugs, ranging from aspirin to synthetic heparin, are anticoagulants that help to prevent embolisms in susceptible patients. Blood banks prevent clotting in donated blood by adding substances that remove calcium, which is needed for the sequence of reactions that causes clotting.

Bleeding disorders represent the opposite problem: too little clotting. Bone marrow diseases or damage, such as that brought on by radiation, can result in too few platelets

BLOOD TYPES: GIVING AND RECEIVING

BLOOD GROUP	ANTIGENS	ANTIBODIES	CAN DONATE TO	CAN RECEIVE FROM
AB	A & B	None	AB	AB, A, B, O
A	A	B	A & AB	A & O
B	B	A	B & AB	B & O
O	None	A & B	AB, A, B, O	O

being created. Without enough platelets, the blood vessels soon begin to leak blood from innumerable tiny breaks throughout the body. Bruises and tiny purplish blotches caused by the bleeding show up on the skin. Liver damage—such as that caused by cirrhosis or vitamin K deficiency—can also bring on bleeding because the liver makes clotting factors that are released into plasma. Perhaps the most familiar example of a blood clotting disorder is hemophilia, a genetic condition in which even a minor trauma can become a life-threatening event.

Many bleeding disorders have a seemingly simple solution: a blood transfusion. And in fact, transfusions of whole blood and packed blood cells are common, lifesaving procedures, but they are not as straightforward as they seem because of the existence of incompatible blood types. Before the discovery of blood types (see "Breakthrough," opposite),

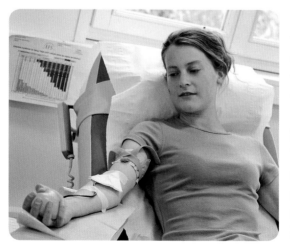

Donated blood can save lives, but the donor's and recipient's blood types must be compatible.

a few physicians had experimented with animal-to-human and human-to-human blood transfusions, but with little success. They attributed their patients' deaths to the rapid clotting of the transfused blood.

Like most other cells in the body, red blood cells have specialized proteins—called antigens—on their surfaces. These antigens are like ID tags, marking the cells as belonging to the individual that produced them. When someone receives a blood transfusion from an incompatible donor, the body's immune system detects the foreign antigens and attacks the cells, making them clump together and eventually burst. The ruptured cells release hemoglobin into the bloodstream, which can carry the protein to the kidneys and shut them down.

Blood Groups

There are at least 30 common varieties of antigens on red blood cells, resulting in at least 30 different blood groups within the human population; about 100 other blood groups exist in small groups of related people. The two most important blood groups from a transfusion standpoint—the ones most likely to cause a major reaction—are the ABO group and the Rh group. The body typically doesn't react strongly to the other blood groups, although some may cause a reaction if a person is repeatedly transfused with them.

The ABO groups consists of four types—A, B, AB, and O—identified by their antigens. Type A blood has antigen A, type B has antigen B, type AB has both antigens A and B, and type O has neither.

The Rh blood group (so named because the first of its antigens, or factors, was discovered in rhesus monkeys) has at least eight types of antigens, but only one, antigen D, is important in transfusions. Blood that contains the Rh antigen D is called Rh positive (Rh+); about 85 percent of the U.S. population is this type. Blood lacking the antigen is Rh negative (Rh-). ABO type and Rh factor are not linked—they are inherited separately—but they are typically reported together, such as A+ or O-.

Everyone can receive a transfusion of his own blood type as well as type O, which is called the universal donor. Those with AB blood, called the universal recipient, can receive all types of donated blood. The Rh factor must also be matched between donor and recipient so that an Rh- patient doesn't receive an Rh+ transfusion. An adverse reaction doesn't happen the first time Rh+ is transfused into an Rh- person, but after it has occurred, the recipient's body builds up antibodies to the foreign Rh type. The next time he is transfused with that type, the antibodies will attack the transfused blood cells and rupture them.

Rh blood types are not only important for blood transfusions; they are also important for expectant mothers and can cause serious problems in fetuses or newborns. If an Rh- woman is carrying a fetus with the Rh+ blood type, leakage of only a small amount of fetal blood into the mother's circulation can cause the mother to form Rh antibodies. While these Rh antibodies do not develop quickly enough to affect this fetus, they cause miscarriage, stillbirth, or serious health problems for subsequent Rh+ children. In the United States, blood typing is a routine part of prenatal care. Rh- pregnant women typically receive injections of Rh immune globulin before and after delivery to prevent formation of Rh antibodies.

BREAKTHROUGH BLOOD TYPES

In 1901, Austrian biologist Karl Landsteiner (1868–1943) discovered that the blood serum of some people would attack the blood of other people, and that this transfusion reaction was a natural, immunological property. He and his colleagues soon identified four basic blood types—A, B, AB, and O—based on this reaction. In 1907, the first transfusion cross-matching donor and patient blood was performed at Mount Sinai Hospital in New York. In 1937, Landsteiner, now an American, and colleague Alexander Weiner discovered the Rh factor, the other important blood type in transfusion reactions.

Blood group test

THE HEART

A Powerful Beat

Press two fingers firmly between the fifth and sixth ribs (from the top) on your left side. The steady thump-thump-thump you feel is your beating heart. Contracting and relaxing 100,000 times a day, 35 million times a year, almost 3 billion times in the average lifetime, the heart tirelessly pumps blood throughout the body, keeping every cell oxygenated, fed, protected, and functioning.

Around the Heart

The heart is lodged between the lungs and on top of the diaphragm in a central cavity in the chest, called the mediastinum. It sits behind and just left of the breastbone, its flattened top below the second rib and its pointed apex, at the bottom, behind the fifth and sixth ribs on the left. The size of a large fist, it is hollow and light, weighing about 8 ounces in adult women, 10 ounces in adult men.

Surrounding and protecting this vital organ is the pericardium, a double-walled membrane.

DID YOU KNOW?

HEART FACTS

+ In the course of an average lifetime, the heart will pump about **48 million gallons** (182 million L) of blood through the body.

+ Anger and fear increase the heart rate by **30 to 40 beats per minute.**

+ Grab a **tennis ball and squeeze it tightly:** That's how hard the beating heart works to expel blood.

The outer wall is a tough, inelastic sac that attaches to the diaphragm at the bottom and wraps around the heart, fusing to the arteries where they exit the heart. The inner wall is a thin, double-layered membrane called the serous pericardium that clings tightly to the heart. A slippery fluid between its two delicate layers eases friction as the heart expands and contracts. If the membrane becomes inflamed, a condition known as pericarditis, the layers may rub together painfully, and fluid buildup may occur.

Heart Parts

The heart itself is a muscular little four-chambered machine. Its walls have three layers. The outermost, the epicardium, is the same membrane that makes up the inner layer of the serous pericardium; it consists of thin connective tissue. In most adults, it contains quite a bit of yellowish fat.

The middle layer, called the myocardium, is where the hard work happens. It is made primarily of cardiac muscle, a specialized kind

THE HEART

Superior vena cava

Right pulmonary artery branches

Aorta

Pulmonary trunk

Right pulmonary veins

Pulmonary semilunar valve

Right atrium

Tricuspid valve

Chordae tendineae

Inferior vena cava

Right ventricle

Left pulmonary artery

Left pulmonary veins

Left atrium

Aortic semilunar valve

Mitral valve

Left ventricle

Papillary muscle

Myocardium

Interventricular septum

Trabeculae carneae

Endocardium

Myocardium

Epicardium

of involuntary muscle tissue. The branching cells of cardiac muscle are connected at spots called intercalated disks. There electrical signals pass from one cell to another. Held together by connective tissues, the cells form muscle bundles that spiral around the heart.

Among the cardiac muscle cells are autorhythmic fibers, specialized cells that are self-excitable, which means they can contract rhythmically without outside stimulation. In fact, individual fibers grown in a cell culture will beat spontaneously. The heart's inner layer

is called the endocardium. This white connective tissue lines the chambers of the heart and covers its valves.

The Heart at Work

The four-chambered heart is an efficient and durable piece of muscular machinery. It can be thought of as two side-by-side pumps: The left side collects oxygen-rich blood from the lungs and pumps it to the rest of the body, while the right side collects oxygen-poor blood from the body and pumps it to the lungs for oxygenation.

BREAKTHROUGH OBSERVING THE HEART

Three inventions improved the ability to detect the heart's actions without opening the chest. In 1816, physician René Laennec (1781–1826) invented the first: the stethoscope. Reluctant to put his ear to the chest of a female patient, he rolled papers into a tube and "perceived the action of the heart . . . more clear and distinct." He built a monaural stethoscope and published the design. In 1903 came the electrocardiograph, right. Physiologist Willem Einthoven's (1860–1927) device measured electrical current in the heart. A silver-coated quartz string was attached to the skin with an electrode; as current passed through, it quivered. A photographic plate registered its movements. By 1911, readings from the "string galvanometer" were used to diagnose heart disease. The third advance, the echocardiogram, was the first medical use of ultrasound. It was invented in the 1950s by cardiologist Inge Edler (1911–2001) and physicist Carl Hellmuth Hertz (1920–1990). Using a sonar

machine from a shipyard, they recorded the echoes of Hertz's heart. In 1954, their "reflectoscope" recorded movement of the heart walls. By 1956 their reflectoscope could be used to diagnose valvular diseases.

The chambers consist of two atria on top (atrium meaning "entry hall") and two ventricles below (ventricle meaning "little belly"). The atria are separated by a thin wall, the interatrial septum. The atria are the collecting rooms for returning blood; with each beat, they pump this blood through valves into the larger and more muscular ventricles, which pump it out of the heart. The ventricles are separated by a partition as well, the interventricular septum.

Three veins—the superior vena cava, the inferior vena cava, and the coronary sinus—feed oxygen-poor blood into the right atrium. As the heart contracts, it squeezes this blood into the right ventricle through the tricuspid valve (so called for its three leaflike parts). When the heart relaxes, the valve closes, preventing the blood from flowing back into the right atrium.

The right ventricle, which makes up most of the visible surface of the heart when seen from the front, pumps the blood it receives into the pulmonary trunk, which divides into two pulmonary arteries that lead to the lungs. A pulmonary valve between the right ventricle and the pulmonary trunk guards against backflow of the blood into the heart.

Bright-red blood returning from its refreshing trip to the lungs enters the left atrium through four pulmonary veins. As the heart contracts, this blood passes into the left ventricle through the two-leaved mitral valve—so called because its two parts look like a bishop's miter.

The left ventricle forms the apex, the heart's pointy bottom, located toward the back of the body. Its half-inch-thick, muscular walls are three times as thick as those of the right ventricle. The hardest-working part of the organ, the left ventricle is responsible for pumping blood through all of the body's blood vessels except those leading to the lungs. With a powerful contraction, it expels oxygen-rich blood through the aortic valve into the aorta, the body's largest artery, which branches off to carry it into the rest of the body. (A little blood from this ventricle also feeds into the coronary arteries to supply the heart muscle itself.)

Valves

The heart's four valves are the guardians of its circulation. The two between the atria and the ventricles are aptly called the atrioventricular valves. When the contracting muscle of the heart presses the blood against them, it forces them open; as the heart relaxes, the valves return to a tightly closed position. Their leaves typically don't get everted, or pulled inside-out into the atria, because they are connected to the ventricle by slender cords called chordae tendineae (the "heart strings" of popular song).

Between the ventricles and the big arteries—the pulmonary trunk and aorta—are the semilunar valves, named for their three crescent moon–shaped leaves. Pressed open by each ejection of blood from the ventricles, they close when the heart relaxes, preventing blood from reentering the ventricles. No valves stand between the incoming veins and the atria; as the heart contracts, the entrance to those veins is compressed so that very little blood flows back into them.

Typically, blood flow through the heart is quiet and smooth, producing the characteristic heartbeat sounds (see "Did You Know?," page 146). Sometimes, however, doctors will hear additional sounds, like clicking or whooshing

The aortic valve regulates the flow of blood to the aorta, which supplies blood to the rest of the body.

noises. These murmurs may be "innocent," but in many cases, particularly in adults, they are the sound of blood flowing through malfunctioning valves. Diseases, congenital defects, or scarring can damage the valves so that they don't open completely (stenosis) or close completely (incompetence). Either condition makes the heart work harder as it pumps blood through narrowed openings (in stenosis) or pumps some backflowing blood again and again (in incompetence). Mitral valve prolapse, in which the mitral valve leading into the left ventricle is weakened, is a common example, and its effects can range from benign and unnoticeable to severe. In serious cases of valve damage, doctors can put in a new valve; some replacement valves are taken from pig hearts (and treated to prevent rejection) or human cadavers, and some are synthetic.

Electricity

The heart is powered by electricity. The signal begins in the sinoatrial (SA) node, also known as the heart's natural pacemaker, located near the top of the right atrium. This tiny piece of tissue is made of autorhythmic fibers, specialized heart cells that take in and expel calcium and other electrolytes to regularly change their electric charge. The electrical impulse produced by the cells of the SA node spreads through the walls of the atria, moving from one cardiac cell to the next via gap junctions, passageways that connect two cells. The signal makes the muscle cells of the atria contract

BREAKTHROUGH HEART-LUNG MACHINE

Before the 1950s, open-heart surgery was an extremely tricky business as it involved cutting into a still-beating heart. Prolonged operations to repair heart valves or to bypass blocked coronary arteries were virtually impossible to perform. After the death of a young patient with blocked lung circulation in 1930, American surgeon John Heysham Gibbon (1903–1973) became determined to build a machine that could completely take over the work of the heart and lungs during surgery. Working with technician Mary Hopkinson, who later became his wife, Gibbon pursued the development of the machine for two decades, funded in the 1940s by Thomas J. Watson, chairman of IBM. The Gibbons found that blood cells passing through a machine were easily damaged, and the blood was liable to clot. The researchers therefore passed the blood in a thin sheet over a layer of film to oxygenate it and added screens and filters to block air bubbles and clots. By 1952, the machine was working successfully in dogs; in 1953, Cecelia Bavolek became the first patient to successfully undergo heart surgery while her blood was being shunted through the heart-lung machine. It was still a risky endeavor in its early days—three of the first four patients to use the machine did not survive the surgery—but with further improvement, the heart-lung machine, right, became a standard addition to heart surgery and a piece of equipment in every hospital's cardiac surgery unit.

An artificial heart pacemaker stands out in this colored x-ray.

in a sequenced fashion, pressing blood into the ventricles.

As the blood enters the ventricles, the electrical impulses from the atria converge on a second junction, the atrioventricular (AV) node, located in the interatrial septum just above the tricuspid valve. The electrical signal pauses for a tenth of a second at the AV node while the ventricles fill with blood. Then it shoots out through the atrioventricular bundle (AV bundle, or bundle of His), autorhythmic cells that run down through the interventricular

DID YOU KNOW?

SHOCK TO THE HEART

In 1952, Dr. Paul Zoll (1911–1999) attached electrodes to the chest of a man whose heart had stopped because the natural pacemaker signal was blocked at the AV node. Two-millisecond electric shocks from the electrodes passed into the heart, restarting it. Zoll's innovation led to the development of the portable defibrillators widely used today.

septum. The AV bundle splits into right and left branches, which in turn lead into strands of specialized cells called Purkinje fibers that fan out through the ventricle walls. Electrical currents from these bundles and fibers spread through the muscular ventricles, causing them to contract and pump blood into the arteries. The whole signaling process lasts less than a second.

Current Detection

The electrical currents generated by the heart can be detected all over the body and convey a surprising amount of information. Electrodes placed on the arms, legs, and chest can pick up the signals and transmit them to an electrocardiograph machine, which translates them into the rounded and spiky waves of an electrocardiogram, or ECG (sometimes called EKG from the German *Electrokardiogram*).

Three waves appear, one after the other, with each beat of the heart. The first, the P wave, represents the atria contracting. The second wave, the tall peaks and valleys of the QRS complex, represents the ventricles contracting. The third wave, the T wave, shows the signal repolarizing as the ventricles relax.

In a normal ECG, these waves repeat in a predictable pattern. Abnormal ECGs can yield clues to the location of a heart problem. Large P waves, for instance, might mean an enlarged atrium; an unusually tall R peak in the QRS complex could point to enlarged ventricles, while a flattened T wave could mean insufficient oxygen to the heart (as in coronary artery disease). Loose, random ups and downs indicate ventricular fibrillation, a serious condition in which the SA node loses control of the heart rhythm and the organ is taken over by rapid, irregular contractions. The heart can

WHEN HEARTS ATTACK

Heart attacks don't begin in the heart: They start in the slender coronary arteries that feed the heart muscle. When those arteries become stiff, thickened, and clogged—a condition known as arteriosclerosis—they can no longer deliver enough blood to the heart. Starved of oxygen, the heart tissue begins to die; the weakened heart may falter and stop beating normally. All too often, the first sign of this disease is its last one—a fatal heart attack. Coronary artery disease is the leading killer of adults in the United States, accounting for more than 650,000 deaths a year.

Although other diseases or traumas can stop the heart from beating properly, the artery-clogging condition known as atherosclerosis (one form of arteriosclerosis) is the most common cause of a classic heart attack. Doctors aren't sure how atherosclerosis starts, but it seems to begin when the inner lining of an artery develops a rough or injured spot, maybe due to inflammation, high blood pressure, carbon monoxide from cigarette smoke, or high LDL (low-density lipoprotein) cholesterol, among other factors. Cholesterol, fats, calcium, cellular waste products, and platelets gather at the injured spot, forming a plaque that narrows the blood vessel.

When this happens in coronary arteries, the reduced blood flow that results can starve and weaken the heart muscle, a condition called myocardial ischemia. It can happen "silently," without symptoms, or it can bring on the squeezing sensation of angina, especially during exercise, when the heart needs more blood than usual.

Even more dangerous is a complete blockage of a blood vessel. Sometimes a blood clot forms at the site of the plaque; if the clot breaks loose, it can be moved by the bloodstream into a narrower vessel, walling it off completely. In a blood vessel feeding the brain, it causes a stroke; in the arms or legs, it results in tissue damage; in coronary arteries, it brings on a myocardial infarction, or heart attack.

The narrowing of the artery on the center left leaves this patient at risk for heart attack.

SIGNS & SYMPTOMS

Some heart attacks strike silently, leaving victims (if alive) unaware that they even occurred. More often, there is some kind of chest discomfort: The symptoms can simulate bad indigestion, a tight band around the chest, or pressure so intense that it feels like an elephant is sitting on the chest. Pain can radiate into the arms, shoulder, jaw (usually the left side more than the right) or torso. Other symptoms—somewhat more common in women than in men—include nausea, shortness of breath, profuse sweating, and dizziness. The stereotypical scene in which someone clutches his chest, drops to the floor, and dies is unusual, but because they are expecting similarly dramatic symptoms, many patients wait too long to call for emergency assistance. Those who don't wait—who reach a hospital within an hour of the first symptoms—have the best chances for survival.

An infarction kills heart muscle and can trigger ventricular fibrillation and death. Hospital personnel who suspect a heart attack will give the patient clot-dissolving drugs and may perform emergency surgery, such as angioplasty, to open up clogged arteries, or coronary bypass surgery, which uses blood vessels from other parts of the body to bypass blocked coronary arteries. After the infarction, the damaged heart muscle will be replaced with scar tissue within six or eight weeks, leaving the heart permanently weakened. This can lead to heart failure.

TREATMENT

Treatment for the disease can mean drugs or surgery, or both. Cholesterol-lowering drugs will, as their name indicates, drop blood cholesterol levels substantially, helping to keep plaque from building up. Beta blockers are medicines that ease strain on the heart by preventing epinephrine from attaching to beta receptors and pumping up the heart rate. Nitroglycerine dilates coronary blood vessels, while digitalis increases cardiac contractibility and, therefore, cardiac efficiency.

If exercise, diet, and medication are not enough to keep blood vessels open, surgery might be needed. Heart surgeons might bypass blocked coronary arteries by attaching a blood vessel, taken from elsewhere in the patient's body, to the aorta at one end and an unblocked portion of the coronary artery at the other. They may also keep blood flowing by inserting a balloon into the blocked artery, inflating it temporarily to squash plaque against the artery walls, and then deflating it. Or they may insert a stent, a latticelike tube of steel, into the artery to keep it open. Some stents are coated with drugs that reduce the odds that the artery will clog up again; questions have arisen about whether these stents lead to an increased risk of clotting, so studies are underway to check on their efficacy and safety. At least one recent study suggests that, in many cases, blood-thinning drugs can be as effective as stents in keeping arteries open. Decisions regarding the specific choices listed above are best made by an informed patient and physician. Ongoing studies may well change current thinking.

RISKS & PREVENTION

The risk factors for coronary artery disease include some that can be controlled and others that cannot. Controllable factors include weight, blood pressure, smoking, diabetes, blood cholesterol (particularly, high levels of LDL), and exercise. With regard to weight, the distribution is significant, with excess pounds around the belly being more dangerous. Factors that cannot be changed include age (risk increases for men over age 45 and for women over age 55) and a family history of early heart disease.

Preventing coronary artery disease, as much as it is possible, involves removing the controllable risks—keeping weight within normal limits, exercising, refraining from smoking, and controlling blood pressure and cholesterol levels. Although the human heart itself is durable, it is as dependent as the rest of the body on the health of the blood vessels that nurture it.

During the cardiac cycle, the heart's chambers pass through two alternating periods: systole when the chambers contract, and diastole, when they relax. When the atria are in systole, the ventricles are in diastole, and vice versa.

Atrial systole begins when the SA node sends out an electrical pulse that makes the walls of both atria contract, compressing the blood inside the chambers. Blood pressure in the atria rises, and the blood presses against the AV valves, opening them. The blood flows into the ventricles, which are already about 70 percent full of blood that entered while they were relaxed. Pressure drops in the atria and they relax, entering diastole.

The full ventricles enter systole and begin to contract. Increasing pressure pushes the AV valves closed, preventing backflow into the atria. Briefly, all four valves around the ventricles—the AV valves and the semilunar (SL) valves guarding the major arteries—are closed. Then, as the pressure in the ventricles is rising, the SL valves open and the ventricles eject their blood into the arteries. The muscular left ventricle shoots the blood into the aorta at a pressure of 120 mmHg (normal systolic blood pressure), while the smaller

Surgeons replace a patient's aortic valve.

sometimes be restored to normal rhythm by shocking it with a defibrillator.

The Cardiac Cycle

Blood flows continuously through a healthy heart, always moving from regions of higher pressure to regions of lower pressure. The sequence of contraction and relaxation associated with one heartbeat is called the cardiac cycle. It typically lasts about 0.8 seconds, pumping about 2 ounces (70 ml) of blood out of each ventricle.

DID YOU KNOW?

HEAR THE BEAT

The steady *lub-dup*, pause, *lub-dup* noise of successive heartbeats coincides with the closing of two sets of heart valves. The first (*lub*) sound occurs when the AV valves close as the ventricles begin to contract. The second, quieter (*dup*) sound is made by the closing of the SL valves when the ventricles begin to relax.

right ventricle sends blood to the lungs at 20 to 30 mmHg.

For about 0.4 seconds, all the chambers are in diastole, or relaxed. This is the quiet period of ventricular filling. Blood flows into the atria from the higher-pressure veins leading back from the rest of the body and then from the atria into the relaxed ventricles. The backflow of the blood into the ventricles closes off the valves leading to the aorta and pulmonary arteries, and the cycle begins again as the atria start to contract.

If heartbeats were controlled only by the SA node, the heart would go through about 100 cardiac cycles a minute. But the heart at rest is inhibited by the parasympathetic nervous system—specifically, by the vagus nerves, which release acetylcholine to slow the heart rate to about 75 beats a minute. During more strenuous activity, the sympathetic nervous system activates and triggers the release of epinephrine and norepinephrine from the adrenal glands and nerve endings. These chemicals prompt the SA node to accelerate and increase the contractility of the heart's chambers, pushing them to eject more blood with every contraction. Other factors can also affect heart rate. Increased body temperature will speed up the heart, while hypothermia, an unusually cold body temperature, will slow it; surgeons take advantage of this tendency and chill some patients during heart surgery to slow their heart rate.

Strenuous activity like mountain biking activates the sympathetic nervous system and results in an adrenaline rush.

BLOOD VESSELS

Pathways and Passages

Within its closed loop, the entire circulatory system encompasses thousands of miles of blood vessels, reaching virtually every cell in every tissue in the body. Three types of blood vessels—arteries, veins, and capillaries—make up this intricate road map. Arteries carry blood away from the heart; capillaries, the tiniest pathways, carry it into the tissues; and veins take it back to the heart.

All blood vessels except the smallest are made of three layers of tissue, called tunics, surrounding the blood-containing space known as the lumen. The innermost layer, the tunica interna, is lined with the endothelium that coats all the inner surfaces of the cardiovascular system,

DID YOU KNOW?

SMALL VESSELS

Capillaries, the smallest blood vessels in the body have an average diameter of 9 µm (about 4 millionths of an inch), much finer than a human hair. A capillary's lumen, the open space within it, is just large enough for individual red blood cells to slip through in single file.

including the heart's. The middle layer, the tunica media, consists of elastic and muscular tissue; the outer layer, the tunica externa, is made mainly of elastic and collagen fibers. The larger blood vessels are fed by their own vessels within their walls.

Arteries

Arteries come in two varieties. Elastic arteries are the biggest; ranging from about one-third of an inch to an inch in diameter, these seven arteries include the three parts of the aorta, the two common carotids, and the two pulmonaries. Their elastic walls expand and

Greatly magnified, a cross section of an arteriole reveals the red blood cells inside.

The body contains thousands of miles of blood vessels.

recoil as blood is propelled out of the ventricles. Because elastic arteries carry blood to the medium-size arteries, they are sometimes called conducting arteries.

The blood then enters the medium-size arteries with their thick, muscular walls. These vessels carry blood to the various organs and limbs; the smooth muscles in their walls are capable of constricting or dilating to restrict or increase blood flow as needed. Blood flows into smaller and smaller arteries, reaching the arterioles, tiny vessels with thin walls, before entering the capillaries. Arterioles can also constrict or dilate to temporarily feed or starve the capillary beds.

Capillaries

As tiny as they are, the microscopic capillaries are arguably the most important parts of the circulatory system, for they are the vessels that do the work for which the system was built. Known as exchange vessels, the capillaries transport nutrients, wastes, gases, chemicals, and even heat between the blood and the interstitial fluid that fills the spaces between the cells of tissues.

The narrow, open space within the capillary is only wide enough for individual red blood cells to pass. Capillaries also have very thin walls, consisting only of one layer of endothelial cells and a basement membrane.

KEEPING HEALTHY

PREVENT HIGH BLOOD PRESSURE

Rush hour traffic can raise anyone's blood pressure for a short time, but about a third of adults have hypertension (or high blood pressure) in which blood pressure stays high throughout the day. This damages blood vessels, weakens the heart, and increases the risk of heart disease and stroke. Even a slight increase over the normal blood pressure (120/80) can be risky.

High blood pressure can be treated with medication, but there are also other steps people can take to reduce blood pressure or prevent the condition altogether.

+ Follow a heart-healthy diet: Healthy choices include fruits, vegetables, whole grains, and low-fat dairy products, which are high in potassium and other nutrients. Avoid foods high in sodium, cholesterol, fat, and sugar.

+ Get regular exercise: Even a 30-minute walk five times a week can help lower blood pressure.

+ Quit smoking and limit alcohol: Women should drink no more than one serving of alcohol a day; men should drink no more than two servings.

Taking any of these steps is progress, but combining several will go the furthest toward keeping high blood pressure under control.

Blood pressure gauge

Some, called fenestrated capillaries, have pores or windows in their endothelium that allow for rapid entrance and exit of nutrients and hormones.

Even more porous are the sinusoidal capillaries found only in specialized tissues such as the bone marrow; their Swiss cheese–like walls allow large molecules and even blood cells to pass through. The capillaries branch through the tissues in capillary beds, networks of ten to one hundred capillaries apiece. Little sphincters control the flow of blood into the capillary beds, contracting and relaxing five to ten times a minute in response to the needs of the area.

Diffusion from higher to lower pressure pulls oxygen and carbon dioxide, glucose, amino acids, hormones, and many other substances through the gaps between the cells in the capillary walls. A few substances (such as insulin) squeeze through the walls using a process called transcytosis, in which they are first enclosed within little vesicles. At times, large volumes of fluid and solutes are driven by bulk flow pressure from capillaries into tissue or vice versa. If too much fluid floods into the interstitial space, the tissues will swell, a condition known as edema.

Veins

After releasing their oxygen, nutrients, and other chemicals and picking up wastes and other cargo from the tissues, the capillaries merge to form small veins called venules, which

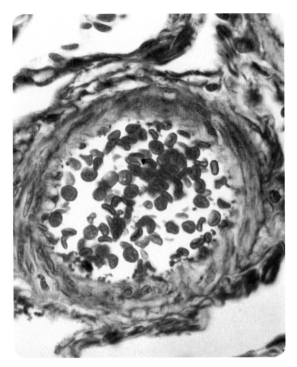

A cross section of the veins shows the layers in the blood vessel walls.

physical, blood pressure is measured in the brachial artery in the upper left arm, with the arm held level with the heart. The first reading an examiner takes after relaxing a blood pressure cuff (which corresponds to the first sound heard through the stethoscope) is the systolic pressure, the pressure that the blood exerts on the big arteries as it leaves the contracting ventricles; the reading at which the sound suddenly becomes faint is the diastolic pressure, the pressure in the arteries as the ventricles relax.

Blood pumping through the elastic arteries creates a visible wave with each heartbeat—the pulse. The pulse can be felt in any artery close to the body's surface (see chart below) by pressing lightly on the skin. The pulse beats at the same rate as the heart, about 70 to 80 times a minute at rest.

drain into larger veins that carry the deoxygenated blood back to the heart.

Blood flowing through veins is at a relatively low pressure and veins are accordingly thinner walled and less muscular than arteries. Some veins, particularly in the arms and legs, contain valves that prevent the blood from flowing backward as it is pumped upward toward the heart. Skeletal muscles in the legs help to pump the blood toward the heart as well by constricting around the veins and milking the blood upward through the valves. Leaky valves, sometimes damaged by stress, can lead to swollen varicose veins that hold pools of backflowing blood.

Pressure & Pulse

Blood pressure varies greatly from place to place within the circulatory system. During a

FINDING YOUR PULSE

ARTERY	LOCATION
Superficial temporal artery	Temple, next to eye socket
Facial artery	Lower jaw on a line with the corners of the mouth
Common carotid artery	Neck, next to larynx
Brachial artery	Upper arm, underside of biceps
Femoral artery	Upper, inner thigh
Popliteal artery	Behind the knee
Radial artery	Bottom half of wrist
Dorsal artery of foot	On top of the instep

Chapter 6

VITAL EXCHANGE

RESPIRATION

The Breath of Life

The human body's essential link to the outside world is the respiratory system. Immersed in a bath of air, the body takes in oxygen with every breath. Each of its trillions of cells depends on this infusion for its survival: Without oxygen, brain cells begin to die within five minutes, and other tissues expire within hours.

In biological terms, however, respiration is more complex. It encompasses the whole process by which oxygen is delivered to the cells and used to break down glucose from food, releasing energy. Respiration also removes the toxic waste product of that process, carbon dioxide, with each exhalation. Other respiratory tasks include moving air across the vocal cords for speech and pulling smells into the nose.

The respiratory system can be divided into two zones, conducting and respiratory, according to their function. The conducting portion includes all the tubes and channels that carry gases in and out of the body. It begins with the nose and nasal cavity, the space within and behind the nose. Below the nasal cavity is the oral cavity, the inside of the mouth; the nasal and oral cavities are connected by the pharynx (throat). At the base of the pharynx lies the larynx (voice box), which leads into the trachea (windpipe). This passageway branches like an upside-down tree, with one branch, or bronchus, extending into each lung. These primary bronchi subdivide into secondary bronchi, one for each lobe of the lungs. The secondary

DID YOU KNOW?

BREATHING

+ At rest, most people breathe about 8 to 16 times a minute.

+ An inhalation lasts about two seconds; an exhalation, about three.

+ Stress can cause hyperventilation: abnormally fast or deep breathing. This action reduces carbon dioxide and increases oxygen in the blood, lowering blood acidity and making the person feel light-headed.

THE RESPIRATORY SYSTEM

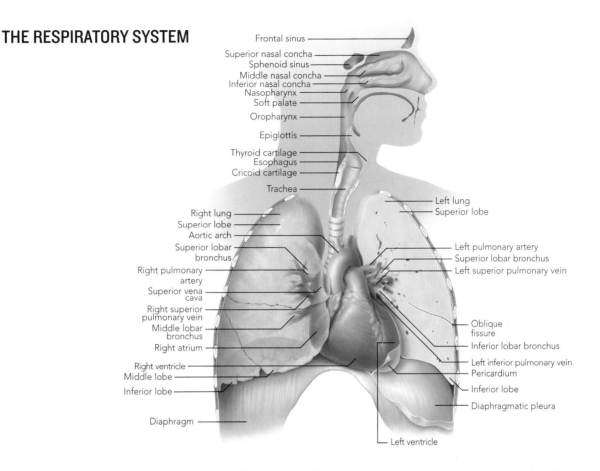

Frontal sinus
Superior nasal concha
Sphenoid sinus
Middle nasal concha
Inferior nasal concha
Nasopharynx
Soft palate
Oropharynx
Epiglottis
Thyroid cartilage
Esophagus
Cricoid cartilage
Trachea
Right lung
Superior lobe
Aortic arch
Superior lobar bronchus
Right pulmonary artery
Superior vena cava
Right superior pulmonary vein
Middle lobar bronchus
Right atrium
Right ventricle
Middle lobe
Inferior lobe
Diaphragm
Left lung
Superior lobe
Left pulmonary artery
Superior lobar bronchus
Left superior pulmonary vein
Oblique fissure
Inferior lobar bronchus
Left inferior pulmonary vein
Pericardium
Inferior lobe
Diaphragmatic pleura
Left ventricle

bronchi branch into tertiary bronchi, which split repeatedly into ever smaller tubes, or bronchioles, some no larger than 0.02 inch (about half a millimeter) across.

Respiratory Zone

The respiratory zone is made up of the tissues in the lungs where gas exchange occurs. It consists of the smallest, respiratory bronchioles and the alveoli, delicate air cavities that are attached—like clusters of grapes—to them via alveolar ducts. Capillaries cover the alveolar sacs, picking up oxygen and delivering carbon dioxide through the porous alveolar walls.

The lungs themselves are made up of both conducting and respiratory structures. These

elastic, spongy organs consist mainly of air spaces, connective tissue, and blood vessels. They occupy most of the chest cavity, with the heart nestled next to the left lung and connected to both lungs via the pulmonary arteries and veins.

Both lungs are covered by a fluid-filled double membrane, the pleura. Beneath them, a dome-shaped muscle called the diaphragm contracts and relaxes to expand and relax the lungs. Several areas in the brain, known collectively as the respiratory center, collaborate to control the motions of breathing (pulmonary ventilation), relieving us of the task of consciously breathing in (inhalation or inspiration) and out (exhalation or expiration) every moment.

Running in cold weather, this athlete can see his exhaled breath,
which contains warm carbon dioxide and water vapor, by-products of respiration.

Conducting Zone

The most visible parts of the respiratory system, the nose and mouth, are the entryways for the conducting zone. The nose plays a key role in drawing in air, moistening it, and filtering out foreign (and possibly harmful) substances. It also houses receptors that enable the sense of smell and adds resonance to the voice.

Within and behind the external nose (which is made of bone and flexible cartilage under the skin) is the nasal cavity. The area just inside the nostrils is called the vestibule; lined with coarse hairs (vibrissae) that trap incoming particles, it is the body's first line of defense against dust, pollen, and microbes. The nasal cavity is divided into right and left sides by the nasal septum, made of cartilage; at the top, just behind the bridge of the nose, is the olfactory epithelium, a membrane holding receptors for smell.

Mucous membrane containing cells with hairlike projections (cilia) lines the nasal cavity. Mucous glands under the membrane secrete about a quart of sticky mucus a day. The mucus traps dust and attacks invading germs with antibacterial enzymes. The cilia then sweep the mucus and contaminants into the pharynx, where they can be swallowed into the stomach and destroyed by stomach acid. Occasionally,

DID YOU KNOW?

WHEN IT'S COLD

Cold, crisp weather can cause an unpleasant side effect: a runny nose. The cold air slows the thousands of moving cilia within the nose, allowing mucus to drain out the nostrils and not down the throat. The tendency of moist breath to condense in cold temperatures compounds the effect, adding droplets of water to the whole mess.

irritating particles tickle nerve endings within the nose, triggering a sneeze that expels the invaders.

Above and around the nasal cavity are the sinuses, cavities in the skull that are lined with mucous membrane and are thought to aid in warming and moistening inhaled air and to resonate during speech. Most people become aware of them when they are inflamed from allergy or infection, a painful condition called sinusitis.

Beneath the nasal cavity is the oral cavity, the space between the lips and the throat. More important for digestion than for respiration, it serves as an alternate passageway for air into the lungs. It is lined by mucous membrane but is not as efficient at moistening and filtering air as the nose. The pharynx connects both cavities to the larynx and trachea below. Muscular and funnel-shaped, it conducts air and food, houses the tonsils, and creates resonance for speech.

This x-ray shows the nasal cavity and sinuses.

Throat Zones

The pharynx is divided into three sections: the nasopharynx, behind the nasal cavity; the oropharynx, behind the oral cavity; and the laryngopharynx, which connects to the larynx and the esophagus, the tube that leads to the stomach. Two auditory, or Eustachian, tubes lead from the nasopharynx to each middle ear, helping to equalize air pressure. Ciliated cells in the nasopharynx propel mucus (containing dust and microbes) down the throat. The back of the nasopharynx also contains the adenoids (pharyngeal tonsil). This oval lump of tissue, like tonsils in the throat, contains lymphoid tissues and, as part of the immune system, helps to trap and fight bacteria.

Below the nasopharynx, the oral cavity opens into the oropharynx through the fauces, the opening seen at the back of your opened mouth when you look at it in the mirror. The oropharynx holds two pairs of tonsils: the lingual tonsils at the base of the tongue and the palatine tonsils visible at the back of the mouth, one on each side. Occasionally, and especially in children, the palatine tonsils are overwhelmed by a bacterial or viral infection and become inflamed, swollen, and sore. This condition is known as tonsillitis; the tonsils may be removed in intractable cases.

The laryngopharynx, the lowest section of the throat, leads to the larynx, where the airways and digestive passages take separate paths. The esophagus carries food to the stomach, whereas the larynx takes air to the trachea and lungs.

Just two inches long, the larynx plays a large role in social interactions. Within the larynx are the vocal cords; with the mouth and other parts of the respiratory system, they make speech possible. The larynx is more than a voice box, however. It also serves as a passageway for air and as a guardian mechanism that directs food

THE LARYNX

Hyoid bone
Epiglottis
Thyroid cartilage
Corniculate cartilage
Arytenoid cartilage
Cricoid cartilage

Hypoepiglottic ligament
Thyrohyoid membrane
Vestibular ligament
Vocal ligament
Elastic cone

Cricoid cartilage
Elastic cone
Arytenoid cartilage
Glottis
Vocal ligaments
Thyroid cartilage

down the esophagus. The entire larynx rises when you swallow, as you can feel if you put your hand on the front of your throat.

Short as it is, the larynx is relatively complex. Its wall consists of nine pieces of cartilage connected by muscles and ligaments. The epiglottis, a large, spoon-shaped piece of cartilage, drops down over the top of the larynx when you swallow, keeping food out of the airways. Below it is the thyroid cartilage, whose prominent midline ridge is visible as the

Adam's apple—larger in men due to hormones released during puberty. The ring of cricoid cartilage below the thyroid cartilage attaches to the top of the trachea; it is the point of entry for an emergency tracheostomy.

Most important for speech is the pair of arytenoid cartilages at the top and back of the cricoid cartilage. These wedge-shaped structures attach to the vocal cords and to the muscles of the pharynx that move them.

Creating Speech

The vocal cords (or vocal folds) are two bands of tight, elastic ligament covered with white mucous membrane and attached to either side of the larynx. When relaxed, they form a V-shaped opening, allowing air to pass freely into the windpipe. When muscles in the larynx contract, the V closes, leaving a narrow space between the cords. Exhaled air rushing past the tensed cords makes them vibrate, producing "voiced" sounds.

The pitch of those sounds depends on the tension and speed of vibration, which ranges from about 80 to 200 times a second in adults and can exceed 300 times per second in children; the faster the vibration, the higher the pitch. Larger vocal cords,

The versatile larynx can produce sounds ranging from a whisper to the train whistle scream, like the one immortalized by Janet Leigh in the movie *Psycho*.

such as those in men, produce deeper sounds. If the cords are relaxed, air passing through them produces little to no sound, and at a very low pitch. The throat, mouth, nasal cavity, and sinuses all contribute to the voice's resonance. Muscles of the face, tongue, and lips help to articulate the sounds of speech like vowels and consonants.

Most people don't appreciate the complex mechanics of speech until they lose it; laryngitis, an inflammation of the larynx, can make the vocal cords swell until they cannot vibrate, resulting in hoarseness or complete voicelessness. Professional vocalists spend years studying and working with their vocal cords, the resonating chambers in the head, the lift of the soft palate, and other areas of voice production to achieve their high degree of vocal control.

Below the larynx is the trachea, or windpipe. About four to five inches long, it carries air in and out of the bronchi, the two main passageways leading to the lungs. Like the nose and larynx, it is lined with mucous membrane and ciliated cells, which sweep dust and other particles up toward the pharynx and away from the lungs. Highly irritating particles can be expelled at speeds of up to 100 miles per hour in a cough.

Sometimes, though, a piece of food or other foreign object gets stuck in the trachea, blocking the airway and cutting off air supply. In the Heimlich maneuver, an energetic thrust up into the diaphragm forces air out of the lungs and through the trachea, dislodging the blockage. When an immovable obstruction occurs above the larynx, doctors can perform a tracheostomy, cutting into the trachea just below the cricoid cartilage to create a new opening that allows air to flow in and out of the lungs.

KEEPING HEALTHY

VOICE LESSONS

Think of how many times a day you use your voice to communicate ideas to a colleague, talk on the phone, sing in a community choir, or say good night to your family.

A raspy voice or a tremble in the voice can be a sign of illness, fatigue, or other problems. Other signs of voice damage are sore throat, decreased singing range, difficulty talking, and repeated throat clearing. Voice damage can occur for many reasons: a cold or other infection, gastric reflux (heartburn), overuse, or a miscommunication between the brain and the voice-producing muscles. Some types of head and neck cancer can also cause changes in the voice.

The good news is that most voice problems can be treated. See your doctor if you notice a change in your voice that continues for more than two weeks, especially if you have a history of tobacco use.

Even if you don't have a medical problem, you can still be kind to your voice by following these tips:

+ Drink plenty of water and eat foods high in vitamins A, E, and C to help keep the mucous membranes in the throat healthy.

+ Avoid throat irritants, like smoke, alcohol, and spicy foods.

+ Don't overuse your voice by shouting or raising your voice in a noisy room.

+ Keep the air moist with a humidifier.

+ Exercise regularly to increase stamina and muscle tone, and to improve posture and breathing techniques.

THE LUNGS

A Hardworking Pair

The lungs are the central organs of respiration. They are so light that they float in water. In fact, the word *lung* comes from an early Germanic word for "light." In older British usage, *lights* is a synonym for *lungs,* as in the expression "punch your lights out." The well-known lightness of the lungs reflects what is perhaps their key characteristic: They are filled with millions of air pouches, or alveoli, that provide an immense amount of surface area for moving gases both in and out of the bloodstream.

Pale pink at birth, the lungs gradually darken to gray or even black with age as they pick up carbon particles. These two cone-shaped, spongy organs occupy most of the chest cavity within the ribs: The tip, or apex, of each can be found just above and behind the collarbone, while the base rests on the diaphragm above the lowest pair of ribs. Between the lungs, in an area called the mediastinum, are nestled the heart and its attendant veins and arteries, along with the esophagus, trachea, bronchi, and some lymph nodes and nerves.

The "root" of each lung is the bundle of bronchi, lymphatic and pulmonary vessels, and nerves that enters through the hilum, an indentation in the central portion of the lung. The left lung is about 10 percent smaller than the right because it contains the cardiac notch that curves around the heart. The right lung, however, is shorter than the left, because the liver presses the diaphragm up under its base.

The lungs are divided into sections, or lobes: three in the right lung and two in the left. Each lobe is supplied with air by its own bronchus and each has its own artery and vein. The lobes

Each lung is a bundle of bronchi, lymphatic and pulmonary vessels, and nerves.

DID YOU KNOW?

+ Together, the lungs weigh just over 2.5 pounds.

+ More than 2,600 gallons (10,000 L) of air pass through the lungs each day.

+ The body loses about 11 ounces of water every day in the form of exhaled water vapor; people may lose a small amount of weight in their sleep due to this water loss.

are subdivided into segments and then further divided into lobules, about 130,000 in each lung. These polyhedral structures give lungs their slightly lumpy, tiled appearance.

Defense System

The lungs are versatile defenders against airborne invaders: dust, pollen, viruses, bacteria, and other unwanted substances. Specialized cells in the lung tissue form an important part of the immune system. Hairlike cilia on some cells sweep invaders out of the airways; others detect and pick up foreign matter in the air sacs (see "Breathing," page 166).

Encasing each lung is a thin, double-walled covering called the pleura or pleural membrane. The outer wall of the membrane, called the parietal pleura, clings to the inside of the chest (thoracic) wall and to the surface of the diaphragm. The inner layer, the visceral pleura, covers the lung's surface. The pleurae produce a

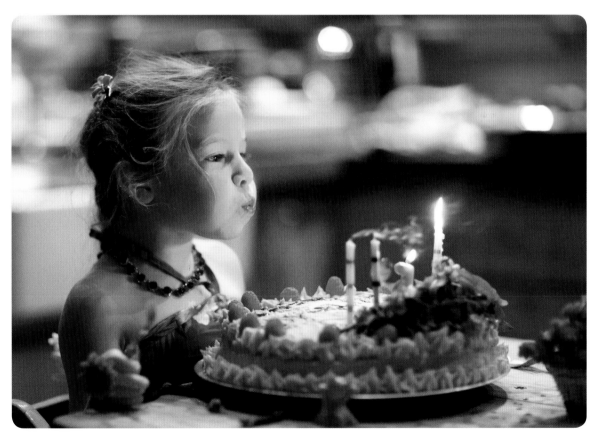

This girl puts her lungs to good use as she blows out her birthday candles.

BRONCHIAL TREE

(Anterior view)

Key of broncho-
pulmonary segments

Right Lung
Superior lobe
Apical
Posterior
Anterior

Middle lobe
Lateral
Medial

Inferior lobe
Superior
Medial basal
Anterior basal
Lateral basal
Posterior basal

Left lung
Superior lobe
Apicoposterior
Anterior
Superior lingular
Inferior lingular

Inferior lobe
Superior
Medial basal
Anterior basal
Lateral basal
Posterior basal

slippery pleural fluid that fills the narrow cavity between the two layers, allowing the membranes to slide easily over each other. The surface tension of the fluid makes the two membranes stick together. When the diaphragm contracts, it pulls the outer pleural membrane with it, which in turn pulls the inner membrane and with it, the lungs, so that they expand.

The pressure within the pleural cavity is less than atmospheric pressure, which also helps to keep the lungs expanded. If air gets into the pleural cavity (say, if it is pierced by a broken rib), the pressure within it equalizes with the atmosphere and the lung collapses. Doctors treat this condition, called a pneumothorax, by inserting chest tubes to suck the air out, reinflating the lung.

Excess pleural fluid is constantly being pumped from the membranes into the lymphatic system. Occasionally the pleural membranes become inflamed, a condition known as pleurisy. In pleurisy's early stages, the two membranes rub together painfully; in its later stages, they may produce too much fluid, which presses on the lungs.

Tree of Life

The lungs enclose a branching network of airways known, appropriately, as the bronchial tree. The trunk of that tree is the trachea, the central airway leading from the larynx. At the trachea's final ring of cartilage, it divides into two large branches, the primary bronchi, one leading into each lung.

Where the two bronchi divide at the trachea is an internal ridge, called the carina, lined with sensitive mucous membrane; any tickling or irritation of the carina leads to violent coughing. The right primary bronchus is shorter, wider, and more upright than the left; inhaled objects—such as food when choking—are more likely to take that path. They typically end up lodged in the right lung rather than the left.

The larger bronchi, like the trachea, are ridged with cartilage and lined with ciliated mucous membrane. The cilia move ceaselessly to propel foreign particles up the airways and eventually out the mouth or down the esophagus into the stomach. As the bronchi become smaller, stiff cartilage gives way to elastic, smooth muscle and the tissues change from ciliated, mucus-secreting cells to a simpler, single layer of cells without cilia.

The primary bronchi divide again in the lungs into secondary (lobar) bronchi, one for each lobe of the lungs. They branch again into tertiary (segmental) bronchi and then into smaller and smaller bronchi and bronchioles until, after about 16 divisions, they taper into the thousands of so-called terminal bronchioles, each one not much wider than a needle.

During exercise, hormones relax and dilate these small airways, or bronchioles, allowing more air to flow through them. But their elasticity has a negative side, as well: Allergic reactions and asthma attacks can close off the bronchioles, choking off the body's supply of oxygen (see "Gasping for Air," page 170).

Blood Connection

Surrounding the entire bronchial tree and covering it like a fine webbing of vines are the veins and arteries that feed into and out of the lungs. No two systems in the body are more closely wedded than the respiratory and the circulatory. Pulmonary arteries from the right side of the heart pump oxygen-poor blood into the lungs, where it flows through branching blood vessels along the bronchi until it reaches the capillaries that cover the alveoli. Carbon dioxide then diffuses from the capillaries into the alveoli as oxygen diffuses from the alveoli into the capillaries. The oxygenated blood returns to the left side of the heart via four pulmonary veins.

Gas Station

Reaching out from the terminal bronchioles are the tiniest twigs on the bronchial tree, the

DID YOU KNOW?

BLACK LUNG

+ **Coal workers' pneumoconiosis** is a scourge of miners, who inhale coal dust into their lungs. Dust particles scar and thicken tissues, causing shortness of breath and, in some cases, emphysema or even heart failure. There is no cure. The disease is less common since the Federal Coal Mine Safety and Health Act of 1969 reduced the coal dust in mines.

DID YOU KNOW?

ALVEOLI

+ **Lungs contain about 170 alveoli** (in purple above) per cubic millimeter.

+ **If stretched flat**, the surface of the alveoli would cover about 861 square feet (80 sq m), roughly the area of a handball court.

+ **Humans are born** with about one-fifth their adult alveoli; they usually reach the full number by age 8.

respiratory bronchioles. Slender tubes called alveolar ducts connect these bronchioles to bumpy alveolar sacs containing a cluster of delicate, cuplike alveoli, each about 0.008 inch (0.02 cm) wide. These are the central respiratory structures of the lungs, the site of gas exchange between the respiratory and circulatory systems. The estimated 300 million alveoli that fill the lungs—each one barely visible to the naked eye—provide a vast amount of surface area for supplying oxygen to the bloodstream and removing carbon dioxide.

Far thinner than tissue paper, the wall of each alveolus consists of a single layer of cells. Slender capillaries surround the alveoli, each one fused to the surface of an alveolus, forming a shared wall known as the respiratory membrane. Only one-sixteenth the diameter of a red blood cell, the membrane allows gases to diffuse easily between the air sacs and the bloodstream.

Surfactant

On the alveolar side of the respiratory membrane, the wall consists of two types of cells. Type I cells, the most common, are the main portals for gas exchange between the alveoli and the capillaries.

The other kind, called Type II, releases alveolar fluid, which coats the surface between the cells and the air. In this fluid is surfactant, a slippery, detergent-like substance that lowers the surface tension of alveolar fluid. Without it, the water molecules in the fluid would cling

A premature baby is treated with artificial surfactants to aid breathing.

together, pulling away from the air and ultimately collapsing the delicate alveoli.

Surfactant is not made in the lungs until the final two months of fetal development. Its absence can be deadly because babies born without surfactant, as with infants delivered prematurely, struggle to breathe. Until the discovery and analysis of surfactants in the 1950s, respiratory distress syndrome (RDS) was the leading cause of death in premature infants. It caused the deaths of some 10,000 babies a year in the United States alone. Today, pregnant women going into premature labor can be injected with steroids to encourage the fetus to begin developing surfactant on its own. Premature infants with RDS are treated with positive-pressure respirators and natural or artificial surfactants to aid their breathing. Since the development of these treatments, deaths from RDS have decreased to about 1,000 a year.

Within the fluid along the lungs' alveolar walls are millions of macrophages, cells that eat and destroy invading microbes and foreign particles. These vigilant guard dogs of the immune system help keep the alveoli remarkably clean. When their task is accomplished, the macrophages themselves are swept up into the bronchial tubes and out of the lungs.

The airways, with their easy access to the bloodstream, can provide a direct pathway for medicine that can be inhaled in a fine mist of droplets. Many asthmatics are familiar with nebulizers that deliver medicine to dilate the bronchial tubes; nebulization can also carry antibiotics directly into the lungs. Medicines delivered in this way may be more quickly absorbed into the bloodstream, which can result in faster onset of action and a lower risk of systemic side effects.

BREAKTHROUGH DANGERS OF SMOKING

On Saturday, January 11, 1964, reporters were ushered into the auditorium of the State Department in Washington, D.C., shut in without telephones, and handed a 387-page report entitled "Smoking and Health: Report of the Advisory Committee to the Surgeon General of the Public Health Service." For the first time, the U.S. government officially acknowledged what scientists had long known: Cigarette smoking damaged lungs, right, and caused lung and laryngeal cancer and chronic bronchitis. After 90 minutes, reporters were allowed out of the auditorium to spread the news to the public. It became one of the leading stories of the year. Despite a vigorous defense of smoking from tobacco companies, the U.S. Congress quickly passed laws to require health warnings on cigarette packs, ban cigarette advertising on television, and require further annual reports on cigarettes and health. The campaign to end smoking, jump-started by the 1964 report, dropped smoking rates among U.S. adults from 42.4 percent in 1965 to 20.9 percent in 2005, a new low. Increased awareness of the dangers of secondhand smoke has prompted bans on smoking in many workplaces and public spaces.

BREATHING

The Mechanics of Inhaling and Exhaling

The steady, even in-and-out motion of breathing—called pulmonary ventilation, in medical terms—relies on an elegantly simple mechanism of pressure changes within the chest. It also depends on a complex feedback system between the body and the brain.

The chest, or thoracic cavity, is enclosed by the rib cage and the intercostal muscles, which are located between the ribs. The diaphragm, a dome-shaped muscle, forms the floor of the thoracic cavity. When relaxed, the diaphragm curves up under the lungs. When it contracts, it pulls down and flattens out.

Within the thoracic cavity, the lungs are attached to the chest wall and to the diaphragm by the double-layered pleural membrane. Pleural fluid fills the thin cavity between the layers; the fluid holds the layers together while letting them glide freely over each as the lungs expand and contract. Also helping the layers hold together is the low pressure inside the pleural cavity: Normally at about 756 mmHg (millimeters of mercury), it is below the atmospheric pressure of 760 mmHg.

Balancing Act

The air pressure inside the lungs themselves varies, depending on whether a person is inhaling (breathing in) or exhaling (breathing out). Just before each inhalation, air pressure inside the lungs is the same as air pressure outside. The lungs are, essentially, at rest.

RESPIRATORY ACTS	
YAWN	**A deep inhalation, with the jaws wide open; cause unknown**
CRY	**Very similar to laughing, with different sounds and facial expressions**
SNEEZE	**Exhalation muscles contract, forcing air through the nasal passages**
LAUGH	**An inhalation followed by a number of short exhalations; the glottis stays open and the vocal folds vibrate**
HICCUP	**Spasms of the diaphragm followed by short, sharp inhalations; the glottis closes, making a sharp sound with each inhalation**
COUGH	**A deep breath followed by a closing of the glottis; then a sharp blast of air from the lungs forcing the glottis open and leaving through the mouth**

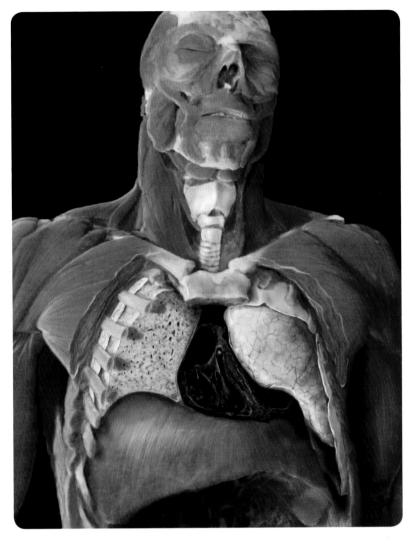

A cutaway model of the muscles used in respiration exposes the intercostals around the ribs and the dome-shaped diaphragm.

the pressure. Thus, as the lungs expand, the air pressure within them drops until it is about 1 to 3 mmHg lower than the air pressure outside. This pressure difference is enough to trigger an inflow of air to equalize the pressure; typically, about 500 milliliters (a little more than 2 cups' worth) restores the balance.

During an exhalation, the respiratory muscles relax. The diaphragm regains its rounded shape and the rib cage returns to its narrower contours. The lungs recoil, the volume within them dropping, which increases the air pressure inside until it surpasses the pressure outside the body. Air from the lungs (now containing carbon dioxide that has diffused from the capillaries) is then expelled as the lungs again equalize the pressure inside and out. Not all the air leaves the lungs during exhalation, however; a residual amount remains, keeping the alveoli open and the lungs inflated.

During an inhalation, the diaphragm contracts, pulling the floor of the chest down. At the same time, the external intercostals pull the ribs up and out. This increases the volume within the chest cavity and, with it, the volume of the lungs, which are attached to the chest wall.

A basic rule of physics, Boyle's law, states that at a constant temperature the pressure of a gas varies inversely with its volume: The greater the volume for a given amount of gas, the lower

This is the process of quiet inhalation and exhalation, the sort of breathing one does at rest. During exercise, while singing, or sometimes when a person is suffering from an obstructive pulmonary disease (such as asthma, emphysema, or pneumonia), respiration becomes a more active process and uses additional respiratory muscles. In forced

ALVEOLI

Alveolar ducts, sacs, and alveoli

Respiratory bronchiole

Terminal bronchiole

Smooth muscle

Alveolar duct

Alveolar sac

Capillaries

Alveolus

inhalation, muscles in the neck, in the front of the chest, and around the spine help to pull the rib cage out. Abdominal and internal intercostal muscles work together to force exhalation.

Ease of breathing is affected both by airway resistance and by lung compliance, a measure of flexibility. Air flowing through bronchi and bronchioles faces resistance from the walls of the passageways, just as liquid flowing through plumbing meets resistance in the walls of the pipes. The narrower the passageway, the greater the resistance. Some diseases— asthma, for instance (see "Gasping for Air," page 170)—thicken the walls of the airways and

BREAKTHROUGH THE IRON LUNG

The iron lung, right, was once a lifeline for people with respiratory paralysis from polio, and other causes. The 1885 prototype, the "pneumatic cabinet," was a body-size box enclosing the patient. An attendant moved a rubber membrane, and changes in air pressure inflated and deflated the patient's lungs. A vacuum-pump-operated iron chamber that left the patient's head outside followed by 1929. "It looked more like a torture chamber than anything else," noted an observer. In the late 1950s, small positive-pressure ventilators began replacing the big machines, signaling the end of an era.

increase resistance to airflow, making breathing more difficult. Some illnesses, such as tuberculosis or pulmonary fibrosis, scar or stiffen the lung tissue, decreasing compliance and making breathing more difficult. A few, such as emphysema, weaken the tissue, which also makes it less elastic and reduces the flow of oxygen into the blood.

Each breath draws air down through the bronchi and bronchioles into the alveoli. Here, in the air sacs, occurs the central process of external respiration, in which oxygen enters the bloodstream and carbon dioxide leaves it. Internal respiration describes the exchange of gases between the body's bloodstream and its cells and tissues.

In internal respiration, blood pumping through the body gives up its oxygen to the cells, which use it to break down glucose for energy, releasing carbon dioxide as a waste product. The oxygen-poor, carbon-dioxide-rich blood returns to the heart and is pumped through the right atrium and ventricle and then into the lungs. There, blood vessels in the presence of low oxygen concentrations constrict and drive the deoxygenated blood toward oxygen-rich areas, so that there is always a close match between blood that needs oxygen and alveoli that have it. (The opposite happens in most of the body's other tissues; arterial blood is directed toward, not away from, oxygen-poor areas.)

Once the oxygen-starved blood has reached the capillaries around the alveoli, oxygen and carbon dioxide diffuse in opposite directions through the respiratory membrane. The gases are driven to do this by differences in pressure: Every gas exerts its own pressure, and each will move from an area of greater pressure to an area of lower pressure until it reaches equilibrium. Normally, oxygen in the alveoli exerts a pressure

DID YOU KNOW?

SNEEZING

Sneezes have generated their share of "urban legends."

+ Your heart stops for a moment when you sneeze. **Not true: Your heart rate might change briefly, but your heart does not stop.**

+ You can't keep your eyes open during a sneeze. If you do, your eyeballs will pop out. **Wrong: It's difficult to do, but some people can manage it—and keep their eyeballs safely in their heads.**

+ A woman won't get pregnant if she sneezes immediately after sex. **Not so: One sneeze, or several, will not discourage or dislodge 300 million determined sperm.**

of 104 mmHg, while that in the pulmonary arteries is only 40 mmHg, so alveolar oxygen diffuses rapidly through the respiratory membrane between the alveoli and the capillaries until both sides have equal oxygenation—a process that takes about three-quarters of a second.

Meanwhile, carbon dioxide travels in the opposite direction from the blood to the alveoli. Although the difference between carbon dioxide's pressure in the blood and its pressure in the alveoli (45 mmHg to 40 mmHg, respectively) is

GASPING FOR AIR

Asthma is a common, chronic, and, for now, incurable respiratory disease. About 20 million Americans are afflicted with asthma, including about 1 in 11 children and 1 in 12 adults. Every year, nearly 4,000 Americans die from it, and the prevalence in the United States continues to grow. Although the exact cause of the disease is unknown, recent research has produced some promising discoveries that may lead to new ways to treat and even prevent asthma in the future.

The symptoms of asthma wax and wane in most people, but a typical episode is marked by wheezing, whistling, or noisy breaths; coughing; tightness in the chest; and shortness of breath. In the most severe attacks, the untreated asthma sufferer will become so short of oxygen that he or she may lose consciousness.

HISTORY

The word *asthma* comes from the Greek word meaning "to gasp." Until the 18th century, Western medicine typically treated asthma attacks as an imbalance of the four humors (yellow bile, black bile, blood, and phlegm) with asthmatics considered to be excessively phlegmatic. In the 18th and 19th centuries, physicians were able to study the disorder more closely in hospitals and apply their new understanding of anatomy and lung function to the disease. They began to understand the link between asthma and allergens and pollution: Many asthma sufferers found relief in moving to clean, dry, desert environments. By 1900, physicians also began to prescribe epinephrine to relax the airways during an asthma attack.

Asthma's symptoms are caused by a narrowing of the bronchi, brought on both by inflammation—swelling—of the tissue in the airways and by a tightening of the smooth muscles in the bronchial walls. Bronchial cells may also produce more mucus than usual, further clogging the passageways. Air struggling to flow through the narrow airways will make a whistling sound. The airways also become more sensitive and are likely to tighten up quickly in reaction to small quantities of an irritant.

TRIGGERS

Many things can trigger an asthma attack, and they vary from person to person and episode to episode. Among the triggers are exercise (particularly in cold weather); allergens, including pollen (from trees and grass), mold, animal dander (from the skin, hair, or feathers of animals), dust mites in house dust and cockroach excretions; indoor and outdoor pollutants such as cigarette smoke, car exhaust and other air pollution, industrial pollutants, and paint fumes; medicines, including aspirin, NSAIDs (nonsteroidal anti-inflammatory drugs, such as ibuprofen), or beta-blockers (medications used to treat high blood pressure and certain heart conditions); respiratory infections such as a cold or the flu; and stress.

The underlying causes of asthma—the reasons why the airways become inflamed in the first place—are still being investigated. Genes play a key role. A child with one asthmatic parent has about a 20 to 30 percent chance of developing the condition based on heredity alone; with two asthmatic parents, the odds rise to 60 to 70 percent. But environmental conditions also play a role. Studies show that people in urban areas have a higher risk while children raised on farms have a lower risk. Some researchers believe that early exposure to the diversity of microbes (germs) found on farms may protect against asthma.

Asthma may begin with an immune reaction either to an infectious agent such as a virus, or to a foreign body such as a pollutant or mold. The immune system mobilizes lymphocytes, which react to these invaders in two ways. They may call up increased numbers of allergy cells called mast cells, which release irritating chemicals that cause inflammation. Alternatively, they may stimulate the production of antibodies, particularly Immunoglobulin E (IgE), an antibody that also produces chemical irritants.

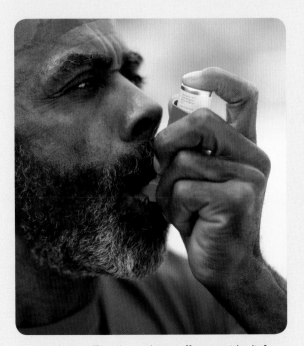

An inhaler offers this asthma sufferer rapid relief.

Until recently, researchers believed there was one basic type of asthma that varied by time of onset (childhood or adulthood) and severity (mild, moderate, or severe). New evidence points to several different variations—or phenotypes—of the disease, based on the amount of inflammation, the severity of symptoms, timing of onset and other factors, such as obesity. This information may lead to new targeted treatments. For example, people with obesity-associated asthma may benefit from weight loss as well as treatment with bronchodilators.

TREATMENTS

In addition to avoiding environmental triggers, medical treatment for asthma usually combines quick-relief drugs taken when wheezing begins and controllers, which are taken every day to prevent attacks. The task of rapid relief falls to bronchodilators, which relax the muscles of the airways. Many of these are derived from synthetic epinephrine, a hormone that in its natural form is produced by the adrenal glands. A bronchodilator is typically delivered as a spray from a handheld inhaler, in measured doses. Patients may also use a nebulizer, which delivers the medicine in a fine mist through a mouthpiece or mask. Long-term controllers typically mean corticosteroids, which are anti-inflammatory medicines. In recent years, the ranks of controllers have grown to include leukotriene modifiers, which help block inflammation. Another new treatment that involves the use of antibodies to counteract the misdirected immune response shows promise in severe cases.

The good news about asthma is that it is usually controllable by following the recommended treatment. Most people with asthma learn how to manage the disease and find that it does not get in the way of an active and healthy life.

KEEPING HEALTHY

IS LAUGHTER THE BEST MEDICINE?

In his 1979 memoir *Anatomy of an Illness*, Norman Cousins wrote that he could alleviate a painful spine condition for two hours ten minutes by laughing at sitcoms and Marx Brothers films. Through the years, researchers and patients alike have made other claims for laughter as medicine.

In the traditionally lighthearted but well-researched 2013 Christmas issue of *BMJ*, the *British Medical Journal*, researchers published a review of the literature on the health effects of laughter. Scouring the medical literature from 1946 to June 2013, they found a variety of articles outlining the benefits—and harms—from laughter.

Among the findings: Laughter burns about 40 calories every 15 minutes. It may also reduce the risk of heart attack, increase pain thresholds, and improve lung function. However, the benefits do not come without risks. Laughter may also induce urinary incontinence, abnormal heart rhythms, and fainting.

On the whole, the researchers reported that "laughter in any form carries a low risk of harm and may be beneficial." Laughter may not be the best prescription, but it can still be good medicine.

much less than that of oxygen, it diffuses more rapidly because it's much more soluble in plasma (the noncellular fluid portion of circulating blood) and alveolar fluid.

A little (about 1.5 percent) of the oxygen entering the bloodstream from the lungs dissolves in the plasma. The rest latches on to hemoglobin, a protein found in red blood cells, turning it bright red. Carbon dioxide, meanwhile, enters the lungs from the bloodstream, where some (about 20 percent) was bound to a different portion of the hemoglobin molecules, and the rest was either dissolved in the plasma or riding along in the form of bicarbonate ions made from carbon dioxide and water.

Other gases can occasionally hijack this efficient mechanism. For example, hemoglobin has an affinity for carbon monoxide, and if exposed to this colorless, odorless gas will bind to it quickly, blocking out oxygen. People who breathe in carbon monoxide (say, in gases from a fire) become starved of oxygen and can die without treatment. In many hospitals, treatment involves hyperbaric therapy, in which the patient is held in a sealed chamber and breathes 100 percent oxygen at high pressure, forcing oxygen back into the tissues.

The Brain's Role

The many patterns of breathing—the quiet in-and-out of the resting body, the vigorous panting of exercise, sobbing, laughter, gasping with cold—are controlled by a network of respiratory centers in the brain stem and receptors elsewhere in the body. Although it is possible to consciously regulate breathing for short periods, control soon reverts to the respiratory centers, which are responsible for sending the right amount of oxygen to the right place at the right time.

Two closely related groups of neurons in the brain stem—the inspiratory center and the expiratory center—control the basic rhythm of inhalation and exhalation. In the typical cycle of quiet breathing, the inspiratory center sends a series of signals, lasting two seconds, to the diaphragm and external intercostals (rib muscles), prompting them to contract, expanding the chest cavity and drawing air into the lungs. Then the center becomes inactive for three seconds; the diaphragm and intercostals relax and the lungs recoil, exhaling air.

During forceful breathing—while exercising or singing—the inspiratory center also activates the expiratory center, which prompts the internal intercostal and abdominal muscles to contract during an exhalation, adding oomph to each expelled breath. Other sites in the brain stem—the pneumotaxic area and the apneustic area—help to coordinate the exact timing of inhalations and exhalations as well.

The brain's cerebral cortex controls voluntary breathing, though it can be overruled by the respiratory centers if oxygen runs low. Emotional and pain stimuli travel from the hypothalamus, at the base of the cerebrum, to the respiratory centers, prompting rapid breathing when angry, for example, or gasping when in pain.

The job of monitoring oxygen and carbon dioxide levels in the blood falls to chemical receptors both in the brain stem and on the walls of the aorta and the carotid arteries. If the amount of carbon dioxide in the blood rises, or

During singing the inspiratory center activates the expiratory center, which causes the muscles around the ribs and abdomen to contract.

This extreme skier's breathing rate is likely affected by several factors, including high altitude, exercise, and excitement.

if the amount of oxygen falls, these receptors signal the inspiratory center to increase the breathing rate, bringing in more oxygen and expelling more carbon dioxide until the blood gases return to normal levels.

Hyperventilating—breathing unusually rapidly and deeply—can build up oxygen levels and drop carbon dioxide in the blood to the point where the chemical receptors no longer send signals to the respiratory area. People who hyperventilate may be able to hold their breath an unusually long time, which sometimes lowers their oxygen level until they faint.

None of the finely tuned mechanisms that regulate breathing is needed until the moment of birth. Until it takes its first breath outside the womb, the fetus does not use its lungs at all; they are collapsed and filled with amniotic fluid (quickly drained at birth). Instead, the fetus picks up all the oxygen it needs through the umbilical cord from the mother's placenta. The moment of birth sees dramatic changes in the newborn baby. With oxygen from the mother cut off, carbon dioxide levels build up rapidly in the baby's blood. This activates the respiratory center in the brain stem, which signals the respiratory muscles. The baby takes a deep, gasping breath and cries: the first sign of a pair of healthy lungs.

Exercise Effects

Several systems come into play at once during exercise. Just the anticipation of movement appears to activate the limbic system, a part of the brain that deals with emotions and behavior. The limbic system then signals the respiratory center to speed up breathing. As soon as a person begins to exercise, proprioceptors (specialized groups of nerve endings within tendons, joints, and muscles that monitor movements) and neurons in the brain's motor cortex also signal the respiratory area that it's time to breath more deeply.

As increased activity continues, the respiratory center responds to the decreased oxygen and increased carbon dioxide in the blood, as monitored by chemical receptors. All these signals together prompt deeper, faster breathing; the heart, beating rapidly, pumps more blood into the lungs to pick up the additional oxygen. These mechanisms can increase the body's ventilation by 10 to 20 times its resting rate during really vigorous exercise.

Other stimuli can speed up or stop breathing, as well. Body temperature, for instance, has a direct effect on the rate of respiration, speeding it up when the body is hot (as in a fever) and slowing it when the body is cold. A rise in blood pressure can slow respiration, while a drop can accelerate it. Sudden pain can make a person stop breathing briefly, but prolonged pain may speed up respiration. Anxiety, fear, or anger can also ratchet up breathing, perhaps as part of a flight or fight response that triggers the body to prepare for self-preservation.

Exposure to high altitudes, where oxygen is thin, can leave a person short of breath after even modest exercise. After a few days, the body becomes acclimatized to the altitude; oxygen uptake increases slightly but will not return to the amounts found at sea level. Skiers and mountaineers often suffer a mild form of altitude sickness as the body adjusts. Astronomers living at 13,780 feet (4,200 meters) on Hawaii's Mauna Kea have a blood oxygen content similar to that of patients with severe emphysema.

DID YOU KNOW?

HOLDING YOUR BREATH

You cannot hold your breath until you die. Respiratory centers in the brain stem are constantly monitoring the level of carbon dioxide in your blood, a level that rises steadily as you hold your breath. When carbon dioxide reaches a critical amount, the brain overrides your conscious control and forces your body to take a breath.

Most people can hold their breath for a minute to a minute and a half, on average. The world record for breath-holding belongs to Dane Stig Severinsen, who held his breath underwater for 22 minutes while floating in water. Don't try this at home: Swimmers who hyperventilate before submerging, to lower their blood carbon dioxide and hold their breath longer, risk drowning by passing out in the water from lack of oxygen.

Chapter 7

BODY FUEL

DIGESTIVE SYSTEM

From Food to Energy

Though some people live to eat, everyone must eat to live. Food nourishes the body by providing nutrients that are essential for the functioning of the heart, nerves, muscles, and glands. It also fuels growth, tissue repair, and the maintenance of organs and systems. In short, without food the body would soon die. The habit of eating good food is so important that food plays a major role in holidays and rituals. We celebrate life by partaking of that which gives us life, enjoying the sight, smell, and taste of food.

Fueling the Body

What does the digestive system do? We stoke the furnace by eating, but then the digestive system automatically takes over, transforming food into chemical compounds that can go through the membranes of a cell. From milk, for example, the digestive system extracts molecules of nutrients from every category it needs—proteins, carbohydrates, fats, vitamins, minerals, and water. From bread we get carbohydrates and fats, as well as some proteins and minerals.

The journey from slice of bread to molecule of carbohydrate takes place in the gastrointestinal (GI) tract, or alimentary canal, a 30-foot-long tube that begins with the lips and ends with the anus. The teeth begin the breakdown process by tearing, mashing, and grinding the food. While the tongue is helping distribute the food around, salivary glands located beside and beneath the mouth cavity secrete saliva and mucus to moisten it into a soft paste.

Chewing is primarily a voluntary activity; but once the food is swallowed, the reflexes take over. The food slides down the throat and esophagus toward the stomach. The chemical transformation of food begins there. Acidic

The digestion process starts at the first bite of food.

DIGESTIVE SYSTEM

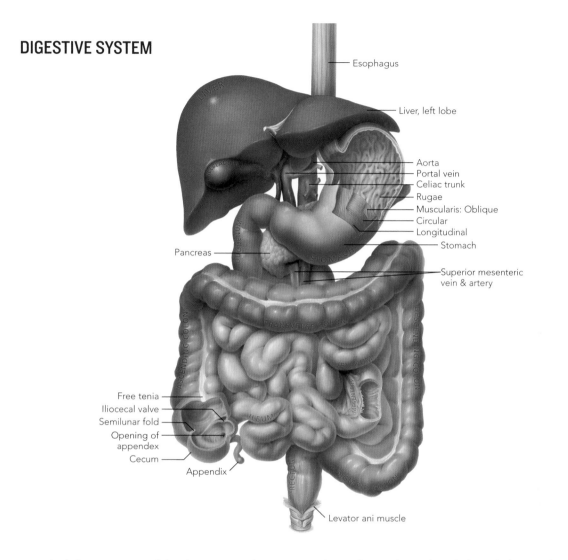

Esophagus

Liver, left lobe

Aorta
Portal vein
Celiac trunk
Rugae
Muscularis: Oblique
Circular
Longitudinal
Stomach

Superior mesenteric vein & artery

Pancreas

Free tenia
Iliocecal valve
Semilunar fold
Opening of appendex
Cecum

Appendix

Levator ani muscle

gastric juices secreted in the stomach turn the food into a soupy mixture, which is then churned until it passes into the small intestine. Hydrochloric acid secreted by the stomach is strong enough to burn through a thick rug, but a lining of mucus protects the stomach from a similar fate.

Once the food passes into the small intestine, the liver and the pancreas aid the chemical breakdown. These two glandular organs outside the GI tract are connected to it by small-caliber tubes (ducts). Situated on about stomach level, the triangular-shaped liver fits

into the right side of the abdominal cavity beneath the right diaphragm. Among the liver's many functions are decomposition of old red blood cells, detoxification of alcohol and other poisons, production of clotting proteins, and storage of carbohydrates and some proteins and fats. Its main role is to produce bile salts, which help absorb fats in the small intestine. Bile is concentrated and stored in the gallbladder.

The pancreas is a large gland that secretes digestive enzymes into the small intestine. These enzymes break down proteins, carbohydrates, and fats into compounds that can be further processed within the small intestine.

The longest part of the digestive system, the small intestine measures about 21 feet in a cadaver, but about 10 feet in a living person because of a normal state of partial muscle contraction. This coiled organ folds into the lower part of the abdominal cavity. Most digestion occurs within the small intestine, aided by the liver, pancreas, and glands within the small intestine itself, which produces about one quart of intestinal juices a day. As the food is traveling through the small intestine, it is absorbed through the cells lining the intestine and delivered to the cells of the bloodstream and lymph system. To increase the surface area of the small intestine some eight times to about 2,000 square feet, it is lined with millions of tiny projections called villi. From pores in the villi, digested nutrients pass into the capillaries and thus into the bloodstream.

Like a widening river, the small intestine joins with the large intestine, or colon. This final section of the bowels frames the small intestine, by ascending on the right side of the body, traversing the body just above the coils of the small intestine, and dropping down the left side. It then takes an S-like turn to the right as the sigmoid colon and joins with the muscular rectum, which connects to the anus, the posterior opening of the GI tract. Though it absorbs water, the main function of the large intestine is to move waste out of the body.

A miraculous processing factory, the digestive system is constantly at work. Each step in the process and each organ will be detailed in the rest of the chapter.

Digestive Processes

The process of digestion has five main divisions: ingestion, propulsion and peristalsis, mechanical and chemical digestion, absorption, and defecation. Digestion begins with ingestion, the taking in of food—in other words, eating and drinking. Once food is in the GI tract, propulsion (with the help of gravity) moves it along. Propulsion begins with the act of swallowing, a voluntary or involuntary activity. From then on propulsion is accomplished by peristalsis, the involuntary contraction and relaxing of muscle walls. Through peristalsis food reaches your stomach even if you stand on your head. In addition to transporting the food along the GI

This color-enhanced reconstruction shows the intestines in white.

tract, propulsion serves to mix the food with secretions from within the cells of the tract and the accessory organs. Some seven quarts of water, acid, enzymes, and buffers are daily secreted into the tract.

Mechanical digestion occurs inside the mouth, when the teeth break up food into smaller units that, when mixed with watery saliva, may be swallowed. While the first part of ingestion obviously involves conscious effort—eating food—the tongue and cheeks work deliberately and automatically to move it around until it is ready for swallowing. Though chewing is not strictly necessary for digestion, it helps: Careful chewing of the food is beneficial to maintaining a trouble-free digestive system. The churning of food within the stomach and intestines by peristalsis also comes under the heading of mechanical digestion.

In chemical digestion large molecules of carbohydrates, fats, and proteins are broken into smaller molecules by hydrolysis (the splitting of bonds resulting in the addition of the elements of water). Chemical digestion starts in the mouth with the secretion of enzymes by the tongue and salivary glands. In the stomach and small intestine, more enzymes are introduced, some from the pancreas. A few substances within food need no chemical digestion, including vitamins, minerals, glucose, cholesterol, amino acids, and water.

Absorption—the passing of molecules, ions, and ingested and secreted fluids through the plasma membranes of epithelial cells lining the GI tract—is the ultimate goal of the digestive system. These end products of digestion move into the blood or lymph system and circulate to cells in the body.

Unabsorbed material (indigestible substances, bacteria, and cells sloughed off from

DID YOU KNOW?

HERBAL REMEDIES

Herbal treatments have long been used for a variety of ailments. Health professionals caution that pregnant women, the very young or old, or anyone on pharmaceuticals should consult their doctor before taking herbal medicines. Some useful herbs include ginger as a treatment for motion-sickness nausea and garlic for lowering cholesterol.

the lining of the GI tract) gets shunted out of the body through defecation. All this waste exits the body through the anus and is called feces.

Food & Nutrition

To stay healthy, people need nutrients—chemical compounds used for energy and for cell growth, repair, and maintenance. The most essential nutrient is water: Humans require two to three quarts a day. The organic nutrients—carbohydrates, fats, and proteins—compose the major part of the diet. Carbohydrates are compounds made of carbon, hydrogen, and oxygen; they take the form of sugars and starches. Simple carbohydrates (sugars) come from fruits and vegetables; complex carbohydrates (starches) derive from breads and cereals.

Of the sugars, the simple monosaccharide glucose is the one that fuels the cells. Neurons and red blood cells depend almost exclusively on glucose for energy. A temporary shortfall of glucose can lower brain function and cause body systems to shut down. Hence carbohydrates are the body's workhorses. Although the body can rely on fats and proteins, carbohydrates provide a ready source of glucose. Current thinking calls for five daily servings of fruits and vegetables, and 6 to 11 servings of bread, cereal, or pasta.

Lipids (organic compounds not soluble in water) provide the most concentrated form of potential cell energy. Saturated fats come from meat and dairy products, unsaturated fats from nuts, seeds, and vegetable oils, and cholesterol from egg yolks, liver, and milk. Though "fat" has become the bad guy of the diet, the body needs it for many reasons: for the absorption of fat-soluble vitamins; as a cushion around organs; as insulation beneath the skin; and as an easily stored energy source.

DID YOU KNOW?

CALORIES

A calorie is a unit of heat energy—the heat needed to raise the temperature of one gram of water by one degree Celsius. In dietary terms, large calories (or kilocalories) are used to measure the energy-producing potential contained in food. Depending on size and exercise routine, about 1,700 to 5,000 large calories are burned per day.

Essential proteins come in high doses in eggs, milk, and meat. Nuts, legumes (beans and peas), and cereals also contain protein but lack important amino acids, the building

Fruits and vegetables are packed with a variety of vitamins, which are an essential part of the diet.

blocks of protein. Leafy vegetables have low amounts of the most essential amino acids. Proteins are used in forming skin and muscle tissue and regulating enzymes, hormones, and other functions. Nutritionists suggest consuming protein equivalent to about 0.08 percent of a person's body weight. A 150-pound (68 kg) person should eat about 2 ounces (56 grams) of protein a day.

The body needs a variety of foods to get the nutrients it needs. No one food can provide all the necessary vitamins and minerals. Some vitamins, such as B, D, and K, are synthesized within the body. Since vitamins assist the other nutrients in doing their jobs, they are an essential part of the diet. Vitamins A, C, and E act as antioxidants that neutralize free radicals and thus help prevent some cancers. Though vitamin supplements may be of some help, the best source of all essential vitamins is a well-balanced diet. Finally, the human body requires moderate doses of seven minerals—calcium, phosphorus, potassium, sulfur, sodium, chloride, and magnesium—as well as trace amounts of some 12 others. Vegetables, legumes, and milk are good sources. Like vitamins, minerals work in concert with other nutrients. Iron binds to red blood cells so that they can carry oxygen. Calcium, phosphorus, and magnesium provide strength and hardness to bones and teeth. Ions of sodium and chloride in the blood help to keep body fluids balanced. And potassium maintains osmotic pressure in cells and helps conduct nerve impulses.

VITAMINS

How vitamins help the body

VITAMIN A	Maintains skin cells, helps form light-sensitive pigments, used in bone growth; found in liver and milk, also formed in GI tract from substances found in yellow and green vegetables
VITAMIN D	Necessary for absorption of calcium and phosphorus; UV light converts 7-dehydrocholesterol in the skin to vitamin D_3, which is then converted to active form by enzymes in the liver and kidneys; food sources include fish liver oils, egg yolk, and fortified milk
VITAMIN E	Inhibits breakdown of unsaturated fats, involved in forming DNA, RNA, and red blood cells; dietary sources include nuts, wheat germ, seed oils, and leafy vegetables
VITAMIN K	Essential for synthesis of clotting proteins and others produced in the liver; synthesized by bacteria in large intestine, also gleaned from spinach, cauliflower, cabbage, and liver
VITAMIN C	Used in formation of connective tissues, healing of wounds and creation of antibodies, and as an antioxidant; sources include citrus fruits, tomatoes, and green vegetables
VITAMIN B_1	Aids in the metabolism of carbohydrates, essential in production of neurotransmitter acetylcholine; found in eggs, whole grains, lean meat, and liver
VITAMIN B_2	Functions as coenzyme in carbohydrate and protein metabolism, especially in eye cells, mucosa of intestine, and blood; sources include milk, yeast, liver, egg whites, whole grains, and legumes

INGESTION

The First Step of Digestion

Putting food in your mouth is the first step in digestion. At that point accessory organs—the teeth, tongue, and salivary glands—take over. The boundaries of the mouth, formed by the lips and cheeks, help keep food in position while you chew. The palate forms the roof of the mouth. In the anterior, or front part of the mouth, the hard palate is composed of bone covered by a mucous membrane. In the posterior part of the roof of the mouth, the soft palate is formed of muscle; at the back of the soft palate hangs a muscular process called the uvula. When we swallow, the soft palate and uvula rise to prevent food from entering the nasal cavity. To see how this works, try breathing and swallowing at the same time.

Defining the floor of the mouth, the tongue is composed of skeletal muscles covered by a mucous membrane. The top and sides of the tongue are covered with little projections called papillae, many of which contain taste buds. While we're enjoying the taste of the food, the tongue sees to it that food becomes thoroughly chewed. Lingual glands on the tongue also chip in by secreting mucus and an enzyme called lingual lipase which helps break down lipids.

When the tongue has maneuvered the food around and mixed it with saliva, the food becomes shaped into a wet lump called a bolus. Once the bolus is positioned at the back of the tongue, it can be swallowed and projected into the pharynx and then to the esophagus. Muscles within the tongue allow it to change shape—it can be thick or thin, rounded or flat—to accommodate the needs of speech and eating. A membrane called the lingual frenulum attaches the tongue to the floor of the mouth and restricts its movement.

To moisten the food and start the chemical digestion process, salivary glands excrete saliva into the mouth. Even with no food, saliva is

DID YOU KNOW?

SALIVA

About 97 to 99.5 percent of saliva is water. The rest is an assortment of useful chemicals. Some buffer acidic foods. The antibody immunoglobulin A keeps microbes from getting into the epithelium (covering and lining tissue). Contributed mostly by the parotid glands, salivary amylase is a digestive enzyme that helps break down starch.

present in small amounts to keep the mouth wet and the teeth clean. With eating, saliva production increases. A person produces about 1 to 1.5 quarts (1,000 to 1,500 mL) of saliva a day.

The three main pairs of salivary glands lie outside the oral cavity, though some saliva is produced by glands inside the mouth. The largest glands, the parotid glands, are situated along the cheek and jawline in front of each ear, tucked between the skin and muscles. The walnut-size submandibular glands lie within the jaw. The smaller sublingual glands are farther forward, just under the tongue. Mumps is a virus that attacks the parotid glands, causing fever, malaise, sore throat, and swelling of the cheeks.

The sight or smell of food is often enough to trigger salivary flow. Just the thought of a hot waffle with butter and maple syrup might make a hungry person's mouth water. Chemical receptors in the mouth send impulses via the nervous system to salivatory nuclei in the brain stem, which then signal the salivary glands to begin production in anticipation of a meal.

Chewing & Swallowing

Chemical digestion begins in the mouth, when salivary amylase and lingual lipase, released from the salivary glands and tongue, interact with the food. Assisting the tongue and the salivary glands, the teeth are also considered digestive organs because they participate in

The tongue gets the digestive process started by secreting an enzyme that helps break down fats.

DID YOU KNOW?

When you swallow, your larynx rises, and the epiglottis pops down to seal off the respiratory passages. If you talk and eat at the same time, food can get stuck in the larynx. Without air getting to the lungs, the coughing reflex does no good. The Heimlich maneuver, below, has saved many chokers from death through a jab just above their navels.

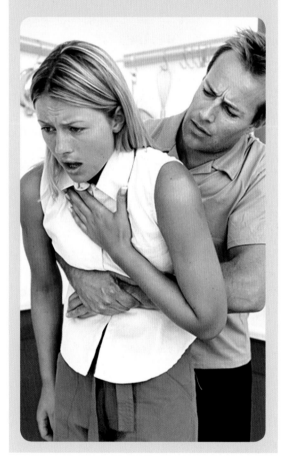

absorb monosaccharides, thus the starches must be simplified to be of use. Salivary amylase can break bonds between glucose units and reduce starches to di- and trisaccharides, a process that can continue—once the saliva is mixed in—for an hour after swallowing. The lingual lipase does not even begin working until it is in the stomach. There, stimulated by stomach acids, it breaks triglycerides into diglycerides and fatty acids.

Down the hatch the food goes, the hatch being a funnel called the pharynx (throat). The pharynx is made of skeletal muscles lined with mucosa and is employed for breathing and digestion. During the act of swallowing, or deglutition, the bolus of food moves into the oropharynx and then involuntarily into the laryngopharynx. Repeated constriction of muscles in the pharynx moves the food down to the esophagus, which is positioned behind the trachea (windpipe).

The act of swallowing not only makes the soft palate and uvula seal off the nasopharynx; it also signals the larynx to rise and a large piece of cartilage called the epiglottis to cover the larynx (voice box) and block access to the trachea. Without this action, food would end up choking you. With the epiglottis covered, the vocal cords constrict, widening the tunnel to the esophagus. In seconds the food is through, and breathing can resume.

The esophagus is a 10-inch-long (25 cm) tube that collapses when not in use so that its curving walls can almost touch each other. From the end of the laryngopharynx, the esophagus passes through the diaphragm (the muscle between the thoracic and abdominal cavities) at an opening called the esophageal hiatus. A hiatal hernia occurs when a portion of the stomach pushes up through the esophageal hiatus,

mechanical digestion by chopping up the food during chewing, also called mastication.

In the saliva, salivary amylase goes to work on starches, which account for most of the carbohydrates we consume. Starches are complex polysaccharides, but the bloodstream can only sometimes contributing to heartburn. The innermost layer of the esophagus is a mucosa made up of squamous (scaly) epithethial cells, connective tissue, and smooth muscle. There are also mucous glands in this layer near the stomach. The cells protect the esophagus from the wear and tear from the daily stream of foods and liquids. Beneath this layer is a submucosa that holds connective tissue, blood vessels, and mucous glands. The outer layer is called the adventitia and is composed of connective tissue that attaches the esophagus to the surrounding muscles and dividing structures.

The esophagus acts as an expressway, transporting food from the mouth to the stomach. Nothing is absorbed, and only mucus is excreted. Like a tollgate, a band of muscle called the upper esophageal sphincter, normally contracted, relaxes during swallowing and allows the food to move from the pharynx into the esophagus.

Peristalsis then causes a sequence of contractions and relaxations, which push the food along. The circular muscle layers above the food contract, squeezing the food down, and the longitudinal muscle fibers beneath it also contract, shortening that section and pulling the food down. The net result is a wave of contractions that both push and pull the food toward the stomach. Mucus secretions in the esophagus keep the bolus sliding along. The whole commute from the top of the esophagus to the stomach is a mere four to eight seconds.

This cutaway of the esophagus shows the layers lining the digestive tract.

THE STOMACH

What Happens to the Food

After the esophagus, the GI tract enlarges into a flexible reservoir called the stomach. Lying below the diaphragm, the stomach is situated in the left portion of the upper abdomen. Inhalation pushes the stomach down slightly; exhalation raises it. In shorter, stouter people, the stomach tends to lie horizontally ("steer-horn shaped"); in taller, thinner people it lies more vertically ("J shaped"). In the stomach food is churned and liquefied into a creamy substance called chyme. It is then held there until ready for release into the small intestine.

Anatomy

The adult stomach measures from about six to ten inches in length, but it is greatly expandable. An empty stomach has a volume of nearly 2 ounces (50 ml); a completely distended one can hold a gallon (about 4 liters). The empty stomach looks like a collapsed pouch, with lengthwise folds called rugae. Of the four major regions, the cardia (near the heart) lies at the entrance from the esophagus; the fundus is the large rounded portion; the body is in the middle; and the antrum is the funnel-shaped end of the stomach, which connects with the duodenum (first part of small intestine) at the pyloric sphincter. In terms of cellular anatomy, the stomach walls have four layers, or tunics. The serosa (outside layer) is composed of simple squamous epithelial cells and connective tissue. The muscularis externa contains within itself three layers of smooth muscles. Next is the submucosa, made of loose connective

DID YOU KNOW?

HUNGER

Although a growling stomach can signal hunger, it can also be a sign of digestion. Peristaltic waves rippling through the stomach and small intestine vibrate gas and fluids in the GI tract and cause "rumbling." Hunger contractions themselves can make noise—from 10 to 20 minutes at a time. The contractions then repeat within an hour or two.

Food enters the stomach via the esophagus, upper left, and exits through the duodenum, lower left, into the small intestine.

hormones that instead of going into the stomach, circulate via the blood vessels to trigger responses in other digestive organs.

Inside the Stomach

Once it has food, the stomach becomes a churning, spurting, bubbling plant of digestion. All of the digestive processes except ingestion and defecation occur here—mechanical and chemical digestion, secretion, peristalsis, and absorption. Every 15 to 25 seconds a series of peristaltic waves ripples across the stomach, sloshing its warm brew. During digestion, waves are more intense, with the strongest waves hitting the chyme as its enters the pylorus. Back and forth waves spill a little chyme into the pyloric sphincter until eventually all the chyme makes its way out of the stomach.

The stomach continues the job of starch digestion and begins work on protein and triglyceride digestion. Food stays in the fundus for about an hour, continuing to mix with salivary amylase. Then gastric juices kick in and activate lingual lipase, which dissolves triglycerides into fatty acids and diglycerides. In children the enzyme rennin acts on milk. Gastric lipase also breaks apart triglycerides of milk into fatty acids and monoglycerides. Parietal cells are pumping out ions of hydrogen (H+) and chloride (Cl-), which form hydrochloric acid. Chief cells secrete pepsinogen, which is not converted to the protein-digesting enzyme pepsin until it comes in contact with hydrochloric acid, but the stomach lining stays protected by a thin layer of alkaline mucus. Though most absorption takes place in the small intestine,

tissue. The mucosa (lining) is the most complex layer, composed of a number of different kinds of cells. The stomach lining is speckled with millions of gastric pits, leading deep into gastric glands that secrete gastric juice. The cells lining these pits are mostly goblet cells, which produce mucus or the stimulatory hormone gastrin. Within the fundus and body of the stomach there are more kinds of secretory cells. Mucous neck cells in the higher parts of a gland produce an acidic mucus. Parietal cells in the middle area of stomach glands secrete hydrochloric acid (HCl), an intrinsic factor (needed for vitamin B_{12} absorption in the small intestine). Then there are the chief cells, which release an inactive form of the protein-splitting enzyme pepsin. Finally, enteroendocrine cells produce

Found in the lining of the stomach's inner layer, these pits produce digestive enzymes to break down food.

the stomach does absorb water, ions, short-chain fatty acids, alcohol, and aspirin, and the latter two can add to problems with gastric ulcers (see "Breakthrough," opposite).

Phases

Neural and hormonal mechanisms regulate gastric contractions and secretions. This stage can be divided into three overlapping phases: cephalic, gastric, and intestinal. Every day gastric glands can unleash up to three quarts of stomach juice—an acid strong enough to dissolve nails. The cephalic phase occurs before food enters the stomach. Like Pavlov's dogs, we're conditioned to respond to sensory stimuli—the aroma, sight, taste, or even thought of food triggers the stomach to prepare itself. Olfactory nerves and taste buds relay signals to the brain's hypothalamus, which transmits the message to the medulla oblongata and thence to neurons in the cranial nerves. The signal stimulates gastric glands to release hydrochloric acid, pepsinogen, mucus, and gastrin. The same impulses also stimulate stomach motility. Depression, anger, and fear can inhibit this neural activity.

In the gastric phase, about two-thirds of the gastric juice is secreted. Both neural signals and hormones control this phase. The stretching of the stomach, for example, sends messages to the medulla, which in turn leads to the release of more gastric juice. The hormone gastrin plays an even larger role in secretion. Undigested proteins, caffeine, and an increasing pH level trigger the output of gastrin, which then causes parietal cells to

DID YOU KNOW?

EATING DISORDERS

When a person's appetite controls are out of balance, with an obsessive fear of weight gain, he or she may have an eating disorder. They often develop during adolescence and occur most often in women. Anorexia nervosa sufferers have a mortality rate of 0.6 percent. Bulimia nervosa patients typically binge eat and then purge by vomiting.

BREAKTHROUGH UNDERSTANDING ULCERS

Some 5 to 15 percent of Americans suffer from peptic ulcers, lesions in the GI tract exposed to gastric juices. The mucosa can become damaged and cannot produce mucus to protect the stomach lining from hydrochloric acid. Symptoms include a gripping pain that starts shortly after eating and often subsides with the next meal. Worldwide, about 90 percent of duodenal ulcers and 80 percent of gastric (stomach) ulcers are caused by *Helicobacter pylori* bacteria, which are found in 50 percent of the population. The 2005 Nobel Prize for medicine went to two Australian scientists, Barry J. Marshall and J. Robin Warren, who proved that *H. pylori* was the leading cause of gastric ulcers, right. Marshall went so far as to deliberately infect himself with the bacterium to study its effects. The bacteria can often be treated with antibiotics. Gastric ulcers are also caused by nonsteroidal anti-inflammatory drugs like aspirin, as well as by oversecretion of hydrochloric acid. Avoiding aspirin, alcohol, caffeine, and cigarette smoking can help keep the mucosa healthy and less likely to be hurt by HCl. To cut down on hydrochloric acid, medications, such as H2-blockers and proton-pump inhibitors, have proved effective.

generate more hydrochloric acid. The higher in protein content the meal, the greater will be the output of HCl and gastrin. HCl release is triggered by two other chemicals in addition to gastrin: acetylcholine (released by parasympathetic nerve fibers) and histamine (released by mucosal cells). When all three chemicals bind onto parietal cells, hydrochloric acid fountains out.

The intestinal phase is activated by the small intestine. Instead of causing secretion and movement, the small intestine slows things down so that it does not become overloaded with chyme. Chyme rich in fatty acids and glucose triggers the small intestine to release into the blood two hormones that will have an effect on the stomach: secretin and cholecystokinin (CCK). Secretin decreases the secretion of gastric juices. CCK inhibits the stomach from emptying too quickly into the small intestine. These hormones also affect the liver, pancreas, and gallbladder and their digestive operations.

After a meal, the stomach's three layers of muscles mix the chyme. Near the cardiac sphincter, peristalsis is gentle, but closer to the pylorus the waves are more vigorous. This churning is the last chance for chyme to become completely blended before entering the small intestine.

The antrum of the stomach holds only about 1 ounce (30 mL) of chyme. As the peristaltic waves wash chyme toward the partially-open pyloric valve, they end up spilling about one-tenth ounce (3 mL) of liquid and small particles out into the small intestine. Secretin and cholecystokinin inhibit motility to make sure that the small intestine does not receive more food than it can handle. It takes up to four hours for the stomach to process an entire meal.

TOO MUCH WEIGHT

A health problem on the rise in the united States, obesity afflicts one-third of the nation's adult population. Simply put, an obese person is too fat. That is, he or she has too much adipose (fatty) tissue relative to body mass. In general, a person who is 20 percent over normal weight is considered obese. To find out if you are overweight, multiply your weight in pounds by 705 and divide by your height in inches squared. The resulting number is your body mass index (BMI). A BMI of 25 to 30 indicates overweight; more than 30 is obese. Keep in mind, though, that weight is not always an accurate measurement for obesity, since a very muscular person may weigh more than a less-fit person of the same height. For an older person, who has lost muscle mass, the index may underestimate body fat. A body fat content of about 20 percent of your total weight is normal; if your fat content is more than 22 percent, however, you are carrying too much adipose tissue.

If Rubenesque figures—a term that refers to the full-bodied women portrayed by Dutch artist Peter Paul Rubens—were a modern ideal, obesity might not be the problem it is, were it not associated with a host of serious maladies: hypertension, arteriosclerosis, coronary artery disease, diabetes, pulmonary disease, arthritis, colon cancer, and varicose veins, to name but a few.

CAUSES

Why are some people too fat? Though there is no one overriding factor, there are many reasons. In general the problem lies with an out-of-balance food intake mechanism. Several influences can cause morbid obesity (obesity as a disease), including genetics, excess food intake, or bad eating habits learned young, tension-induced overeating, and trauma to the food-control center of the hypothalamus. Appetite regulation and fat-processing within the body are both contributing factors.

Among causes, genetics plays a definite role—those who inherit two obesity genes are nearly certain of having the disease. In the United States, though, only about 5 percent of obese people can blame genetics. These unfortunate people will store excess calories as fat, while others store some of them as muscle tissue.

Another main cause of obesity is childhood eating habits. Children who learn to overeat and who eat a lot of fast food, junk food, and other high-fat-content food add fat cells to their body, which are able to hold more fat as the person ages. Normal signals that tell the body that it is full are overridden in some way, possibly by stimulation from the fat cells themselves.

Another current theory holds that fat people are better at storing fat than others. Many obese people actually eat less than normal people but end up fatter because they convert food to fat and then do not easily burn it off. Another theory holds that thin people are fidgeting more than fat people and thus use more energy.

High-fat diets are particularly bad for people, because fat foods are the slowest to be exhausted as fuel. Carbohydrates are the most rapidly used. About 23 percent of carbohydrate calories eaten are tapped for metabolism, with the rest being stored, as opposed to a measly 3 percent of fat used and a walloping 97 percent of fat stored. Fat cells also beget more fat cells by growing alpha receptors, which help accumulate fat. In obese people, the lipoprotein lipase enzyme, which draws fat from the blood, is heavily produced and highly efficient.

TREATMENT

Ranging from pills to surgery, treatments vary widely in type and efficacy. Diet plans themselves come in a multitude of flavors, a new one or two popping up every year in popular books and magazines. While some of them may work temporarily, the problem is that the pounds shed too often return when the dieter begins slipping back into his or her old lifestyle. High-protein (low-carbohydrate) diets may have some benefits, but they can also be deficient in vital nutrients. Diuretics, or water pills, simply cause the dieter to lose water weight; and they can also cause electrolyte imbalance and dehydration.

Obesity is associated with a variety of health problems, including diabetes, hypertension, and cancer.

Gastric bypass surgery can be effective, but the best solution is to cut calories and increase exercise.

Some anti-obesity drugs, created since the 1970s, have achieved moderate success. The fenphen regimen involved the use of both fenfluramine to increase serotonin secretion (and thus increase feelings of satiety) and phentermine (an amphetamine) to stimulate metabolism. However, in 1997 fenfluramine was recalled when researchers discovered a link to problems with heart-valve closure.

Another drug, Orlistat, was approved in 1997. Orlistat blocks the action of pancreatic lipase, an enzyme that breaks down triglycerides in the intestines, so that ingested fat is never broken down and absorbed in the first place and so is moved out undigested. New drugs are being developed that bind to fat cells and increase the rate in which they are used up in the bloodstream. But all such drugs have side effects and should not be taken without the supervision of a physician.

Surgical options, such as stomach stapling and gastric bypass surgery, are sometimes effective in weight loss by giving food less chance for absorption. In one radical technique, biliopancreatic diversion, two-thirds of the stomach and half the small intestine are removed and pancreatic juice and bile are redirected so that essentially no fat is processed. The fat is simply eliminated. With liposuction, excess fat is vacuumed out. The problem is that these are major surgeries, with all the attendant risks.

The best solution to losing weight remains simply cutting back on fat intake and calories in general. A very low-calorie diet, coupled with a doctor-supervised exercise program (often starting with about 30 minutes of walking per day, five to seven days per week, and a gradual increase in intensity) is the best method for weight loss. After a few months, caloric intake can be increased to an optimal level.

ABSORPTION

Nutrients in the Intestines

After food leaves the stomach, it enters the small intestine, where most of digestion and absorption take place. Some 90 percent of nutrients and water are absorbed in the small intestine. The small intestine measures about 1 inch (2.5 cm) in diameter and 10 feet (3 m) in length in a living person, or 21 feet (6.4 m) in a cadaver because of lack of muscle tone. The circular folds, villi, and microvilli of the small intestine give it a large surface area for the absorption of digested nutrients.

Small Intestine

The small intestine has three sections. The duodenum (from the Latin for "twelve") extends about the width of 12 fingers, or some 10 inches (25 cm), starting at the pyloric sphincter. Shortest in length, the duodenum has the most complex architecture—the bile duct from the liver joins here, as does the pancreatic duct. The jejunum (meaning "empty," as at death) is 3 feet long and is thicker walled and contains more circular folds than the ileum. The final section, the ileum, extends about 6 feet, joining with the large intestine at the ileocecal sphincter, or valve. The circular folds are like rifling in a gun barrel—they spiral around the small intestine, from the end of the duodenum to the middle of the ileum, providing extra surface area and forcing the chyme to spin and thus slow its advance. The jejunum and ileum hang in sausage-shaped coils, supported by a membrane called the

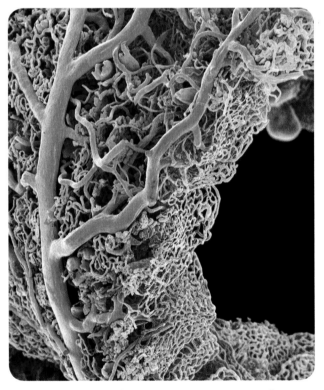

The inner wall of the small intestine contains villi that aid the body's absorption of nutrients.

mesentery. And, of course, a network of nerves and blood vessels service the small intestine.

The histology, or small-scale structure, of the small intestine reveals a number of amazing adaptations for carrying out digestion and absorption. With its circular folds, villi, and microvilli, the intestine has about the surface area of a two-story house (2,150 sq ft/200 sq m). These features are more prominent in the first part of the small intestine; then they decrease nearer to the end.

The villi are tiny finger-like projections of the mucosa, about 1 mm in length. Muscle contractions can lengthen the villi for greater absorption. The villi number about 20–40 per square

Cells in the small intestine control peristalsis, which moves liquefied food down the intestinal tract.

millimeter and account for the velvety look of the small intestine. Each villus is covered with absorptive cells and equipped with blood capillaries and a lymph capillary called a lacteal. Nutrients are absorbed into these capillaries. Other cells within the mucosa include goblet cells (which secrete mucus), enteroendocrine cells (which secrete hormones that regulate digestion), and Paneth cells (which secrete the bactericidal enzyme lysozyme).

Lining the tops of the absorptive cells are microvilli, each 1 micron (a millionth of a meter) long. These microscopic structures form a "brush border" with the inside, or lumen, of the intestine—the term comes from their bottlebrush appearance. There are more than 200 million microvilli per square millimeter of small intestine. Enzymes from the microvilli complete the job of breaking down carbohydrates and proteins.

Mechanical digestion in the small intestine is accomplished by segmentation and peristalsis. In segmentation, one localized portion of the intestine contracts to pinch off the chyme's access to the rest of the intestine. At the midpoint of this sealed off segment, circular muscles contract to form two smaller sections. Then the original contracting muscles relax, allowing the chyme to slosh to and fro and be exposed to absorption cells. These repeated contractions occur about 12 times a minute in the duodenum, but only about 8 times a minute in the ileum. When most of the food has been absorbed, segmentation stops and peristalsis takes over. Starting in the lower part of the stomach, a wave of peristalsis advances, slowly pushing the chyme along. One peristaltic wave can take up to two hours to reach the end of the small intestine. It takes three to five hours for one meal to clear the small intestine.

The chemical breakdown of carbohydrates, proteins, and lipids is completed in the small intestine. Bile, pancreatic juice, and intestinal juice all contribute to this final stage of digestion. Glands in the small intestine secrete about one quart of juice per day.

Starches (complex carbohydrates) are already partially broken down by salivary

KEEPING HEALTHY

ABSORB YOUR FOOD

The digestive system has a big job to do, breaking food down into the components humans need to feel healthy and energized. Some simple tricks can help your body get more out of food.

+ **Cook vegetables—but not too much:** Lightly steaming vegetables helps soften the food and release disease-fighting antioxidants and other nutrients more easily during the digestion process.

+ **Add a touch of oil:** Add a small amount of dressing or some avocado slices to your salad to help the body absorb the healthy carotenoids in the raw vegetables.

+ **Take time to chew:** A study showed that people get more energy and nutrients from almonds that have been chewed into tinier pieces. This increases the surface area and makes it easier for the body to access the nutrients.

+ **Make winning combinations:** Vegetarians will get more iron from plant sources such as beans, tofu, or spinach when they eat them with foods rich in vitamin C, such as citrus fruits, strawberries, or red peppers.

amylase in the mouth and stomach. In the small intestine, pancreatic amylase goes to work, splitting both starches and glycogen (a glucose polymer). It does not work on cellulose; in fact, this polysaccharide plant fiber is indigestible by humans, but a diet high in fiber, or roughage, is considered a good hedge against constipation and cancer-causing agents in the colon. The disaccharides sucrose, maltose, and lactose are broken down only in the small intestine. The monosaccharides are then absorbed. Lactose-intolerant people do not produce enough lactase in their small intestine—dairy products can produce gas, diarrhea, and abdominal cramps.

The breakdown of proteins, which began in the stomach, is completed in the small intestine. Pancreatic enzymes convert proteins into smaller units, called peptides, each enzyme splitting the peptide bonds at a different point. Lipids are mostly digested in the small intestine, when pancreatic lipase hydrolyzes triglycerides into fatty acids and monoglycerides. The nucleic acids DNA and RNA are also digested there.

Absorption in the small intestine is the culminating process of digestion. After the long journey from mouth to stomach, the food gets drawn into the body through the epithelial cells of the small intestine. Here 90 percent of all absorption takes place. The carbohydrates have been broken down into the monosaccharides glucose, fructose, and galactose and are ready to be absorbed in the blood and lymph systems. Likewise, the proteins have been chopped up into dipeptides, tripeptides, and amino acids. The fats are now fatty acids, glycerol, and monoglycerides.

The vast chemistry lab of the small intestine has several ways of absorbing these materials:

diffusion, facilitated diffusion, osmosis, and active transport. Each will be explained as we examine how the different nutrients are taken up.

The small intestine is capable of absorbing 4 ounces (118 mL) of monosaccharides every hour. Under normal conditions every carbohydrate eaten (except cellulose) will be absorbed in the body. Molecules of fructose pass into the bloodstream by facilitated diffusion, which means that, since they are too large to simply pass through membrane pores, they have to be escorted through by a transporter protein. Glucose and galactose are absorbed via active transport—they expend energy going across a concentration gradient

DID YOU KNOW?

CELIAC DISEASE

Caused by inadequate absorption in the small intestine, celiac disease stems from a sensitivity to gluten. Consumption of gluten leads to damage of the villi, which impairs the body's ability to absorb nutrients. Symptoms include weight loss, diarrhea, abdominal cramps, weakness, and stunted growth. The only cure is to eliminate gluten from the diet.

against the normal path of diffusion; in other words, they go against the flow. Likewise, amino acids and the end products of proteins are taken in by an active transport mechanism. About 95 to 98 percent of proteins in the small intestine end up being absorbed; the rest is passed out. Half the amino acids absorbed come from food; the rest come from digestive juices and mucosal cells.

Simple diffusion takes lipids across the boundary between the small intestine and blood vessels. Adults absorb some 95 percent of all lipids; the figure is about 10 percent lower in newborns, as they don't produce as much bile. Short-chain fatty acids diffuse into the villi of the small intestine and move into the lacteals. Long-chain fatty acids and monoglycerides need bile salts for dissolving before they can be taken up by the villi. About 95 percent of the bile salts then get recycled through the blood and into the liver. People who do not produce enough bile salts—usually because of blocked bile ducts or liver disease—may

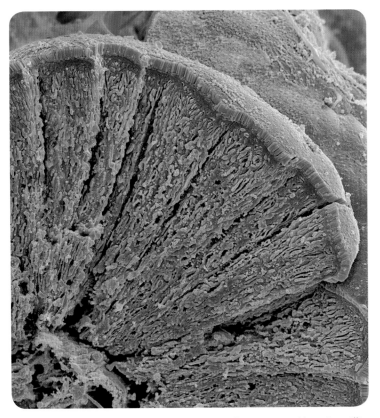

Lining the small intestine, the velvety villi (pink) are topped by microvilli, which are instrumental in breaking down carbohydrates and proteins.

end up passing 40 percent of their ingested fats along as waste. The fat-soluble vitamins A, D, E, and K then are not sufficiently absorbed.

The small intestine also absorbs water—about 2 gallons a day—through osmosis into the bloodstream. Only about 2 quarts come from ingested water, the rest from GI tract secretions (1 quart from saliva, 2 quarts from gastric juice, 1 quart from bile, 2 quarts from pancreatic juice, and 1 quart from intestinal juice). This leaves 1 quart, most of which is absorbed by the colon; 0.1 is excreted in feces.

The Liver

The versatile liver is the body's largest gland, weighing in at 3 pounds (1.3 kg). Among its many regulatory and metabolic functions are synthesis of plasma proteins, detoxification, storage of glycogen and iron, destruction of old blood cells, and synthesis of vitamin D. It produces bile, which breaks down fats.

The blood-rich, wedge-shaped liver is mostly tucked in the right rib cage, just below the nipple. It occupies the right hypochondriac and epigastric regions. The liver is divided into two main lobes: the right and left lobe. From the visceral, or back (posterior),

HEPATITIS A
Tiny viruses, in red, cause the disease, which is transmitted through contaminated water and food.

HEPATITIS B
The Hepatitis B virus is transmitted through the blood. In severe cases, it may lead to cirrhosis and death.

HEPATITIS C
This virus also spreads via the blood. Currently, there is no vaccine.

side of the liver, the left lobe shows two additional, smaller lobes: the caudate and quadrate lobes. The right and left lobes are divided by the falciform (sickle-shaped) ligament, a part of the peritoneum, which helps suspend the liver. At the bottom, or free, end of this ligament runs the ligamentum teres (round ligament), a remnant of the fetal umbilical vein, which connects to the umbilicus, or navel. The liver is hooked up to the circulatory system via a portal between the two lobes containing the hepatic artery and hepatic portal vein. Bile leaves the liver by the common hepatic duct and then the common bile duct, which carries bile to the duodenum.

The liver is made up of sesame seed–size granules called lobules. Hexagonal in shape, the lobules consist of liver cells (hepatocytes) arranged around a central vein. Liver cells are divided into "plates" by leaky capillaries called sinusoids. At each of the six corners of the lobule runs a portal triad—a branch of the hepatic artery (bringing in oxygen-rich blood), a branch of the hepatic portal vein (ferrying off nutrient-enriched blood from the GI tract), and a bile duct. Blood from both the hepatic artery and portal vein seeps via

Tucked underneath the liver, the green gallbladder stores and concentrates bile produced by the liver.

the sinusoids into the cell's central vein, and these veins collectively drain the liver. Kupffer cells within the sinusoids destroy bacteria and worn-out blood cells.

Within the liver cells, the liver's work is done, including the production of bile, which flows from the cells in narrow canals called bile canaliculi. These canals empty into bile ducts at the edges of the cells. The many bile ducts merge to form the left and right hepatic ducts, which themselves join to form the common hepatic duct. This duct joins with the cystic duct and empties into the gallbladder, where bile is stored.

Bile is a yellow-green alkaline liquid composed of water, bile salts, bile acids, phospholipids, cholesterol, pigments, and electrolytes. The liver cells secrete about half to 1 quart (500 to 1,000 mL) of bile per day. The bile salts

and phospholipids (including lecithin) are what get used in digestion. The bile salts, derivatives of cholesterol, emulsify fats—that is, they break large lipid globules into millions of tiny droplets suspended in water. Broken up, the lipids thus have more surface area for enzymes to act on. Bile salts also help make cholesterol soluble and aid in fat and cholesterol absorption. The bile salts are then recycled back into the blood in the ileum of the small intestine.

Bilirubin is the main pigment in bile. It derives from heme, liberated from the destruction of worn-out red blood cells. One of the by-products of bilirubin imparts a brown color to feces. Feces with a grayish white color and streaks of fat indicate that insufficient bile is being secreted and thus fats are not being absorbed.

When there is no digestion going on, bile cannot get into the duodenum because the pancreaticobiliary sphincter remains closed like a dam. During these times, the bile simply backs up the cystic duct and is stored in the gallbladder. Then, when the gallbladder contracts, bile is squeezed through the cystic duct and common bile duct into the duodenum. The principal stimulus for gallbladder contraction is the intestinal hormone cholecystokinin (CCK), the release of which itself is triggered by acidic, fatty chyme in the duodenum. CCK does several things—it stimulates secretion of pancreatic juice, causes the gallbladder to contract,

The cells of the pancreas secrete digestive enzymes.

opens the pancreaticobiliary sphincter, and produces a feeling of satiety. It also slows gastric emptying and regulates maintenance of the pancreas. Although CCK is the chief stimulus for the release of bile, impulses from the vagus nerve (a cranial nerve) also cause gallbladder contraction and thus bile release.

Gallbladder & Pancreas

The gallbladder is a plum-size, green, muscular sac hanging from a depression in the posterior surface of the liver and peeking out below the liver's right lobe. Its job is to store and concentrate bile from the liver. By removing water and ions from the bile, the gallbladder can make bile up to ten times as strong as that coming from the liver.

The bulging part of the gallbladder is called the fundus, the middle the body, and the tapered upper portion the neck. Inside, the gallbladder has a mucosa that pleats into rugae (as in the stomach) when the gallbladder is empty. The middle layer of the gallbladder wall is composed of smooth muscle fibers. When these muscles contract, the gallbladder squirts bile into the cystic duct. The outer, and final, layer of the gallbladder is formed by part of the visceral peritoneum.

A more complex organ, the pancreas (from "all" plus "flesh") produces digestive enzymes and secretes them into the small intestine. The pancreas is about 5–6 inches (12–15 cm) long and consists of a tail that abuts the spleen, a body, and a head that nestles into the curve of the duodenum.

Pancreatic juices, rich in digestive enzymes, pass into the pancreatic duct, which is joined by the bile duct just before it enters the duodenum at the ampulla of Vater and surrounds a valve, called the sphincter of Oddi, that controls the flow of juice. The pancreas also has an accessory duct located about an inch above, or superior to, the sphincter that empties directly into the duodenum.

Surrounding the ducts are acini, little clusters of enzyme-secreting epithelial cells, called acinar cells. Pancreatic islets (or islets of Langerhans) among the acini have an endocrine function: They produce hormones, including glucagon, insulin, and somatostatin. The pancreas is both an exocrine gland producing enzymes and an endocrine gland producing hormones.

The pancreas secretes approximately 1.2–1.5 quarts (1,200–1,500 mL) of pancreatic juices a day. A clear, colorless fluid, the juice is made of water, salts, sodium bicarbonate, and enzymes. The sodium bicarbonate helps create the proper pH for the enzymes in the intestine. Since the pancreas has no mucosa, it secretes enzymes in an inactive state so that the pancreas will not digest itself. Trypsinogen, for example, is converted to trypsin only when it is in the small intestine. Trypsin then can activate the enzymes carboxypeptidase and chymotrypsin. In addition to these protein-digesting enzymes, the pancreas secretes the carbohydrate-digesting enzyme pancreatic amylase and the chief triglyceride-digesting enzyme, pancreatic lipase. It also secretes the nucleic acid–digesting enzymes ribonuclease and deoxyribonuclease.

The secretion of pancreatic juice is regulated by both nerve impulses and hormones. Before food enters the small intestine, impulses from the vagus nerves tell the pancreas to secrete enzymes. When chyme enters the small intestine, the latter secretes the enzymes cholecystokinin (CCK) and secretin into the bloodstream; CCK makes its way back to the pancreas and stimulates the production of bicarbonate ions, while the secretin stimulates enzyme secretion. (Yes, this is the same hormone that inhibits gastric secretion.)

Just as alcohol abuse can damage the liver, it can also lead to a condition called pancreatitis.

DID YOU KNOW?

JAUNDICE

A yellowing of the skin, whites of the eyes, and mucous membranes are signs of jaundice, caused by too much bilirubin in the blood. Newborns often have jaundice because their livers have not matured; usually it clears without treatment. In adults it could be caused by bile duct obstruction from a gallstone, a tumor, hepatitis, or cirrhosis of the liver.

Instead of releasing the inactive trypsinogen, the pancreatic cells may release trypsin, which then starts eating away at the pancreas. Attacks can continue to occur even after treatment. Pancreatitis may also be triggered by gallstones that lodge at the common Y-connection, where the pancreatic duct and bile duct meet at the ampulla of Vater. Recurrent symptomatic gallstones call for complete surgical removal of the gallbladder, which can be done by laparoscopy.

ELIMINATION

Removing Waste From the Body

After food passes from the small intestine, the final stages of the digestive process can begin. Food moves to the large intestine, making its way through the colon toward the rectum. During these phases, the last bits of nutrients are extracted before the body rids itself of any leftover by-products and indigestible material.

Large Intestine

Wider but shorter than the small intestine, the large intestine measures about 2.5 inches (6.5 cm) in diameter and 5 feet (1.5 m) in length. Its job is to absorb water from indigestible foods and then eliminate the remaining semisolid waste, called feces.

Extending from the ileum to the anus, the large intestine is suspended from the abdominal wall by a double layer of peritoneum called the mesocolon. Undigested material from the small intestine enters the large intestine via the ileocecal sphincter, or valve. The first chamber in the cavern of the large intestine is the cecum. Attached to the cecum is a twisted tube called the vermiform (worm-shaped) appendix (see "Did You Know?," opposite).

The next, and longest, part of the large intestine is the colon. It begins with the ascending colon, which travels up the right side of the abdomen to just below the liver. Here it turns left and continues as the transverse colon all the way to the left side of the abdomen, where it drops down as the descending colon to the sigmoid colon. This final S-shaped section of the colon snakes back to the midline of the body and ends at the rectum, which includes the terminal 4 inches (10 cm) of the GI tract. Three transverse folds in the rectum, the rectal valves, keep feces from escaping when gas is passed. The final 1 inch (2.5 cm) of the rectum is the anal canal. The anal canal ends at the anus, which is opened during defecation.

The large intestine is shown in blue.

Internal and external anal sphincters control the opening and closing of the anus. The internal sphincter is composed of smooth muscle (involuntary) and the external one of skeletal muscle (voluntary).

The mucosa layer of the large intestine has no villi (as in the small intestine) but is dotted with little pits that open into long, tubular intestinal glands. The glands are lined with absorptive cells that absorb water and a huge number of goblet cells that secrete mucus to lubricate the passage of feces and to protect the intestinal wall from noxious gases and acids. Under the mucosa is a submucosa and a muscularis consisting of longitudinal and circular muscles. Three thick longitudinal bands called teniae coli extend along the entire colon. The colon is further divided along its length into little balloon-like pockets called haustra. Attached to the teniae coli, like little tags, are fat-filled pouches of visceral peritoneum called epiploic appendages. Their function is not known.

The final process of digestion happens in the large intestine, not by enzymes but by the action of bacteria that enter the colon via the small intestine or anus. This colony of bacteria (intestinal flora) ferment indigestible carbohydrates such as cellulose, in the process releasing carbon dioxide, dimethyl sulfide, nitrogen, hydrogen, and methane gases.

The colon produces about 17 fluid ounces (500 mL) of flatus (gas) per day. While proteins and fats produce little gas, beans and other carbohydrate-loaded foods produce plenty of it; the dimethyl sulfide contributes heavily to the odor. Besides making gas, the intestinal flora synthesize B vitamins and vitamin K, which the liver requires for blood-clotting proteins. The bacteria also break down

DID YOU KNOW?

APPENDIX

A muscular tube, the worm-shaped appendix, below, opens from the cecum and dead-ends after 3–4 inches (7–10 cm). It has no known physiological purpose but is thought to be a vestige of an organ that helped ancestral species digest cellulose. The appendix occurs in humans, higher apes, wombats, civets, rodents, and some other mammals.

any leftover protein into amino acids and the amino acids into simpler compounds, some of which are excreted. The rest are routed to the liver and later eliminated in the urine.

Movement

Right after a meal, a peristaltic reflex pushes any chyme still left in the ileum from a previous meal into the cecum. In its normal state, the ileocecal sphincter stays partially closed so that food does not pass too rapidly from the small to the large intestine. The hormone gastrin and a distension of the cecum trigger the opening of the ileocecal sphincter. Material then begins steadily accumulating in the cecum and moves up the ascending colon.

The main method of transport in the colon is haustral churning. When a haustrum (pouch) in the colon is filled to a certain point, its muscles contract and squeeze the contents on to the next haustrum. There are also about 3 to 12 peristaltic contractions per minute in the colon—a much slower rate than in the earlier portion of the GI tract. Finally, there are three to four waves of "mass peristalsis" a day, usually during or just after a meal. These are powerful waves that start in the middle of the transverse colon and flush the contents down toward the rectum.

About 90 percent of water absorption has already taken place in the small intestine, but that other 10 percent absorbed in the colon—a pint to a quart—is a crucial part of the body's daily water intake. The vast majority of the colon's water absorption is through osmosis. Vitamins and ions such as sodium and chloride are also absorbed in the colon.

After some 3 to 10 hours, the large intestine has absorbed enough water from the chyme to turn it into a solid or semisolid state, and the chyme is now termed feces. What's in feces? Actually, quite a lot. Water and mucus in the feces help ease its passage; the feces also contains cellulose and other undigested food residue, inorganic salts, sloughed-off epithelial cells, and millions of bacteria.

The feces travel down the descending colon and into the sigmoid colon, where peristalsis pushes them into the funnel-like rectum. Distension of the rectal walls initiates the defecation reflex. Nerve impulses travel to the spinal cord, which sends signals back via parasympathetic nerves to the descending colon, sigmoid colon, rectum, and anus. In response, longitudinal muscles in the rectum contract, shortening the rectum and increasing rectal pressure. Voluntary contractions of the diaphragm and abdominal muscles also help in opening the internal anal sphincter. There is also an external anal sphincter, under voluntary control, that has to be relaxed in order for defecation to occur.

The blood vessels and capillaries of the large intestine supply the glands that secrete mucus to lubricate the passage of feces.

Infants do not have control of the external anal sphincter, hence pressure on the rectum is enough to cause an infant to empty contents out through the anus. Adults also can contract their levator ani muscle to lift the anal canal. If an adult chooses not to defecate right away, the reflex impulses stop, and he or she then may wait until the next mass peristaltic movement.

Complications

Diarrhea (from "through" plus "flow") results when food hurries through the intestines without sufficient water absorption. The feces are then watery and increased in volume and frequency. The excessive motility of the intestine may be caused by microbes, lactose intolerance, or stress. The resulting frequent diarrhea can lead to dehydration and electrolyte imbalance.

The opposite problem, constipation, occurs when the feces remain too long in the colon and so much water is absorbed that defecation becomes increasingly difficult. Causes include failure to heed "nature's call," lack of exercise, emotional stress, and inadequate fluid and fiber in the diet. Mild laxatives may temporarily fix the problem but are not a good long-term solution.

Constipation can lead to another annoying problem called hemorrhoids ("flowing with blood"), or piles. Excessive straining during defecation puts extra pressure on the rectum, which causes rectal veins to swell. If the walls of these varicose veins dilate too much they start oozing blood. Anal itching and bleeding often indicate the presence of hemorrhoids. Over-the-counter creams are often successful in reducing the swelling; in rare cases, hemorrhoids are removed surgically. A diet high in fiber is often the best cure for both hemorrhoids and constipation. A low-fiber diet can narrow the bore of the colon and increase the

KEEPING HEALTHY
EARLY COLON CANCER DETECTION

This x-ray shows colon cancer.

Colorectal cancer is the third most common cancer in the United States and the second highest cause (after lung cancer) of cancer deaths. According to the American Cancer Society, there are 112,340 new cases of colon cancer and 41,420 new cases of rectal cancer per year; they will combine for a total of 52,180 deaths. Presently, some 5 to 6 percent of Americans develop colorectal cancer sometime during their lives; most people diagnosed with the cancer are over 50. Thanks to early detection, the death rate has been on the decline in the past 15 years. The most thorough type of screening is a colonoscopy, which allows a physician to examine the entire colon and remove precancerous polyps. A screening every 10 years starting at age 50 or earlier is the currently preferred strategy.

strength of its contractions, which puts undue pressure on the colon walls. Weaknesses in the walls can balloon into pouches called diverticula (from "to turn aside"). Inflammation of the diverticula, diverticulitis, may cause severe abdominal pain (usually on the lower left side), fever, and vomiting. Fiber is also recommended as a guard against diverticulitis.

Chapter 8
BODY CLEANSING

WASTE DISPOSAL

Efficient Filtration

An intricate and efficient filtration system, the urinary tract is the body's main way of keeping itself clean. The lungs and skin have a small role in ridding the body of metabolic wastes and other toxins, but the urinary system does the majority of the heavy lifting. It accomplishes this by filtering through the body's entire blood plasma more than 60 times a day. This work is carried out by the kidneys. The bladder is simply a storage reservoir, and the ureters and urethra are passageways for urine, the waste product of the kidneys.

Purifying Blood

While the kidneys are making urine, they are also purifying the blood by fine-tuning its chemistry and volume. By regulating the blood's pH level and ion makeup, the kidneys keep the bloodstream and other fluids at optimal condition. The amount of salt (sodium) and water that the kidneys filter, reabsorb, and excrete from the blood regulates the blood's volume. A higher volume of blood raises blood pressure, and a lower volume decreases the pressure. By changing the volume and by secreting the enzyme renin, the kidneys can adjust the body's blood pressure as needed. The kidneys also regulate the osmotic pressure, or concentration of the blood, and all fluids in the body—that is, the amount of dissolved particles within the liquid. They do this by separately regulating the amount of water, in response to changes in the concentration of body fluids (osmolality) and the presence or absence of antidiuretic hormone (secreted by the brain), and the amount of solutes they add to the urine. Among the solutes are calcium, potassium, sodium, phosphorus, and sulfate.

These three interrelated tasks—blood pressure, pH balance, and osmotic pressure—are

Increased water intake can raise urine volume from 1.5 quarts a day to 3 quarts.

THE URINARY SYSTEM

Kidney

Ureter

Bladder

DID YOU KNOW?

CLEANSING

+ The kidneys process about 1 quart of blood every minute, for a total of about 423 gallons a day. This is equal to more than 300 times the body's blood volume.

+ By contrast, the heart pumps about four times the volume of blood daily processed by the kidneys.

+ The kidneys filter about 48 gallons of water from the blood daily. Most is reabsorbed; less than 1 percent, or about 1.5 quarts, is excreted in the form of urine.

+ Increased water intake can raise the urine volume to 3 quarts; heavy sweating can reduce it to a pint.

+ The blood plasma (blood minus the cells) is cleansed by the kidneys at the rate of 60 times per day.

absolutely essential for the functioning of the human body. By maintaining electrolyte and water homeostasis, or equilibrium, within narrow limits, the kidneys make it possible for human bodies to function on dry land, where they must conserve water and salts.

Among the other key tasks performed by the kidneys are the production of hormones, the regulation of the blood's glucose level, and the excretion of waste. The hormone calcitriol is the active form of vitamin D, and the hormone erythropoietin helps stimulate the production of red blood cells. And by creating urine, the kidneys rid the body not only of metabolic wastes such as ammonia and urea but also of foreign toxins—drugs and environmental poisons.

Urine

The waste product produced by the kidneys, urine, is a clear, pale to dark yellow fluid. The yellow tint comes from urochrome, a pigment resulting from the breakdown of hemoglobin. The more concentrated the urine, the darker its color. Occasionally the urine has a pink or brownish tinge, which is usually brought about by consumption of foods such as red beets but can be due to the presence of blood or bile salts in the urine. Vitamin supplements and prescription drugs can also cause discoloration. The frothiness of urine comes from bile salts or proteinuria. Smoky or cloudy urine is often caused by an infection.

Urine has only a slight odor at first, but outside the body it begins to take on the smell of ammonia as its solutes are metabolized by bacteria. Diabetes mellitus (when out of control) lends the urine a fruity odor because of the presence of ketones. Other diseases can change the odor, as can consumption of drugs. Eating certain foods such as asparagus causes sulfur-containing amino acids to break down, lending urine a sulfurous smell.

Although urine is normally slightly acidic, with a pH of about 6 (7 being neutral), it can range

Colored red in this x-ray, the ureters connect the kidneys to the bladder.

into the alkaline (up to around 8) with diet or bacterial infection, or during protracted vomiting—as the stomach loses acid, leaving the blood alkaline. A diet leaning heavily on fruits or vegetables is high in potassium salts of organic acids, which when metabolized leave alkaline residues in the urine. Urine can also veer toward the acidic (down to pH 4.5)—for instance, after eating a lot of protein and whole wheat.

The specific gravity of urine varies from 1.001 to 1.035, depending on the concentration of solutes in the urine. With distilled water having a specific gravity of 1.0, urine concentration thus goes up with more solutes or down with less water. And the higher the amount of solutes per volume of water, the more likely it is that the solutes will precipitate into stones (see "Keeping Healthy," page 219).

DID YOU KNOW?

URINE FACTS

+ Normal urine is nearly odorless. It develops an ammonia smell as bacteria begin to metabolize its solutes.

+ The yellow color derives from the pigment urochrome; the more concentrated the urine, the darker the color.

+ Drinking coffee or tea adds caffeine to the urine; exercise adds lactic acid.

+ The ketone bodies acetone and acetoacetic acid accumulate in the urine of untreated diabetics and those on a low-carbohydrate diet, giving the urine a fruity odor.

If you drink nothing for a full day, your kidneys continue producing urine at a specific gravity of at least 1.025. Without at least 1 quart (946 mL) of water per day, the water for urine will begin coming from tissues, which can lead to dehydration. The kidneys have a limit beyond which they cannot concentrate urine.

Composition

About 95 percent of urine is water; the rest is solutes. The main solute is a nitrogenous compound called urea, which is made from the breakdown of amino acids. Other waste products in urine are ions of chloride, sodium, potassium, phosphate, and sulfate. Other nitrogen-containing substances in urine include creatinine (a white, crystalline, basic compound) and uric acid (a product of nucleic acid metabolism). There can also be found small quantities of calcium, magnesium, and bicarbonate ions.

A higher concentration of any of these urinary solutes often means there is something wrong with the kidneys or some other body system. Another indicator of disease is the presence of glucose, blood proteins, hemoglobin, red blood cells, white blood cells, or bile pigments in the urine.

Each of these abnormal components in urine points to a different probable cause. High amounts of glucose in the urine could mean diabetes. Too much protein in your urine most likely reflects glomerular inflammation (glomerulonephritis), but small amounts of protein can also be seen with high fever or prolonged upright posture. Ketone bodies in the urine point to untreated diabetes but also to fasting or starving.

Excessive bile pigments occur as the result of liver disease or a blockage of bile ducts. Hemoglobin in the urine can result from a reaction to a secondary blood transfusion, anemia, or severe burns.

Red blood cells, also called erythrocytes, in the urine might indicate bleeding in the urinary tract, which could be caused by kidney stones, a tumor, infection, trauma. or glomerulonephritis. Leukocytes, or white blood cells, often indicate a urinary tract infection or an inflammation.

URINE COMPONENTS

WATER	About 95 percent of urine is made up of water
AMINO ACIDS	Urine also has about 3 g of amino acids per day
HIPPURIC ACID	About 0.7 g of hippuric acid per day are excreted through urine
UREA	Of the remaining 5 percent, the most plentiful solute by weight. Urea accounts for more than 50 percent of all solutes and 90 percent of urine's nitrogenous components, with ammonia 2–4 percent, creatinine 3 percent, and uric acid 1–3 percent
OTHER	Of inorganic components, the daily urine contains about 6.3 g of chloride ions, 4 g of sodium ions, 2–4 g of potassium ions, up to 2.5 g of phosphate ions, up to 2.5 g of sulfate ions, and small amounts of calcium, magnesium, bicarbonate, and iron

THE KIDNEY

Controller of Water, Acids, and Salts

The main function of the body's two kidneys is to filter the blood, produce a waste product called urine, and maintain the body's fluid composition. The kidneys also convert vitamin D to its most active form, and they produce an enzyme called renin, which helps maintain blood pressure. The hormone erythropoietin, secreted by the kidneys, stimulates the production of red blood cells.

Situated above the waistline, the kidneys are bean-shaped organs lying behind the abdominal cavity. The liver pushes the right kidney somewhat lower than the left. Both kidneys are partially protected by the lower part of the ribcage. An adult kidney isn't very large; it is approximately the size of a deck of playing cards.

THE KIDNEY

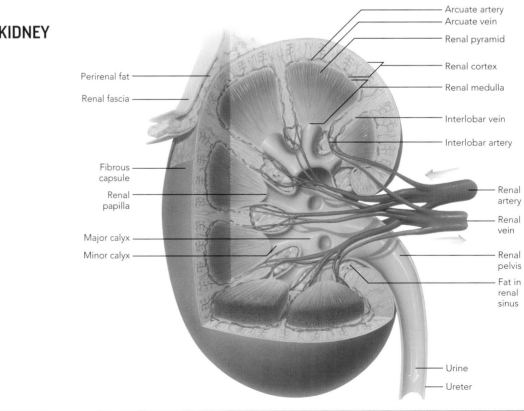

Perirenal fat

Renal fascia

Fibrous capsule

Renal papilla

Major calyx

Minor calyx

Arcuate artery

Arcuate vein

Renal pyramid

Renal cortex

Renal medulla

Interlobar vein

Interlobar artery

Renal artery

Renal vein

Renal pelvis

Fat in renal sinus

Urine

Ureter

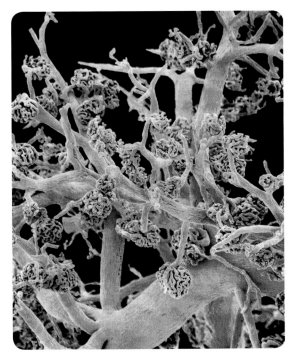

The glomeruli (pictured in red) filter toxins from the blood.

Anatomy

The ureter, nerves, blood, and lymphatic vessels pass through the renal hilus, a deep vertical fissure on the concave side of each kidney. Surrounding each kidney are three layers of tissue. The innermost layer is the renal capsule, a fibrous covering on the kidney itself. Next is a fatty layer called the adipose capsule, which supports and protects the kidney. And on the outside is a thin layer of connective tissue, the renal fascia, which anchors the kidney to surrounding structures.

The adipose capsule is so important that extremely thin people especially run the risk of nephroptosis, or floating kidney, in which the kidney drops and can kink the ureter, causing urine to back up into the kidney, leading to kidney damage and even failure.

The internal anatomy of the kidney breaks into two main regions—a smooth, reddish renal cortex and a darker inner region, the renal medulla. Within the renal medulla are 8 to 18 renal pyramids, tapering to the middle of the kidney. These pyramids are composed of bundles of tiny urine-gathering tubules. Within the cortex are nephrons, the basic processing units of the kidney. These functional units consist of a glomerulus (filtering device composed of a capillary network) and a tubule. At the end of the pyramids are urine-collecting cups called calyces, all of which drain into a collection cavity known as the renal pelvis. From the renal pelvis, the urine flows out through the ureter and to the bladder. Smooth muscles in the calyces, renal pelvis, and ureter move the urine along by peristalsis.

Since the kidneys are the blood's filters, it is no wonder that they are well supplied with blood vessels. In fact, one-quarter of the body's arterial blood is delivered to the kidneys via the renal arteries every minute. Inside the kidney, the renal artery splits into numerous segmental arteries that branch out into the kidney.

Each of the one million nephrons is hooked to an afferent arteriole (a terminal branch of the artery). These arterioles then divide into little tangled capillary balls called glomeruli. Instead

DID YOU KNOW?

KIDNEY FACTS

+ An adult kidney weighs about 4.5 to 5 ounces.

+ The typical kidney is 4 to 5 inches long and about 1 inch thick.

+ The kidneys make up less than 0.5 percent of the body's weight, yet they take in 20 to 25 percent of the resting body's cardiac output and use 20 to 25 percent of the body's oxygen.

DID YOU KNOW?

GOUT

Excess uric acid in the blood can crystallize in a joint—often at the base of the big toe. These crystals, below, cause swelling, redness, and sudden severe pain. Known as gout, this condition was once thought to be due to gluttony. Though it is more likely to affect the obese, gout can hit any person who is producing too much uric acid or not eliminating enough. Gout can be brought on by stress, injury, or diuretics taken for high blood pressure. Ninety percent of cases are in men over the age of 40. The age-old treatment colchicine (an alkaloid from the seeds of meadow saffron) is still used. Sufferers should avoid excessive alcohol and purine-rich foods: sardines, sweetbreads, and liver.

of connecting to a vein, as do the body's other capillary networks, they connect to an efferent (outgoing) arteriole. Some 90 percent of the blood entering the kidney floods the cortex, where most of the glomeruli are. The veins then reverse the route of the arteries, draining each kidney via a renal vein, which connects to the inferior vena cava. The left renal vein is twice as long as the right to reach the inferior vena cava, which lies to the right of the vertebral column.

Development

The kidneys begin to appear during the third week of human fetal development. Not one, but three pairs of kidneys develop in succession. The first pair begins degenerating after the fourth week and disappears by the sixth week; by the eighth week the second pair is almost completely gone as well. The pair farthest down along the fetus's intermediate mesoderm is the one that will survive. By the third month, the fetus is excreting urine into the amniotic fluid; urine then becomes the main component of the amniotic fluid.

On the other hand, renal activity slows with age. Renal blood flow and filtration rate decrease by some 50 percent between the ages of 40 and 70. The kidneys diminish from a mass of about 9 ounces (260 g) in a 20-year-old to 7 ounces (200 g) in an 80-year-old. Kidney stones and inflammation become more common with age, as do urinary tract infections. Other ailments include dysuria (painful urination), hematuria (blood in the urine), and incontinence.

Nephrons

Each kidney has about one million nephrons, the miraculous miniature filtration plants of tubes and capillaries. In each of these microscopic processing units, blood is filtered by glomeruli; the waste product, urine, is created in the tubules, which reabsorb and secrete substances through one of thousands of collecting ducts.

BREAKTHROUGH LIVING WITH ONE KIDNEY

Before Gustav Simon (1824–1876), no one knew the body could still function with just one kidney. Although the kidney's nephrons do not heal if they become injured or diseased, they can enlarge to take on a bigger amount of blood filtering if one kidney is removed or damaged. The healthy one grows big enough to handle 80 percent of the load that two kidneys used to handle. In 1869 the German surgeon treated a woman with a urogenital fistula that could only be remedied by removal of a kidney. Simon had already distinguished himself in experimental surgery—first he proved that animals with only one kidney could still excrete urine at normal levels. His operation on the woman was the first successful removal of a human kidney. The procedure was a medical milestone, ushering in a new age of surgery that would eventually lead to kidney transplants.

The interface end of the nephron is the renal corpuscle, which consists of a glomerulus (a ball of entangled capillaries) surrounded like a helmet by a glomerular capsule (or Bowman's capsule). The glomerulus is a double-layered epithelial cup, which acts as filter through which blood plasma passes. The filtrate from the blood plasma oozes from the glomerulus into the highly porous inner layer of the glomerular capsule, and the resulting filtrate then flows through the renal tubule, where it then becomes urine.

The renal tubule is about 1.25 inches (3 cm) long. The first part of the tubule—the part that exits the glomerular capsule—is a twisting roller coaster called the proximal convoluted tubule. The tubule then straightens out and makes a long hairpin turn called the loop of Henle. Finally, the tubule widens and begins to twist again in the section called the distal convoluted tubule. Fluid leaves the renal tubule as urine and then enters one of the kidney's collecting ducts, which receive urine from many other nephrons.

Some 80 to 85 percent of all nephrons lie primarily in the outermost part of the kidney (the cortex), with the loop of Henle barely entering the medulla before ascending back into the cortex. The rest of the nephrons lie deep within the cortex and contain long loops of Henle that dip far down into the renal medulla.

The kidney is well supplied with blood vessels.

A NEW HOPE

Before the mid-1950s, people with kidney failure faced a grim future. With nothing to stop the deterioration of the kidneys, and no machine on Earth capable of duplicating the kidneys' complex filtering job, patients faced an irreversible death sentence. But the development of artificial dialysis (from the Greek for "separation") and the artificial kidney gave such patients the hope for a much longer and more active life. Although commonplace today, dialysis was an amazing breakthrough that improved the lives of countless patients around the world.

FAILURE

When the kidneys begin to fail, they can no longer sufficiently filter metabolic and other wastes from the blood. As nitrogenous wastes begin to accumulate in the blood, the blood's pH turns dangerously acidic, and severe electrolyte imbalances begin wreaking havoc throughout the body. Symptoms of failure can include nausea, vomiting, edema, weakness, fatigue, uriniferous breath, and labored breathing from lung congestion. Blood tests reveal accumulation of urea (from protein metabolism) and creatinine (from muscle metabolism). The final stages of kidney failure involve convulsions, coma, and death.

Kidney failure occurs when 85 to 90 percent of the kidney's nephrons are destroyed. Nephrons may deteriorate for a number of reasons: diabetes mellitus, high blood pressure, kidney diseases such as glomerulonephritis (inflammation of and sometimes permanent damage to the glomeruli), chronic kidney infections, physical trauma, obstructed blood flow to the nephrons, and exposure to environmental toxins such as mercury or lead that poison renal tubules. Any of these can lead to chronic renal failure, an especially treacherous condition in that it often gives no clear warning until the damage is already done. In fact, in many cases there are no symptoms until renal function is down to less than 25 percent of normal.

Victims of acute renal failure might be luckier because symptoms come on more quickly— fluid retention, weakness, fatigue, and nausea indicate the need for immediate medical attention. The main sign of acute kidney failure is decreased urine output—down to 1 pint (500 mL) or less—and elevated levels of substances normally removed by healthy kidneys, such as blood creatinine and blood urine nitrogen (BUN). Causes for this kind of failure include blockage from an enlarged prostate gland or kidney stones, low blood volume, infections in the bloodstream (called sepsis), as well as toxic effects from drugs and other substances.

INVENTION

The artificial kidney changed everything for patients suffering from these conditions. Dutch physician and medical pioneer Willem Kolff (1911–2009) invented the first dialysis machine in the Netherlands. While a student at the Leiden Medical School, Kolff witnessed the death of a 22-year-old patient from kidney failure. This pivotal event began his research into an artificial kidney.

During World War II, Kolff was working in a small municipal hospital in Kampen in the Netherlands. There in 1941, he developed the first dialysis machine. He made it from coarse materials gathered from a local factory. The machine consisted of a long tube of cellophane sausage wrapped around a slatted wooden drum. An electric motor turned the drum in a tank filled with dialyzing solution. Gravity drew the patient's blood through the tube as the drum spun. The waste materials in the blood filtered out through the tubing into the solution before the clean blood returned to the body.

Over the next few years, Kolff worked to improve on this initial design. By 1945, he had successfully treated his first patient: a 67-year-old woman who lived for 7 more years after treatment. Kolff worked with other physicians to improve the machine's basic design. Kolff's revolutionary machine would become a routine treatment for what was once considered a fatal condition.

Dr. Willem Kolff (center), inventor of the artificial kidney, works with American physicians to improve it.

DIALYSIS TODAY

Currently about 217,000 people in the United States are on some form of dialysis. People with acute renal failure may need the procedure only for a short time until the kidneys recover function. However, in cases of chronic renal failure, patients must have dialysis on a regular basis for the rest of their lives, or undergo a kidney transplant. With end-state renal disease (ESRD), the kidneys function at 10 to 15 percent or less of their normal capacity. In such cases, dialysis or kidney transplant are the only solutions.

The most prevalent form of dialysis, hemodialysis, works in much the same way as Kolff's original machine. It pulls waste out of the blood by filtering it through a membrane. Tubes inserted into the patient's arm or leg form a connection between an artery and a vein and the dialysis machine. The unpurified blood runs through the machine, which filters out impurities and excess fluid. Potassium ions and nitrogenous wastes diffuse out into a solution in the machine, and bicarbonate ions and other substances are added to the returning blood. A typical hemodialysis regimen involves three weekly sessions, each lasting four to six hours.

In continuous ambulatory peritoneal dialysis (CAPD), the patient's own peritoneum (the membrane lining the abdominal cavity) acts as the filter. A catheter is inserted into the abdomen, through which the dialysis solution infuses the abdominal cavity, equilibrates with the blood, and then drains out of it. Tiny blood vessels in the peritoneal membrane do the filtering job the kidneys normally do.

Using this method, it can take an hour or longer for the toxins and fluids to be completely removed, and the procedure must be done four to five times a day. Patients can perform CAPD at home. A continuous cycling peritoneal dialysis machine (CCPD) does the same thing as CAPD—cycling dialysis solution in and out of the peritoneal cavity—while the patient sleeps.

Like little mountain creeks and streams emptying into rivers, the nephrons drain into the collecting ducts, which merge into several hundred papillary ducts. Then the papillary ducts merge into the minor and major calyces, carrying the urine down from the renal cortex to the renal medulla to the renal pelvis.

Zooming in to the cellular level of the nephron, we can see a number of interesting features. While the outside of the glomerular capsule contains a simple layer of epithelial cells, the visceral (inside) layer next to the glomerulus is composed of podocytes, cells with little feet that grip onto the endothelial cells of the capillaries. Between the feet

are filtration slits, or pores, where the filtrate passes into the glomerular capsule.

The renal tubule and collecting duct have different kinds of cells. In the first part of the tubule, the proximal convoluted tubule, there are cuboidal epithelial cells. These multitasking cells both reabsorb materials from the passing filtrate and secrete substances (waste, excess ions, and drugs) into it. A brush border of microvilli (similar to that in the small intestine) increases the surface area of the tubule for the absorption of water and solutes. Remember, the tubules are carting off waste. When they secrete something into the urine, they are at the same time removing

TYPICAL NEPHRON

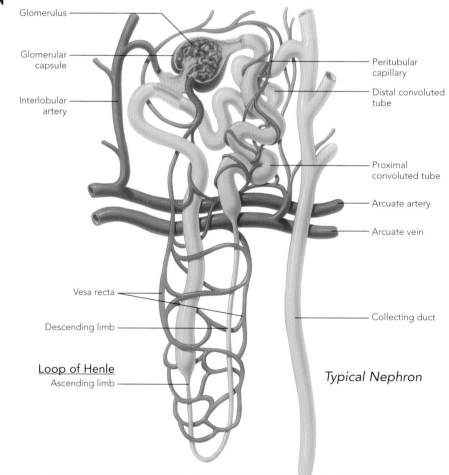

Glomerulus

Glomerular capsule

Interlobular artery

Peritubular capillary

Distal convoluted tube

Proximal convoluted tube

Arcuate artery

Arcuate vein

Vesa recta

Descending limb

Collecting duct

<u>Loop of Henle</u>
Ascending limb

Typical Nephron

it from the blood. In short, their job is to maintain balance (homeostasis) by excreting waste and keeping salt, water, and acid and base in balance.

Simple squamous epithelial cells compose most of the loop of Henle; they are water permeable. Then the thicker, ascending part of the loop contains cuboidal and low columnar epithelial cells. And finally, the distal convoluted tubule has principal and intercalated cells. At this point, much more secretion than absorption is occurring. The principal cells are involved in balancing water and salt, and the intercalated cells help maintain the blood's pH. The urine drains into the collecting and papillary ducts, which are lined by columnar cells.

Kidneys at Work

The three basic processes carried out by the nephrons and collecting ducts are glomerular filtration, tubular reabsorption, and tubular secretion. First, water and solutes filter from the blood plasma into the glomerular capsule. Then, in the renal tubules, some 99 percent of the water, as well as some solutes, are reabsorbed into capillaries. While the filtrate flows, cells lining the tubules secrete wastes from the capillaries, further purifying the blood.

The kidney's nephrons filter wastes from the blood plasma and turn it into urine in several stages. The first is glomerular filtration: The blood plasma filters from the glomerulus into the glomerular (Bowman's) capsule. How much blood plasma actually gets through? Only about 16 to 20 percent of what comes in via the arterioles goes on and becomes filtrate in the nephron. And of that amount, more than 99 percent ends up being reabsorbed by the tubules and ducts. So of the 40 gallons (150 L)

KEEPING HEALTHY

KIDNEY STONES

"A searing pain that comes in waves, moving from the back to the abdomen. The worst pain of my life." Descriptions of having a kidney stone relate what it is like to have a sharp-edged pebble travel down the ureter, into the bladder (where the pain often subsides), and then through the urethra. About one million Americans between the ages of 20 and 40 are treated each year for a kidney stone. Some 10 percent of men and 5 percent of women will have a kidney stone by the age of 70. When the urine becomes too concentrated, solutes such as calcium salts and uric acid can crystallize into little mineral deposits or stones. While 90 percent of stones pass without help, some require surgery or the newer technique of lithotripsy, in which high-energy sound waves crush the stone into fragments that wash out with the urine. The best way to avoid getting them is to keep your urine diluted by drinking three to four quarts (3 to 4 L) of water a day.

Enlarged image of
a kidney stone

of glomerular filtrate that daily passes through a woman's kidneys (more for men), only about 1 to 2 quarts (1 to 2 L) ends up as urine.

The initial filtrate is far closer to blood plasma than urine in its chemistry. The junctions between the capillaries and the podocytes (footlike cells) of the glomerular capsule have a three-layer meshing that does not allow blood cells, platelets, and most plasma proteins to pass through. These "good things" remain in the bloodstream while the rest, the filtrate, enters.

This scan shows a transplanted kidney.

The kidneys maintain a steady glomerular filtration rate (rate at which filtrate is made) of about 3.5 fluid ounces (105 mL) per minute in females and 4.25 fluid ounces (125 mL) per minute in males. They do this by controlling the rate that blood flows into and out of the glomerulus and by adjusting the glomerular capillary surface area.

The filtrate passes from the glomerular capsule into the first part of the renal tubule, the proximal convoluted tubule where reabsorption begins to take place. Remember that 99 percent of the filtrate still has to be reabsorbed before it can be called urine. Most of the reabsorption happens in the proximal convoluted tubule. In addition to water, a number of solutes are reabsorbed, including (in order of quantity) chloride ions, sodium ions, bicarbonate ions, glucose, potassium ions, urea, uric acid, and proteins. Notice that some urea is reabsorbed by the blood. Of all the solutes, though, urea has the lowest percentage of reabsorption—only 0.84 ounces (24 g) out of 2 ounces (54 g) per day (44 percent). The other solutes are absorbed in percentages varying from 91 to 100 percent.

Since 20 percent of the body's entire volume of blood is filtered through the glomeruli into the tubules every 45 minutes, the reabsorption that takes place there keeps it from being drained of fluid. As in the glomerular corpuscle, there are three layers of membranes through which the water and solutes have to pass: the luminal and basolateral membranes of the tubule cells and the endothelium of the peritubular capillaries (the capillaries that return the water and solutes to the bloodstream). Using active or passive transport mechanisms, the substances cross this barrier by a couple of different routes. They can either squeeze between the tightly packed tubule cells, or they can pass right through the tubule cells themselves. Hormones constantly regulate the amounts of water and ions that are reabsorbed in any given time period.

The proximal convoluted tubules absorb about 65 percent of the filtrate's initial water, and the loop of Henle absorbs 15 percent. The distal convoluted tubule then takes out its share—10 to 15 percent—so that when the fluid enters the collecting duct some 90 to 95 percent of its water and solutes has been reabsorbed.

While this reabsorption is occurring, secretion is also happening. Instead of substances moving out of the tubules into the blood, the reverse occurs. The net result of this two-way process, though, is the same—the blood becomes cleaner, the urine is more loaded with waste, and fluid balance and composition is maintained.

Such substances as hydrogen and potassium ions, creatinine, ammonium ions, and organic acids pass from the peritubular capillaries through the tubule cells and into the filtrate. Except for the potassium, most of the substances are secreted in the proximal convoluted tubule, the same place where most of the absorption takes place. Some of the urea and uric acid that have already been passively reabsorbed can, via secretion, make their way through the tubules and then out of the body.

And drugs such as penicillin and phenobarbital are secreted into the renal tubules for elimination.

In addition to removing undesirable substances from the blood, secretion also helps balance the blood's pH. When the blood falls into the acidic range, the tubule cells secrete hydrogen ions into the filtrate and reabsorb more bicarbonate ions. Bicarbonate, for example, which is alkaline (or basic), is retained and helps buffer the blood. When the blood's pH climbs, bicarbonate ions are secreted into the filtrate, and chloride ions (Cl-) are reabsorbed; the alkaline substances are thus passed out of the body in urine, and the blood can return to its proper pH.

Organ transplants offer new hope to those with kidney disease and create a special bond between donor and patient.

URINARY TRACT

Removing Waste From the Body

While the kidneys are the center of action in the urinary system, they need a delivery route for transporting the urine out of the body. That's where the ureters, bladder, and urethra come in. The design and functions of the urinary tract differ somewhat between males and females.

Ureters

After being formed in the kidney, the urine drains from the renal pelvis into the ureter. Each kidney has a ureter, a narrow tube about 10 to 12 inches (25 to 30 cm) long and about 0.04 to 0.4 inches (1 to 10 mm) in diameter that runs from the kidney to the urinary bladder. When the bladder fills, the near end of the ureter closes and prevents urine from backing up to the kidney. Occasionally the ureters do not properly seal off urine from the bladder; microbes can then enter the ureters and cause kidney infections, called urinary reflux.

The walls of the ureters are composed of three tissue layers. The innermost layer, the mucosa, has connective tissue loaded with collagen and elastic fibers that enable the ureter to stretch to accommodate varying amounts of fluid. Mucus secreted by the mucosa keeps the urine from entering the ureter cells, preventing any alteration of their pH and solute concentration.

The intermediate layer of the ureter is the muscularis, which consists of both longitudinal and circular smooth muscle fibers. The distal, or final, third of the ureter has circular muscles sandwiched by layers of longitudinal muscles.

The muscularis propels the urine by peristalsis. Depending on the rate of urine formation, some one to five waves per minute ripple along the ureters. Gravity and hydrostatic pressure also help push the urine through the ureters.

DID YOU KNOW?

URETHRA

+ In both men and women the urethra is the passage through which urine exits the body; in men it also serves as the discharge passageway for semen.

+ The urethra measures 1.5 inches (4 cm) in women and 6 to 8 inches (15 to 20 cm) in men.

+ The flow of urine through the urethra is controlled by both voluntary and involuntary muscles.

+ Newborns void urine 5 to 40 times a day. At two months, a baby voids about 14 fluid ounces (400 mL) of urine a day; adolescents and adults void 1.5 quarts (1,500 mL) a day.

+ Awareness of urination starts at about age 15 months. Control of nighttime urination sometimes takes until age four.

The adventitia is the outer layer of the ureter. Its connective tissue not only secures the ureter into position, but also holds the blood and lymphatic vessels, as well as the nerves that service the muscularis and the mucosa.

Bladder

Urine moves from the ureters into the urinary bladder for temporary storage. The bladder is an expandable muscular sac that is situated on the pelvic floor just in front of the rectum in men and in front of the vaginal canal and uterus in women. In men, the prostate gland surrounds the neck of the bladder at the point where it joins the urethra. The female bladder is smaller because the uterus lies just above.

When empty, the bladder assumes a pyramidal shape. When it begins filling up—which it does continuously—it becomes spherical and then takes on a pear shape as part of it distends up into the abdominal cavity. Like the stomach walls, the bladder walls have folds called rugae that allow for quite a bit of expansion.

The female urinary system—from the kidneys, to the ureters, and the bladder—can be traced in this 3-D visualization.

A medium-full bladder holds about 1 pint (500 mL) of urine and measures 5 inches (12.5 cm) in length. As urine accumulates, the walls of the bladder thin out and internal pressure grows. Fully expanded, the bladder can hold 1 quart (1 L) or more, but it is also capable of bursting. In general, though, the bladder can store urine until it is released through the urethra.

Like the ureters, the bladder walls have three layers—a mucosa, a muscularis, and an adventitia. The mucosa is made up of transitional epithelium and lamina propria (or connective tissue). Under this coat is the muscularis, which in the bladder is called the detrusor ("to thrust out") muscle. The detrusor has three stacked muscle layers—inner longitudinal, circular, and outer longitudinal. The connective tissue of the adventitia makes a smooth connection with the adventitia of the ureters.

The base of the urinary bladder is a triangular area called the trigone, defined by three openings—the two ureters and the urethra. Bladder infections usually occur in the trigone. The muscularis around the urethral orifice here forms an internal urethral sphincter, a valve that, since it is composed of smooth muscles, involuntarily controls the release of urine.

Just beneath this opening is the external urethral sphincter, which wraps around the first part of the urethra. This sphincter is composed of skeletal muscles and is, thus, under voluntary control. This control, in other words, allows a person to decide when he or she will urinate.

Urethra

Urine exits the body through the urethra, a narrow-walled muscular canal that measures about 1.5 inches (4 cm) in females and 6 to 8 inches (15 to 20 cm) in males.

In women, the urethra is attached to the vaginal wall by fibrous connective tissue. Its opening, the external urethral orifice, lies between the clitoris and the vaginal opening. The shorter urethra in females and its location near the vagina and anus make it prone to infection. About 40 percent of women experience urinary tract infections at some point in their lives. Wiping back to front after a bowel movement can cause bacteria to enter the urethra or even colonize the periurethral area. Likewise, sexual

The female urethra is surrounded by two layers of muscle.

DID YOU KNOW?

BLADDER

+ A moderately full urinary bladder holds about 17 fluid ounces. The bladder is capable of holding 1 quart or more of urine.

+ The base of the bladder, the trigone, is defined by the openings for the two ureters and the urethra.

+ Men with enlarged prostate glands that limit bladder outlet often retain more than 5 fluid ounces of urine in their bladder after trying to empty it.

intercourse can push vaginal bacteria up into the urinary tract. Infections can be localized in the urethra or move up into the bladder or kidneys. Fever and frequent and painful urination are typical symptoms; antibiotics are usually successful in treating such infections.

In men, the much longer urethra is more complicated. The male urethra discharges both urine and semen, which contains sperm. The first inch is called the prostatic urethra and is surrounded by the prostate gland, which contributes semen from the vas deferens and ejaculatory duct. Next, the membranous urethra runs for almost an inch from the prostate gland, through the urogenital diaphragm, and on to the penis. Then for about the next 6 inches (15 cm), the spongy, or penile, urethra extends the length of the penis, opening at the external urethral orifice.

The composition of the urethra is similar in females and males. The urethral wall is composed of a mucosa and a superficial muscularis of circular smooth muscles.

In males, the prostatic urethra has ducts that carry secretions from the prostate gland and seminal vesicles and sperm from the vas deferens into the urethra. Some of these secretions neutralize the acid environment of the female reproductive tract and increase sperm motility. Then, along the spongy urethra, ducts of the bulbourethral (Cowper's) glands secrete a lubricating mucus during sexual arousal, as well as an alkaline material to buffer the urethra's acidity before ejaculation.

Urination

The final step in the excretory process is the actual elimination of the waste product, urine. A number of specialized muscles and tissues make this action possible. What happens during urination, or micturition? Put most simply, the bladder is emptied.

What seems the simplest of daily activities is more than a mindless duty for the body. In the absence of any disease, the bladder can hold about 17 fluid ounces (500 mL) or more urine for two to five hours.

But after accumulating about 7 to 14 fluid ounces (200 to 400 mL) of urine in the bladder, stretch receptors there begin to send nerve impulses to the sacral (pelvic) region of the spinal cord to stimulate the micturition reflex. Impulses going to the brain also stimulate the urge to urinate. The more frequent the bladder's contractions occur, the more urgent the signal to start the micturition reflex will be.

The impulses cause contractions of the bladder's detrusor muscle, while relaxing the internal urethral sphincter. Remember that the external urethral sphincter is under voluntary control. When a person decides to urinate, neural signals will inhibit the normally contracted external urethral sphincter and thereby allow it to relax. Then the micturition reflex takes over and urination can happen.

If a person consciously decides not to void the bladder, the contractions should subside within approximately one minute. In addition to the external urethral sphincter, the levator ani muscle on the pelvic floor (the same muscle used in defecation) helps control the urination impulse. When another 7 to 10 fluid ounces (200 to 300 mL) of urine have collected in the bladder, the micturition reflex will begin all over again. This process will keep repeating until the bladder is so full that voluntary control is overridden, and urination occurs.

The male urinary system

Chapter 9

CONTROL CENTER ⊢

NERVOUS SYSTEM

Command Center

Spreading through every tissue in an intricate electrical web, the nervous system is the body's central controller and communicator. Its billions of neurons, or nerve cells, connect with one another in a network of intricate complexity. Along with the endocrine system, the nervous system dictates and coordinates every action the body takes. Its sensory nerves deliver information about the outside world, and its primary organ, the brain, is the seat of the mysterious processes of human consciousness.

What does this all look like, anatomically speaking? It literally starts at the top with the brain—arguably the most important organ in the human body. Enclosed in the skull, the brain is a double handful of pinkish gray, mushy tissue containing billions of neurons and an even greater number of support cells, called glial cells.

Twelve pairs of cranial nerves branch off from the base of the brain to the head and upper body. Many more nerves run from the brain down through the spinal cord in the center of the vertebral column. Thirty-one pairs of spinal nerves extend from the spine to the muscles and glands of the body. Nerves divide as they fan out through the body and, in some places, converge in clusters called plexuses, as in the gastrointestinal tract where they serve as mini-controllers of digestion.

DID YOU KNOW?

DEVELOPMENT

In the womb, the brain takes longer than most organs to develop.

+ **4 weeks:** Neural tube, basis of the central nervous system, has formed.

+ **5 weeks:** Forebrain, midbrain, and hindbrain form at one end of neural tube; spinal cord encompasses rest.

+ **13 weeks:** Cerebrum has begun to form; brain stem, cerebellum appear.

+ **26 weeks:** Cerebrum folds begin.

+ **Birth:** All structures are formed, but neural networks continue to develop throughout childhood.

A Complex System

The nervous system has three connected functions. The first is input: Sensory nerves receive information from outside and inside the body. These nerves gather data about changes in light, sound, touch, and other stimuli in the environment as well as inside information about

THE NERVOUS SYSTEM

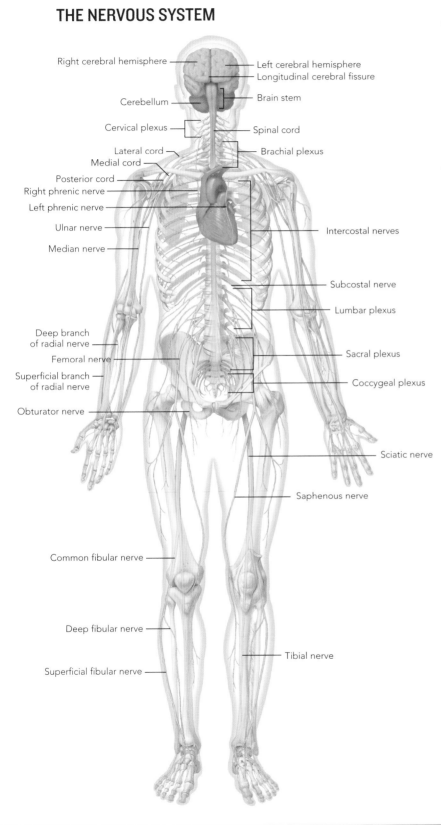

Right cerebral hemisphere

Left cerebral hemisphere
Longitudinal cerebral fissure

Cerebellum

Brain stem

Cervical plexus

Spinal cord

Lateral cord

Brachial plexus

Medial cord

Posterior cord

Right phrenic nerve

Left phrenic nerve

Ulnar nerve

Intercostal nerves

Median nerve

Subcostal nerve

Lumbar plexus

Deep branch
of radial nerve

Femoral nerve

Sacral plexus

Superficial branch
of radial nerve

Coccygeal plexus

Obturator nerve

Sciatic nerve

Saphenous nerve

Common fibular nerve

Deep fibular nerve

Tibial nerve

Superficial fibular nerve

the status of internal organs, muscles, blood chemistry, and myriad other changeable states of the body. They then convert this information into signals sent to the brain or spinal cord.

The second nervous system job is integration: The central portion of the system receives and interprets sensory input and makes decisions about what should be done.

This leads to the nervous system's third function: response. The brain or spinal cord sends a signal—a motor output—that dictates an action in some part of the body. Millions of signals every second tell the body to blink, or increase its heart rate, or utter a syllable, or release epinephrine, or perform countless other actions, most of them unnoticed by the conscious mind.

Neuroscientists typically divide the nervous system into two main parts and a number of subdivisions. The two primary parts are the central nervous system (CNS) and the peripheral nervous system (PNS). The central nervous system consists of the brain and spinal cord—the command and sensory integration centers of the body and the seat of thought, memory, and emotion. A communication network, the peripheral nervous system consists mainly of nerves that carry signals to and from the CNS.

The peripheral nervous system is made up of three parts: motor, sensory, and autonomic. The motor division transmits orders from the CNS to the muscles and glands, telling them to take action. The motor division includes the somatic nervous system, also called the voluntary nervous system because its actions fall under conscious control.

The sensory division carries impulses from the sensory organs and the visceral (or internal) organs to the central nervous system. The autonomic (or involuntary) nervous system carries signals to areas of the body rarely under conscious control, such as the smooth muscle, cardiac muscle, and glands. It regulates processes that we don't actively think about, such as respiration and digestion.

The autonomic nervous system can itself be divided into two parts: the sympathetic and parasympathetic divisions. These motor pathways generally have opposing effects on the body and its systems. The sympathetic division usually mobilizes body systems. For example, the sympathetic division increases the heart rate while the parasympathetic division inhibits and slows it down (see "The Nerves," page 252).

Nerves

Two kinds of cells make up the body's nervous tissue: neurons and neuroglia (also called glial cells). Five to fifty times more common than neurons, the small neuroglia have the unglamorous but essential role of support

Massage therapy may relax the nervous system, slowing heart rate and blood pressure.

NEURON

Dendrites
Nucleus
Cell body
Nissil bodies
(produce neurotransmitters)
Axon
Myelin sheath
(formed by Schwann cells)
Node of Ranvier
Axon terminal
fiber
Synaptic nobs
(for axon terminal of
presynaptic neuron)

and protection in the nervous system. Unable to carry electrical impulses, the six types of neuroglia cells—four in the central nervous system and two in the peripheral nervous system—provide nutrients and a supporting framework of tissue for their neuronal cousins. Some protect brain cells by killing invading microbes and clearing away dead cells. Mature neuroglia can divide and reproduce, filling in areas of the brain where neurons have been destroyed.

The neuron—nerve cell—is the basic functioning unit of the nervous system, notable for its ability to conduct electrochemical messages from one part of the body to another. Like cardiac muscle cells, neurons are long-lived, capable of surviving the length of a human lifetime. Scientists have recently found evidence that neurons do indeed reproduce. This process may play a role in human cognition. However, mature nerve cells do not divide and reproduce, which is why damage to the brain or spinal cord is often irreversible. Neurons are hungry cells, demanding oxygen and glucose for survival. After cardiac arrest, people can suffer brain damage because of the relatively brief loss of oxygen to nerve cells.

The central portion of these highly specialized cells ranges in diameter from 5 to 140 microns, just barely visible to the naked eye. Most neurons consist of three parts: the cell body and two kinds of nerve "processes" or fibers. These fibers are the dendrites—short, branchlike extensions reaching out from the cell body—and an axon—a long, thin, electrically excitable projection.

The cell body contains a nucleus and typical cell organelles. Its surface, the plasma membrane, sometimes serves as a receptive surface for messages from other neurons. Dendrites extend like leafless trees from the cell body; some neurons have just a few thick dendrites, while others have hundreds. These are the main receivers for signals arriving from other nerve cells.

Axons are the most remarkable parts of these remarkable cells. Arising from the cell body, these thin fibers can range in length from just a fraction of an inch to up to several feet long (as from the spine to the toes). Each axon finishes up in a spreading axon terminal with up to 10,000 or more branches, and each branch ends in a rounded bulb or knob.

Most axons are covered with a myelin sheath, a fatty coating that insulates the fiber. It also increases the traveling speed of nerve impulses along its length. Two kinds of neuroglia—Schwann cells in the PNS and oligodendrocytes in the CNS—wrap around the axons to form this protective sheath; between the glial cell wraps are little gaps called nodes of Ranvier.

Nerve impulses travel from one neuron to the next through one-way junctions called synapses, the points where axons' terminal branches reach the dendrites of other neurons. Most neurons are covered with hundreds—or even thousands—of synapses, so that they are able to receive signals from many different neurons. Neurons can also make synaptic connections with a few other kinds of tissue, such as skeletal muscle or glandular tissue, allowing nerves to trigger action in those tissues directly.

Neurons

Neurons come in all sizes and many shapes, but they can be classified into three groups according to the number of dendrites and axons extending from them. Most, including almost all in the brain, are multipolar neurons, with several dendrites and one axon extending from the cell body. A few (such as in the retina) are bipolar, with one main dendrite and one axon. Some are unipolar: A single process extends from the cell body and divides into a T, one end branching out into dendrites and the other into axon terminals. Some unipolar neurons are found in the peripheral nervous system.

Neurons can also be grouped according to their jobs. Sensory neurons carry impulses between the sensory receptors in the skin or

SELECTED NEUROTRANSMITTERS

NEUROTRANSMITTER	WHERE IT IS RELEASED	EFFECTS
Acetylcholine, shown on opposite page	Junctions in skeletal muscles, autonomic motor endings, motor cortex of brain	Muscle contraction; in the brain, involved in memory, sleep, attention
Norepinephrine	Brain, peripheral nervous system	Regulates arousal, moods, heart rate, blood pressure
Dopamine	Brain, peripheral nervous system	Involved in movement, reward, hormones, psychosis
Serotonin	Brain stem, cerebellum, pineal gland, spinal cord	Involved in sleep, mood, depression, anxiety, appetite
GABA (gamma-aminobutyric acid)	Hypothalamus, cerebellum, spinal cord, retina	Inhibits neurons
Glutamate	Brain, spinal cord	Excites neurons; involved in learning and memory
Endorphins	Brain, pituitary, spinal cord	Inhibit pain; natural opiates
Nitric oxide	Brain, spinal cord, adrenal gland, nerves to genitals	Affects digestion, messaging, male sexual response

A color-enhanced microscopic image of acetylcholine, a neurotransmitter that controls muscular contractions and may also play a role in memory and sleep

internal organs and the central nervous system. Motor neurons send signals from the CNS to the muscles and glands. By far the most common are interneurons, which make up much of the brain and spinal cells; these are primarily responsible for integration of signals and for decision-making.

Regardless of shape or place in the nervous system, every neuron can send an electrical impulse down its length and transmit it to another cell. The neuron's plasma membrane contains tiny tunnels called ion channels that allow electrically charged atoms or small molecules to flow in and out of the cell. When the cell receives an impulse from another neuron, the impulse triggers a tiny voltage change (an action potential) on the surface of the receiving

neuron. This voltage change then whizzes down the length of the axon at speeds of up to 65 miles per hour. A neuron can fire repeatedly and rapidly, up to a thousand times per second.

When the signal reaches the terminal bulb at the end of an axon, it usually triggers the release of chemicals called neurotransmitters that are

SYNAPTIC CONNECTIONS

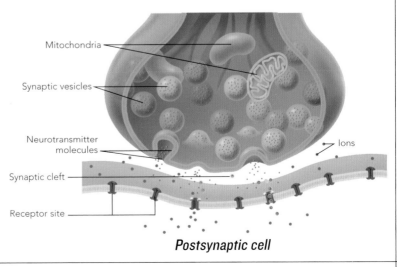

Mitochondria

Synaptic vesicles

Neurotransmitter molecules

Synaptic cleft

Receptor site

Ions

Postsynaptic cell

Serotonin (color-enhanced magnified view, above) is a neurotransmitter involved in regulating sleep, mood, and appetite.

stored there in little bubblelike vesicles. In less than one-thousandth of a second, the chemicals diffuse across the tiny space of the synapse—the gap between the transmitting axon and the receiving target cell—and bind to receptors on the target neuron or effector cell. The receptors then activate their own neuron, prompting it to fire off an impulse in turn or else inhibiting it from firing. After the neurotransmitter has done its job, it is diffused, destroyed by enzymes, or reabsorbed and destroyed by the transmitting terminal (a process called reuptake).

Scientists have discovered dozens of different neurotransmitters at work in the nervous system, with more sure to be found. Many bind to a specific receptor, like a key in a lock, and each has a different, specialized set of effects. Research into neurotransmitters is one of the busiest areas of neuroscience, promising insights into the causes of Alzheimer's disease, Parkinson's disease, depression, schizophrenia, and many other ailments.

The first neurotransmitter was discovered in 1921 by German scientist Otto Loewi. One night Loewi had a dream that outlined an experiment to prove that nervous impulses were transmitted chemically. "I awoke," he wrote later, "turned on the light, and jotted down a few notes on a tiny slip of thin paper. Then I fell asleep again. It occurred to me at six o'clock in the morning that during the night I had written down something most important, but I was unable to decipher the scrawl. That Sunday was the most desperate day in my whole scientific life. During the next night, however, I awoke again, at three o'clock, and I remembered what it was."

In Loewi's experiment, two frog hearts were placed in connected chambers that were filled with a saline solution. Loewi stimulated the vagus nerve connected to the first frog heart, causing it to slow down; soon afterward, the second frog heart also slowed, showing that a substance was flowing from one heart to the other in the saline. This neurotransmitter, later named acetylcholine, is now known to control muscular contractions. It may also be essential to the brain's role in memory and sleep.

Other notable neurotransmitters include dopamine, which controls some movements, can cause psychosis, and actuates during some hormonal responses. The brains of people with Parkinson's disease have a shortage of dopamine; administration of levodopa, from which dopamine is made, can relieve the symptoms of Parkinson's in the short run. Serotonin, found in the brain and throughout the body, may be involved in the experiences of sleep, anxiety, and mood disorders.

Fluoxetine (Prozac) and other drugs that affect serotonin are used to treat depression and similar illnesses. Endorphins, natural opiates that act like morphine, may be released to reduce pain. And even gases can act as neurotransmitters: Nitric oxide and carbon monoxide travel from neuron to neuron by diffusing through their plasma membranes.

KEEPING HEALTHY
EXERCISE YOUR MEMORY

Fitness training may be just as important for our brains as it is for our bodies. People who regularly give their brains a workout may help build and maintain the synaptic connections necessary for learning and memory. How do you keep your brain in tip-top shape?

+ **Get some exercise:** Exercise improves blood flow to the brain. It also results in the release of several different neurotransmitters, including norepinephrine, serotonin, and dopamine, which stimulate synapses in the brain and may help consolidate memories. Exercise may also increase alertness and improve the ability to learn.

+ **Relax:** Stimulation is important, but the brain also benefits from taking time to reduce stress and getting enough sleep. Stress hormones can interrupt the learning process, making it more difficult to absorb information. Sleep, on the other hand, appears to help the brain consolidate memories, making them easier to retrieve later on. REM sleep—which usually takes place late in the sleep cycle—seems to be especially important for memory.

+ **Puzzle it out:** Do crossword puzzles that require word association, list generation, memory, spelling, and vocabulary—all-important cognitive skills. Work with numbers—including solving equations or even just filling in a sudoku magic square—which appears to stimulate brain function, especially when using odd numbers. All this may help strengthen synapses in the brain, making it easier to hold on to memories or feel prepared for new situations. Studies indicate that increasing the difficulty level of the puzzles helps improve results.

THE BRAIN

Communication Hub

"The brain—is wider than the sky," wrote Emily Dickinson. And indeed, not only can the brain encompass the idea of "sky" and more, but it is an organ of astronomical complexity. The total number of synapses in the brain—the connections within its web of neurons—is about 10^{15}, or one thousand trillion, greater than the stars in our galaxy.

The brain is the body's central controller and the coordinator of virtually all its functions, from the automatic in-and-out of breathing to the act of solving an equation. It registers sensations and orders actions based on those sensations; it is the seat of memory, intellect, emotions, behavior, sleep, and wakefulness. It directs communications and creates in humans a sense of self-awareness and identity.

All of these amazing abilities come from a wrinkled, jellylike mass weighing about 3.5 pounds. Composed of about 100 billion neurons and up to 50 trillion neuroglia, the brain sits on top of the spinal cord like a mushroom on a thin stem. Its delicate tissues are protected by layers of bone, membrane, and fluid.

The outermost, and toughest, protective covering is the cranium, the rounded bones that form the top and back of the skull. Beneath the skull are three layers of connective tissue membranes known as meninges. The external layer is the dura mater (meaning "tough mother"), a strong leathery covering. Beneath it is the loose arachnoid ("spidery") mater, connected to the next layer below. This open area is flooded with cerebrospinal fluid that serves as a surprisingly effective liquid cushion for the brain. The subarachnoid space also contains blood vessels that feed the organ. The third layer, the pia

The brain processes learned habits in the cerebellum.

THE BRAIN

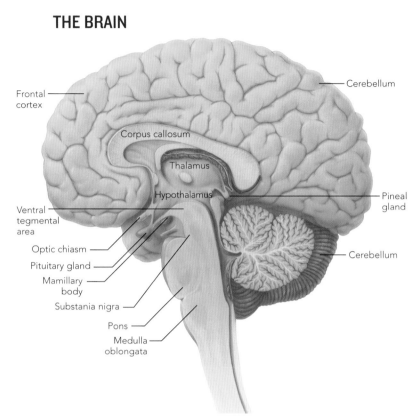

Frontal cortex

Cerebellum

Corpus callosum

Thalamus

Hypothalamus

Pineal gland

Ventral tegmental area

Optic chiasm

Pituitary gland

Mamillary body

Substania nigra

Pons

Medulla oblongata

Cerebellum

As well as serving as a watery buffer for the brain and spinal cord, the liquid nourishes the brain tissue and may carry chemical signals as well. If a tumor or injury blocks the flow of this fluid into the subarachnoid space, it can build up in the ventricles and put pressure on the brain tissue, a condition known as hydrocephalus.

Blood flows into the brain through the internal carotid and vertebral arteries, spreading out into a network of blood vessels. Although the brain makes up only 2 percent of the body's weight, its voracious neurons use fully 20 percent of the body's oxygen and glucose. When parts of the brain become active, blood flow to the regions instantly increases (and can be seen with MRI machines). If blood flow to the brain slows even momentarily, or if glucose levels drop, a person can lose consciousness.

mater ("gentle mother"), is a more delicate membrane, filled with small blood vessels, that clings directly to the brain's surface.

Parts of the brain are curiously wrinkled and grooved. Its ridges are called gyri ("twisters") and its grooves are known as sulci ("furrows"). Deeper grooves, called fissures, separate major areas of the brain; the most noticeable is the longitudinal fissure that separates the cerebrum, the main part of the brain, into two hemispheres.

Curving like horns deep within the brain are fluid-filled ventricles, hollow chambers that connect to the central canal of the spinal cord. Porous capillaries hanging from the walls of each ventricle release cerebrospinal fluid into the ventricles, where it circulates into the subarachnoid space and also into the spinal cord.

The sensitive brain is protected from a host of harmful substances in the blood by a unique blood-brain barrier. The capillaries in the brain, unlike those in the rest of the body, are tightly sealed, their endothelial cells forming particularly tight junctions. A thick basement membrane also surrounds the little blood vessels. As a result, many of the chemicals and hormones that routinely pass through the capillaries in the rest of the body are barred from the brain. Some substances can pass through, including glucose, oxygen, and carbon dioxide. Alcohol, nicotine, and anesthetics can cross into the brain. As useful as it is, the blood-brain barrier frustrates efforts to deliver helpful drugs, such as those that attack brain tumors.

Brain Structure

The brain's four main areas are the brain stem, cerebellum, diencephalon, and cerebrum. The brain stem begins with the medulla oblongata, a 1.2-inch continuation of the spinal cord. The medulla contains all the sensory and motor nerves that run up the spinal cord into the brain. In the medulla, the nerves from the body's left side cross to the brain's right side, and vice versa. The medulla regulates some of the body's most basic functions, including reflex actions such as coughing or swallowing. It contains the cardiovascular center as well as the rhythmicity area of the respiratory center, controlling the pace of heartbeats and breathing.

Directly above the medulla oblongata in the brain stem is the inch-long pons ("bridge"), a relay station that connects the medulla to other regions of the brain, such as the cerebellum.

The third part of the brain stem, the midbrain, also about an inch long, connects the pons to the diencephalon. It holds visual and auditory reflex centers—areas that control pupil size, for instance, and the startle reflex.

Tucked behind the pons, under the cerebrum, is the cerebellum, two rounded hemispheres that are deeply fissured and leaflike in cross section. The cerebellum's primary job is to coordinate movement and balance. It mediates between the muscles and the areas of the cerebrum that dictate movement. Skilled rote activities, such as hitting a baseball or playing the piano, are processed in the cerebellum.

Located in the middle of the brain and cradled by the two hemispheres of the cerebrum, the diencephalon is made of four structures: thalamus, hypothalamus, subthalamus, and epithalamus. Largest of the three is the double-lobed thalamus, which relays sensory

DID YOU KNOW?

BRAIN SURGERY

Brain surgery has a long history. Here are some major milestones:

+ In Europe and South America as early as 5000 B.C., skulls were drilled open and brain parts were cut away.

+ In 1879 British surgeon William Macewen successfully removed a tumor from a young woman's dura mater.

+ In the late 19th century, British neurosurgeon Sir Victor Horsley developed specialized techniques for operating on the brain.

+ In the early 20th century, Harvey Cushing, a pioneering American doctor, removed more than 2,000 brain tumors.

input—perception of pain, temperature, and pressure—to the cerebral cortex and plays a role in emotions, memory, and alertness.

Below it is the hypothalamus, a small but vital part of the brain's structure. Among orchestrating other tasks, it controls the autonomic nervous system, which regulates smooth and cardiac muscles, blood pressure, and other basic body functions; it produces hormones and stimulates the pituitary gland; it regulates body temperature and feelings of hunger; and it is involved in the perception of emotions such as fear and pleasure. Behind the thalamus is the epithalamus, containing the pineal gland, which helps to regulate the sleep-wake cycle.

Curling above is the cerebrum, the structure most people picture as the brain. Divided into two hemispheres, the cerebrum makes up more than 80 percent of the brain's mass. It is the center of intelligence, reason, communication, memory, and imagination.

The cerebrum has three kinds of areas: sensory areas that interpret sensory impulses, motor areas that control movement, and association areas that communicate with the other regions to process information. Its thin, grayish outer layer, the cerebral cortex, is made primarily of billions of neurons; the white matter that makes up much of its interior consists mainly of myelinated axons. Most of the brain's information processing takes place in this region. Specific functions can be mapped to specific parts of the cortex, although many regions connect with one another as well as with other parts of the brain.

Though similar in structure, the right and left hemispheres perform unique functions. Each is divided by fissures into five lobes, named for the cranial bones that overlie them. The frontal lobe is responsible for thought, speech, and skilled movements; the parietal lobe registers

The thalamus region is shown in orange in this rendering of the human brain.

A BRAIN ATTACK

Strokes are the third leading cause of death in the United States and the top cause of adult disabilities. They occur because of the intimate interactions between the brain and the cardiovascular system. Also known as a cerebrovascular accident or, more colloquially, as a brain attack, most strokes occur when the blood supply to part of the brain is cut off and, as a result, that brain tissue dies.

An aneurysm in the carotid artery presages a hemorrhagic stroke.

TYPES

About 80 percent of all strokes are ischemic strokes: They begin when a blood clot or some other material becomes lodged in a cerebral artery and cuts off the flow of blood to the brain. In thrombotic strokes, another kind of ischemic stroke, the blood clot forms directly in the artery of the brain itself, usually because the blood vessel has been damaged by atherosclerosis. An artery in the brain may also shut down because it has become clogged by fatty deposits.

In embolic strokes, likewise a type of ischemic stroke, the clot travels from elsewhere in the body and lodges in the brain's arteries. Most typically, the clot has formed in a heart that suffers from atrial fibrillation, where the irregular beat and blood flow is not strong enough to prevent clotting.

The remaining 20 percent of strokes are hemorrhagic, occurring when a blood vessel in the brain leaks or bursts outright. Intracerebral hemorrhages happen within the brain tissue, while subarachnoid hemorrhages start in an artery and leak into the fluid-filled space between the arachnoid mater membrane and the brain.

When this happens, brain cells may die not because they are starved of their blood supply but because the pooling blood puts pressure on the delicate brain tissue. Most hemorrhagic strokes result from blood vessels weakened by high blood pressure or containing thin, ballooning areas called aneurysms. A few also begin with an arteriovenous malformation, tangled webs of thin-walled blood vessels in the brain that are present from birth.

SYMPTOMS

Strokes come on very quickly. Their symptoms typically include one or more of the following signs: sudden weakness or numbness in the face, arm, or leg, usually on one side of the body; sudden difficulty speaking or understanding speech (called aphasia); sudden blurred, double, or lost vision; sudden dizziness, trouble walking, loss of balance or coordination; sudden severe headache with no known cause; and confusion or sudden problems with memory, perception, or spatial orientation.

Sometimes people experience a few of these symptoms fleetingly, feeling them fade in minutes or a few hours. These may be the signs of a transient ischemic attack (TIA), a ministroke, which by definition lasts less than 24 hours. TIAs are probably caused by clots or plaques that temporarily block a cerebral artery but are then cleared or washed away before serious damage can occur. Harmful in themselves, TIAs are also red flags: About a third of people who experience a TIA will later have a major stroke.

TREATMENT

Whether the symptoms are fading or overwhelming, stroke victims need immediate emergency care. Some stroke experts recommend that a companion perform the following quick test, memorable with the mnemonic FAST:

Face: Ask the person to smile. Does one side of the face droop?

Arms: Ask the person to raise both arms. Does one drift downward?

Speech: Ask the person to repeat a simple sentence. Are the words slurred or repeated incorrectly?

Time: If the person shows any of these symptoms, waste no time. Call 911 immediately.

Cardiologists say, "Time is muscle" when dealing with heart attacks. Neurologists say, "Time is brain" in the case of a stroke: The longer it takes to get treatment, the more brain cells will die.

At the hospital, doctors will scan the patient's brain to locate the region of the stroke and find out what kind it is. If a clot is involved and the patient has come to the hospital promptly, physicians can administer a so-called clot-buster drug, properly called a tissue plasminogen activator (t-PA), which may be able to dissolve the clot to open blood vessels. Used within three hours after the stroke, t-PA can reduce disability by 50 percent. Other surgical treatments can clean out clogged arteries or put a clip at the base of an aneurysm, stopping a hemorrhagic leak.

The severity of a stroke depends on the area of the brain damaged and the extent of that damage. The middle cerebral artery is the most common blood vessel involved; it feeds the language and speech areas of the brain, so difficulty in understanding and forming speech is a frequent result of a stroke. Other areas typically affected by strokes include the facial, arm, and leg areas of the left motor cortex, resulting in paralysis of the right side; Wernicke's or Broca's areas, leading to difficulty in forming and understanding speech, as well as reading, writing, or naming objects; facial and arm areas of the sensory cortex, leading to loss of sensation in the right side of the face and right arm; and the optic tract, causing a loss of vision in one-half of the visual field of both eyes. Physical therapy can often retrain a stroke victim in lost skills: Neurons can grow new connections in the brain, and many stroke victims recover substantially.

About 700,000 Americans suffer new or repeat strokes each year, and of that number, over 150,000 die annually. The death rate from stroke is declining, probably due to better control of some risk factors. Some risk factors—including age (over 55), family history, gender (women more than men), and race (African Americans more than Caucasians)—are uncontrollable. But others are treatable; they include hypertension (high blood pressure), high cholesterol, atrial fibrillation (irregular heartbeats), diabetes, smoking, alcoholism, and obesity. With such circumstances, fully 80 percent of strokes experienced today could have been prevented.

sensations; the occipital lobe processes visual input; and the temporal lobe, sound and music.

The two hemispheres communicate through bridges of nerve fibers. The largest of these is the corpus callosum, consisting of some 200 million nerve fibers in all.

The Mind

All of our understanding of the brain's anatomy does not take us very far toward answering the questions: What is the human mind, and what is the nature of human consciousness?

Clinically speaking, consciousness is awareness and responsiveness to surroundings. But in a larger sense, it is synonymous with mind, more difficult to define. The mind seems to arise from an interplay of processes in the brain: thoughts, perceptions, emotions, memory, imagination, and will. The human mind is aware of itself and, despite the complexity of its sensations, memories, and emotions, feels itself to be a single, perceptive unit.

Scientists cannot yet map out the origins of consciousness in the human brain. Some researchers study how perceptions, which arrive in tiny fragments, are integrated into single images. Others look into how perceptions move into memory and how memories are recalled. A third avenue of investigation is the study of emotions and how they affect both memory and decision-making.

Scientists are learning, for instance, that emotions are mediated through structures and pathways known as the limbic system, found in the cerebral hemispheres. This system, early to evolve, still controls the fight-or-flight reflex. In humans, it is heavily involved in integrating emotions, physical sensations, and memory.

The little almond-shaped amygdala is a key portion of the limbic system. When stimulated,

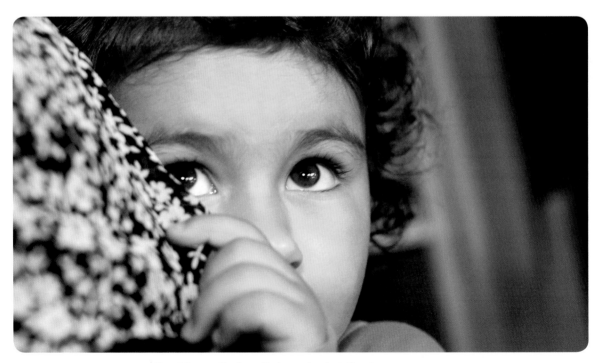

Fear triggers the hypothalamus to start the stress response.

it produces the sensations of fear; it is also activated when a person recognizes angry or fearful expressions on another person's face. In a frightening situation, the amygdala will trigger the hypothalamus to begin a wave of physical responses, including releasing stress hormones and elevating blood pressure. The amygdala and the nearby hippocampus also play a role in integrating emotion with memories. The limbic system connects to the olfactory bulb, responsible for smell, which may account for strong links between odor, emotion, and memory.

Mental Functions

Among the most mysterious functions of the brain are its most important tasks: regulating sleep and wakefulness, producing and understanding language, storing memories, and learning. With new brain-imaging techniques, researchers have begun to trace the pathways involved in these functions (see "Breakthrough," below).

Information in the brain flows through networks of neurons that fire together in exact timing when a job must be carried out. The brain is quite plastic—able to reorganize based on new experiences—so strong connections within these networks are reinforced with new

A girl makes new memories with a pet, which will be stored in multiple locations in the brain

synapses and more powerful signals. Weak ones are pruned away.

Learning is a multistage process of acquiring new knowledge through sensory experience and storing that knowledge in memory. Incoming information is held at first in working memory, a kind of short-term storage in the cells of the prefrontal lobe. Some working memories, perhaps because they are reinforced, then move into long-term memory, a process that seems to be governed by the hippocampus in the limbic system. Patients with damage

BREAKTHROUGH BRAIN MAPPING ⊢

The first inside look at brain function was the electroencephalograph (EEG), invented in 1929. The EEG records electrical activity in the brain by means of electrodes on the scalp, right; abnormal brain waves can indicate disorders. Techniques from recent decades include magnetic resonance imaging (MRI), which records how atoms in the brain resonate in a magnetic field. The patient moves through a cylindrical magnet, and the machine produces a series of cross section images. And magnetoencephalography (MEG), another magnetic imaging technique, can quantify the strength of brain activity in different regions.

to the hippocampus cannot form long-term memories.

Scientists are finding that information is broken down into multiple strands that are organized into categories and stored in scattered locations across the cerebral cortex. The memory of your pet, for instance, is not stored in its own bin in the brain, but more likely in a web of associations such as animal, four legs, small, fur, white, purr, snow, pleasure, and so on. When recalling the cat, all these locations are activated and send their messages on converging pathways to somehow form the idea, my cat Snowball.

Indeed, research now indicates that the brain has at least 20 categories for organizing information, among them plants, animals, body parts, nouns, verbs, proper names, faces, facial expressions, colors, numbers, letters, kinds of sounds, and fruits/vegetables. Studies of people with very localized brain damage show that these categories can have some surprising distinctions: One patient could recognize and name small tools but not large ones; some patients can write words but not numbers; one had difficulty recognizing musical instruments by sight but could name them easily by sound. Memories of learned physical skills are stored elsewhere from memories of concepts or episodes, which is why patients who can no longer recognize a spouse may

A brain scan shows increased brain activity during REM sleep.

still be able to play the piano.

Language, a defining characteristic of the human animal, is dependent on learning and memory. Unlike most functions of the brain, language skills are localized in one hemisphere of the cerebrum, usually the left. Language appears to be broken down into two main components. Speech comprehension involves the prefrontal cortex and Wernicke's area, a region of the temporal and parietal lobes. Speech planning and production occurs in another region, Broca's area, in the frontal lobe. Messages from this region are then passed to a nearby part of the premotor cortex, which controls the larynx and organs of speech.

Far more mysterious even than language or memory is sleep. In fact, even the purpose of sleep is one of the great unknowns of science. We do know that sleep is far more than the absence of wakefulness. It is an active state, and while sleeping an individual passes through several stages, each mediated by various areas of the brain stem and the hypothalamus.

The brain is apparently kept in a state of wakefulness or alertness by nerve impulses sent from the pons and midbrain to the thalamus, which in turn activates the cerebral cortex. It is possible that sleep is induced when neurons in the hypothalamus release inhibitory neurotransmitters that turn off arousal systems.

Typically, a person will pass through four stages of increasingly deep sleep, interspersed with a completely different kind of sleep, called REM—rapid eye movement—for a characteristic bodily experience that occurs during this stage. Brain waves during REM sleep indicate increased brain activity, but all large muscle groups are effectively paralyzed during this stage, possibly to prevent the sleeper from getting up and moving around.

Why do we need to sleep at all? Some evidence shows that long-term memories are consolidated during sleep; sleep may also have ties to proper function of the immune system (see "Keeping Healthy," right). But there are still more questions than answers about this fundamental feature of human life.

Mental Illness

Mental illnesses such as schizophrenia and depression used to be considered diseases of the mind rather than of the brain. Today the distinction between mind and brain is blurred, as is the line between psychiatry and neurology. Most mental illnesses are now viewed as disorders of the communication systems within the brain, the neurotransmitters that speed signals from neuron to neuron. Many illnesses also have a clear genetic component. Some are linked to abnormal brain structure and physical trauma as well. And the effects of a person's environment, particularly psychological traumas and long-term stresses, are also important factors in mental illness.

Mental illness is common, affecting about 20 percent of the American population. Major depression, bipolar disorder (manic-depression), schizophrenia, and obsessive-compulsive disorder are among the most frequent causes of disability in the workforce.

KEEPING HEALTHY
SLEEP'S ROLE IN GOOD HEALTH

Scientists don't know why people sleep, but they do know that we can't do without it. Most adults need from seven and a half to eight and a half hours of sleep each night in order to stay healthy and perform their daily functions on an even keel.

Those who go without sleep for several days or more suffer a wide variety of symptoms, ranging from mild to dangerous: irritability, blurred vision, confusion, and visual or tactile hallucinations. People with shortened sleep schedules, down to three or four hours a night, often do poorly on tasks requiring concentration or memory, like taking tests.

Sleep deprivation is also associated with a drop in the number of immune cells in the body. And, of course, the simple fatigue that accumulates with extended sleep loss is a hazard for drivers, factory workers, and the like, all of whom experience a greatly increased risk for accidents if sleep-deprived.

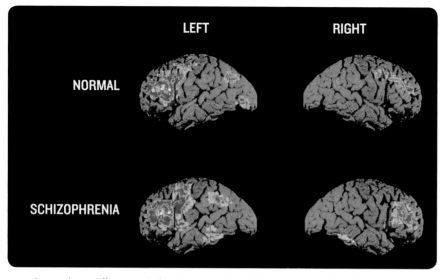

LEFT   RIGHT

NORMAL

SCHIZOPHRENIA

Scans show differences in brain areas activated in a person with schizophrenia.

Schizophrenia is a good example of the complex nature of mental illness. This common disease, which typically shows up during a person's late teens or 20s, is marked by serious disruptions in thinking, perception, and language and in a person's sense of self.

The illness is genetically linked: People with a schizophrenic parent are 12 times more likely to develop the disease, and those with a schizophrenic sibling are 9 times more likely to get it. A person with an affected identical twin has a 46 percent chance of developing the illness. The age of the father plays a part, as well: Men over 50 have three times more risk of fathering a schizophrenic child than those under 25 do. The disease has also been connected to improper brain development during fetal growth, possibly because of viral infection. The forebrain, the hippocampus, and the amygdala of schizophrenics may be smaller than average, while their ventricles—the fluid-filled cavities within the brain—may be larger. And stress is also a major player: A stressful life episode can often trigger schizophrenia.

It is now becoming clear that psychological stress has a powerful effect on the body's hormones, which in turn can reroute the brain's wiring in destructive ways and lead to mental illness. Any external stimulus that threatens the status quo—even just the anticipation of such a threat—can initiate the ancient fight-or-flight reactions within the body.

Under stress, the nervous system signals the endocrine system to release hormones such as epinephrine and cortisol. In the short run, stress hormones pump up the heart, energize the muscles, raise blood pressure, and boost the immune system—all good things if you are facing a charging tiger or fleeing a burning building.

In the long run, however, the hormones released by prolonged stress are damaging,

DID YOU KNOW?

AUTISM

Autism is a developmental disorder—more accurately, a wide spectrum of disorders—marked by difficulty with language, poor communication skills, impaired social relationships, and, sometimes, repetitive body movements or obsessions. Autism occurs more commonly in boys. Its causes are unknown and are the object of intense research. Recently, research has linked it to abnormally rapid growth of the brain during an infant's first year.

leading to hypertension, hypoglycemia (a form of diabetes), and immune system damage. Stress appears to do harm to the hippocampus, a critical site in the brain for learning and memory. Researchers have determined that stress is clearly a factor in some mental illnesses. Some people with serious depression, such as those diagnosed with Cushing's syndrome, for instance, secrete excess amounts of the stress hormone cortisol.

Recognizing that most mental illnesses involve an interplay of internal and external factors, doctors typically treat the illnesses with both medication and psychotherapy or behavioral therapy. Most psychiatric medicines attempt to correct possible imbalances in the brain's neurotransmitters. For instance, many antidepressants block the reabsorption of serotonin and norepinephrine in the brain, allowing those neurotransmitters to remain in the synaptic space longer. Many antipsychotic drugs are dopamine blockers, since schizophrenia seems to involve an imbalance of the neurotransmitters dopamine, glutamate, and GABA, an amino acid.

Neurotransmitters play a key role in drug addiction, which is considered a brain disorder, though not a mental illness. Drugs such as cocaine, alcohol, and nicotine exert their influences by activating the pleasure circuits in the brain and decreasing feelings of stress or pain. Nicotine, for instance, causes a rush of epinephrine and releases dopamine in the brain, increasing pleasure. Alcohol affects the neurotransmitter GABA and has the effect of calming anxiety. Cocaine also increases dopamine in the brain's reward system, relieving pain and causing euphoria. So powerful is this reward system that drug users who become addicted will lose control of much of their lives in order to experience it again and again. Researchers are continually challenged to find medications that can pass the blood-brain barrier to lessen the effects of withdrawal from harmful drugs or, even better, block their effects completely.

Researchers are exploring different approaches to mental illness, including light therapy to relieve depression.

THE SPINAL CORD

Conveying the Message

The spinal cord is the nervous system's central highway connecting the brain to the rest of the body. Information from the body zooms up the cord to the brain, while commands from the brain flash down the cord to muscles and organs. In some cases—such as in the quick reactions of reflexes—signals travel to and from the cord without passing through the brain at all.

THE SPINAL CORD

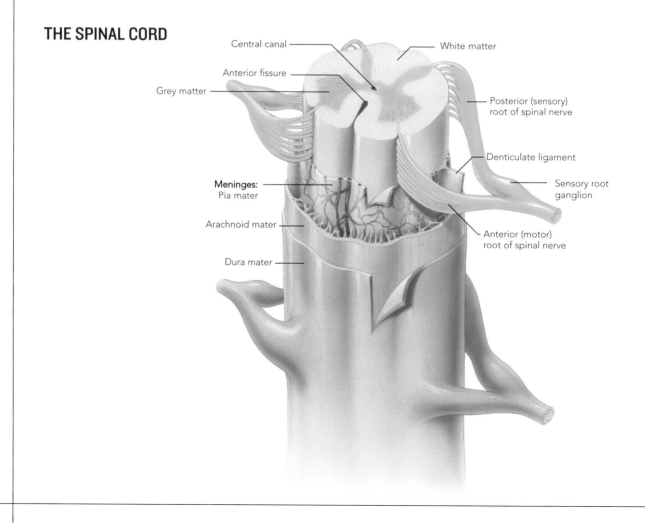

Central canal

White matter

Anterior fissure

Grey matter

Posterior (sensory) root of spinal nerve

Denticulate ligament

Meninges:
Pia mater

Sensory root ganglion

Arachnoid mater

Anterior (motor) root of spinal nerve

Dura mater

The spinal cord and brain together form one continuous entity, the central nervous system. The cord itself extends down from the brain stem and is about 16 to 18 inches long in the average adult. It is held within and protected by the vertebrae, the bones of the spine; the cord runs through the hollow center, called the vertebral foramen, within each stacked vertebra. It is not as long as the spine itself but stops at the first lumbar vertebra, just below the lowest ribs.

The spinal cord is protected by the same three membranes, or meninges, as the brain: the tough, outer dura mater, the weblike arachnoid mater, and the delicate inner pia mater. As in the brain, cerebrospinal fluid fills the space between the arachnoid mater and the pia mater. Between the dura mater and the vertebrae is a padding of fat and connective tissue that gives the cord further cushioning and protection.

The cord, shiny and white, is about three-quarters of an inch thick, but it swells out a bit in two regions. The cervical enlargement, at about shoulder level, is where nerves travel to and from the arms; the lumbar enlargement, in the lower back, is where nerves extend to the legs. In all, 31 pairs of spinal nerves extend from the spinal cord into the rest of the body, each pair emerging from the space between two vertebrae. The lowest pair is part of a fine, hairlike bundle of nerves called the cauda equina ("horse's tail") that drops down from the base of the spinal cord itself.

All of these nerves, part of the peripheral nervous system, connect the brain and spinal cord to glands, muscles, and sensory receptors, some reaching these areas directly but most connecting via other nerves and nerve networks.

The nerve pairs are organized into five groups and named according to the location and level of the vertebral column from which they emerge. Eight pairs of cervical nerves (C1 to C8) extend from the spinal cord in the neck and

BREAKTHROUGH POLIO VACCINES

Poliomyelitis is an infectious disease of the nervous system spread by viruses. Its most notorious outbreak in modern times, the epidemic of the 1940s and 1950s, was brought to an abrupt end by the introduction of two vaccines: the inactivated, or killed, vaccine developed by Dr. Jonas Salk in 1953, and the live, oral vaccine, developed by Dr. Albert Sabin in the early 1960s. The polio virus enters through the mouth and spreads through the bloodstream, causing flulike symptoms such as fever, muscle pain, and nausea. In its most serious form, it invades the cells of the spinal cord and brain stem, causing paralysis in the limbs, throat, or chest. The disease most often infects children. In the 1940s, the National Foundation for Infantile Paralysis sponsored the efforts of Dr. Jonas Salk. Working at the University of Pittsburgh, Salk grew samples of all three virus forms of polio and created a vaccine. By 1961, new cases had dropped by 96 percent. Shortly afterward, Albert Sabin introduced an oral vaccine, right, using weakened live viruses, the

type used today. By the 1990s the disease had been eliminated in the Western Hemisphere, although it still remains in Africa and South Asia, where vaccination efforts are still under way today.

From the central nervous system—the brain and spinal cord—peripheral nerves branch out through the body.

upper back: Among other things, they connect to the nerves of the skin of the ear, neck, chest, and shoulder and to the muscles of the neck, diaphragm, shoulders, and arms. Below them on the cord are 12 pairs of thoracic nerves (T1 to T12), which supply the skin of the arms, chest, and abdomen, and the muscles of the ribs, abdomen, and back. Lumbar nerves come in five pairs (L1 to L5) that connect with the nerves of the legs and groin. Below those are five pairs of sacral nerves (S1 to S5) and a single pair of coccygeal nerves (Co1), which mainly serve the buttocks and legs.

Each spinal and cranial nerve can be mapped to the part of the body it supplies, including an area of skin. These nervous system regions are called dermatomes. Doctors can use this knowledge to test for nerve injury; if sensation is lost in a particular area of skin, they can identify the nerve or nerves that may be damaged or severed. Anesthetists use the dermatome map to block pain in particular nerves.

Spinal Nerves

Each spinal nerve connects to the spinal cord via a two-pronged root, one strand of which consists of sensory axons and the other of motor axons. The cord itself, like the brain, has both gray matter (nerve cell bodies) and white matter (nerve fibers).

Gray matter takes in and integrates information flowing up and down the cord. In cross section, it forms an H or butterfly shape within the white matter. The upright bars of the H (or wings of the butterfly) are called horns. Those that extend toward the front of the cord contain neurons that control movement of the skeletal muscles. Those that reach toward the back of the cord contain groups of sensory nerves. In some segments of the cord, an additional pair of horns holds the autonomic motor neurons that control glands, cardiac muscle, and smooth muscle.

The white matter of the spinal cord is arranged into three columns. Each one consists of bundles, or tracts, of axons—in some cases, over a million of them—with similar origins and destinations. Ascending tracts carry information from sensory receptors up the cord to the brain, while descending tracts convey commands from the brain down motor neurons to muscles and glands.

The names of ascending tracts begin with the prefix *spino-* and end with the region of the brain to which they are connected. Descending tracts begin with the brain region and usually end with the suffix *-spinal*. The anterior spinothalamic tract, for instance, is an ascending tract that carries impulses up the spine to the thalamus. The lateral corticospinal tract brings nerve impulses from the cerebral cortex down the spine.

Different kinds of sensations are transported via different tracts. One set, for instance, carries pain, warmth, itching, tickling, and deep pressure, among other things. Another pair of ascending tracts conveys the sensations of light pressure, vibration, discriminative touch (the awareness of the body part being touched), two-point discrimination (the ability to identify two separate but close points being touched), and proprioception (the awareness of the position of muscles and joints).

DID YOU KNOW?

SPINAL TAP

A spinal tap, or lumbar puncture, is a method of removing cerebrospinal fluid for testing; it is typically used to diagnose nervous system disorders such as meningitis or multiple sclerosis. It's known as a lumbar puncture because a hollow needle is inserted between the lower lumbar vertebrae, below the end of the spinal cord, to extract the fluid.

The brain integrates this sensory information and may, in response, send commands to the muscles and glands down the descending tracts. Some descending tracts carry impulses that control voluntary movements of skeletal muscles—reaching for an apple, for instance—while others control automatic movements—maintaining posture, for instance.

THE NERVES

Communicating Connections

There are millions of nerves that connect the central nervous system—the brain and spinal cord—to the rest of the body. Together, they are known as the peripheral nervous system. Broken down into sensory and motor divisions, the peripheral nervous system registers sensations, carries them to the brain, and then takes commands from the brain back to the rest of the body.

Sensation is not the same thing as perception. Sensation is the body's conscious or unconscious awareness of a stimulus, whereas perception is the conscious awareness and interpretation of the stimulus. The brain can receive sensory information about internal body temperature, for instance, without those sensations becoming perceptions.

The sensory division of the peripheral nervous system includes both general and special senses. General senses pull in sensations from the internal organs (visceral senses) and the skin and muscles (somatic senses). Somatic senses include proprioception—position of body parts—pain, temperature, and tactile sensations such as touch, pressure, and vibration. Special senses encompass smell, taste, hearing, vision, and balance.

Sensory receptors come in three basic forms. They can be free nerve endings, simply exposed nerve fibers that register a stimulus. These are found everywhere in the body, but particularly in the skin and connective tissues, and they sense pain, heat and cold, tickles, itching, and some touch. Encapsulated nerve endings, wrapped in a connective tissue capsule, respond to pressure, discriminative touch, and stretching of muscles. More complex are the separate sensory cells found in the eyes or ears. The highest concentrations of sensory receptors are found in the lips, fingertips, and tip of the tongue.

Sensory receptors can also be identified by the kind of sensation in which they specialize. Many respond to tactile sensations

DID YOU KNOW?

HICCUPS

Hiccups are rarely thought of as a neurological disorder, but they are in fact triggered by an irritation of the phrenic nerve. This nerve, originating high in the spinal cord, controls the motion of the diaphragm. It can send the diaphragm into hiccuping spasms when disturbed. Contrary to popular opinion, standing on your head will not cure hiccups: If this relaxed your diaphragm, you wouldn't be able to breathe in at all.

After a few minutes in a cold swimming pool, the body adapts to the temperature by slowing down messages to the brain.

only—touch, pressure, vibration, itching, tickling. Some of these receptors are free nerve endings wrapped around hair follicles, sensing the hairs' movements; others detect slow or rapid vibrations. Sensations of heat and cold are felt by two different kinds of free nerve endings. Free nerve endings known as nociceptors register pain. Found in every tissue except the brain, nociceptors respond to intense heat and cold, unusual stretching, and chemicals released by injury or irritation. Proprioceptors, little-noticed receptors in muscles, tendons, and the inner ear, tell the position of all the parts of the body. When you close your eyes, stretch out your arm, and touch your nose, your are using your proprioceptive senses.

In response to a constant stimulus, many kinds of receptors will adapt by slowing down the impulses they shoot off until the sensation falls below the threshold of conscious perception. An example: After a few minutes in a cold swimming pool, you no longer notice the temperature. Receptors for touch, smell, and pressure adapt quickly, while other sensations fade slowly, if at all, in our perceptions.

Sensations are not perceived consciously until they reach the brain. Impulses from sensory receptors speed along spinal nerves to the spine or along cranial nerves to the brain stem and are passed on from there first to the thalamus and then to the cerebral cortex of the brain. As they enter the brain, the nerves crisscross, those from the left side of the body entering the right side of the cerebrum and vice versa. An area of the parietal lobes known as the primary somatosensory area takes in the sensory information.

Neurologists have mapped the body onto the surface of the brain in this area, finding that parts of the body with many sensory receptors have, not surprisingly, larger areas of the brain devoted to them. Therefore, a substantial portion of the somatosensory area is devoted to fingers, lips, and tongue; the trunk of the body receives relatively little attention. This brain map is not organized like the body, however: Sensations for the thumb register next to those for the eyes, for instance; the genitals, next to the toes. A leading theory about the phantom pain of amputated limbs says that neurons from adjacent areas may be stimulating the part of the cortex devoted to the absent limb—a touch on the face, for instance, might partly transmit to the nearby area for the hand.

Voluntary Moves

Just in front of the primary somatosensory area of the brain is the primary motor area, where most impulses for voluntary movement begin. Control of complex movements, suppression of unwanted movements, maintenance of muscle tone, and many other subtle balancing acts of the body are coordinated among a variety of brain structures, including the cerebellum, the basal ganglia (nodes of gray matter within each cerebral hemisphere), the thalamus, and motor centers in the brain stem. Commands for movement speed from these areas via the brain stem and spine. Signals travel through cranial nerves (to the face and head) and spinal nerves (to the rest of the body) to reach the muscles. At the end of the journey, the terminals of the motor nerves form junctions with muscle cells, stimulating them to contract.

Although the brain is the executive for most kinds of movements, in some instances sensory impulses and motor commands are integrated in the gray matter of the spinal cord instead. This is the case with spinal reflexes: rapid, involuntary responses to stimuli. Reflexes can be inborn—snatching your hand away from a flame before you're

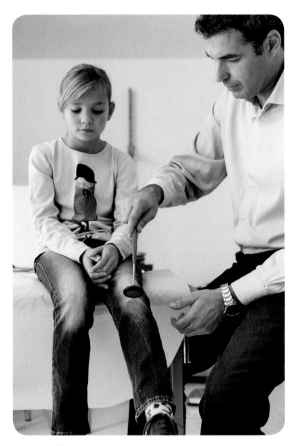

A reflex test shows how quickly a signal travels to the spinal cord.

consciously aware of the heat—or learned—stomping on the brake pedal of your car. Pavlov's dogs, conditioned to salivate at the sound of a bell, are a famous example of a learned reflex.

The input and output pathway for a reflex is known as a reflex arc, and it typically passes through three sets of neurons. A sensation ("Heat!") first speeds along sensory neurons to the spinal cord or brain stem. There it swings through an interneuron, which sends a signal to one or more motor neurons. A command for movement ("Pull back!") then zips back to the relevant area (in this case, the muscles of the arm).

Reflexes can prompt muscles to stretch, to relax, to flex and withdraw (in response to pain, for example), or, in a more complex transaction, to withdraw while balancing on the other side (as in pulling up your foot from a sharp object while balancing on the other). The classic example is the stretch reflex called the patellar reflex, familiar to anyone who has had a physical exam. When a doctor taps on a patient's patellar tendon under the kneecap while the patient's leg is hanging bent over the edge of a table, the sensation triggers a signal to the spinal cord. Motor commands shooting back from the spinal cord command a muscle spindle in the quadriceps (upper thigh) muscle to contract, extending the leg. At the same time, other signals from the spinal cord tell hamstring muscles to relax, so they don't inhibit the contraction. Tests like this one give doctors a quick indication of the health of the spinal cord and nervous system.

By skipping the trip to the brain, reflexes allow the body to react almost instantly in an emergency. The brain is usually involved at the same time, however, registering the sensation of pain, for instance, or modifying the movement according to the surroundings.

Autonomic System

In contrast to the somatic nervous system, which involves conscious perceptions and voluntary movements, the autonomic nervous system controls the unnoticed, involuntary workings of the body's interior. Most of the body's internal housekeeping tasks—heart rate, digestion, sweating, opening or constricting blood vessels, dilation or contraction of the pupils of the eye—belong to the autonomic system. More than any other system, it is responsible for maintaining homeostasis: a constant, stable environment within the body.

The input part of this system, composed of its sensory neurons, is similar to that of the somatic system, except for the deeper location of its receptors. These neurons, found within the blood vessels and organs, monitor and send information to the brain about internal conditions such as blood pressure or oxygen levels. Almost all of the sensations reported by these neurons fall under the radar of conscious perception, although certain kinds of pain will break through into awareness.

It is in its output, its motor activities, that the autonomic system is really distinctive. Commanding smooth muscle, cardiac muscle, and some glands, autonomic motor neurons are

DID YOU KNOW?

BLINK

The average person blinks every two to ten seconds. Regular blinking, which keeps the eyes protected and irrigated, is controlled by a blinking center in the caudate nucleus, deep within the brain.

Blinking also occurs as a reflex when the cornea is touched: The impulse flashes along a cranial nerve to the medulla and back. It is the last reflex to disappear under anesthesia.

typically grouped into sympathetic and parasympathetic divisions. Generally speaking, the sympathetic division stimulates and activates the organs in response to stress, while the parasympathetic division inhibits activity and conserves resources at times of rest. Unlike the somatic nerves, which run more or less directly from the spinal column to the muscles, autonomic nerves pass through a junction first: a chain of ganglia. These ganglia, collections of nerve cell bodies, then pass signals on through a second set of neurons to an organ or to several organs, if necessary. Most organs receive signals from both the sympathetic and parasympathetic divisions.

The parasympathetic division of the nervous system is active when the body is at rest, helping it to conserve energy and rebuild resources. It gives priority to digestive activities, decreases heart rate, and constricts blood vessels. It also promotes sexual response, which is restricted during times of stress. Nerves that serve the parasympathetic system emerge either from the brain stem or from the bottom of the spinal cord and pass through ganglia that are found near or within their target organs.

The sympathetic division is more extensive and complex. It is the activating system that responds to stress and readies the body for exercise; it governs the fight-or-flight responses. Among other things, sympathetic nerves cause the adrenal medullae to secrete epinephrine and norepinephrine; they dilate the airways and blood vessels, increase heart rate and the force of heart contractions, promote sweating, dilate pupils, and decrease the motion of the stomach and intestines. Nerves for the sympathetic division extend from the middle of the spinal cord and pass through one of two groups of ganglia. One group, the sympathetic trunk ganglia, runs in two chains down each side of the spine; nerves extending from this group reach out primarily to organs above the diaphragm. The other group of ganglia, the prevertebral ganglia, run in front of the spine and close to the abdominal arteries. Nerves from these ganglia primarily serve organs below the diaphragm.

The sympathetic division of the nervous system is responsible for activating the fight-or-flight response.

SYMPATHETIC VS. PARASYMPATHETIC

AFFECTED ORGAN	SYMPATHETIC RESPONSE	PARASYMPATHETIC RESPONSE
Heart	Increased heart rate; stronger atrial and ventricular contractions	Decreased heart rate; decreased atrial contractions
Adrenal glands	Secretion of epinephrine and norepinephrine	No effect
Liver	Glycogen converted into glucose; decreased bile secretions	Glycogen created; increased bile secretion
Iris of eye	Pupil dilation	Pupil constriction
Lungs	Airway dilation	Airway constriction
Stomach and intestines	Decreased motion; contraction of sphincters	Increased motion; relaxation of sphincters
Pancreas	Inhibited secretion of digestive enzymes and insulin	Secretion of digestive enzymes and insulin

Physical activity—even the anticipation of physical activity—will trigger the sympathetic system, but so will a variety of emotions, such as fear, embarrassment, or anger. The hypothalamus, found in the center of the brain, above the brain stem, controls most autonomic functions. It communicates through nerve axons that run from the structure down through the brain stem and spinal cord. The hypothalamus takes in sensory input about the state of internal organs, smell, taste, and blood chemistry.

The hypothalamus is greatly influenced by the nearby amygdala, a little almond-shaped structure within the limbic system that processes emotions such as fear. For instance, a person being approached by a snarling dog will first take in sensory information, the sight and sound of the dog, to be processed in the cerebral cortex. Next the cortex relays those memories and information about the snarling

dog to the amygdala, which triggers the emotion of fear and in turn signals the hypothalamus to begin the array of physical phenomena involved in a fight-or-flight response.

Some physiologists identify a third kind of autonomic system: the enteric nervous system. This recognizes the fact that the digestive system is served by an unusually dense, complex network of nerves—as many neurons, in fact, as serve the spinal cord. These nerves control motion and blood flow within the intestines. Although they are influenced by the sympathetic and parasympathetic systems, these nerve networks can also act independently, registering sensations, integrating them, and acting on the intestines without passing through the central nervous system. This so-called brain in the gut is probably a reflection of the importance of the digestive system in human evolution.

THE SENSES

Bringing Information to the Body

Most people, asked to name the senses, will list five: touch, vision, hearing, smell, and taste. But physiologists group them differently. The general senses include the tactile sensations of touch, pressure, pain, and vibration, picked up by sensory receptors in the skin, muscles, and internal organs. The special senses—vision, smell, taste, hearing, and balance—are associated with special receptor cells in the head.

Sight

Of all the special senses, vision is probably the most important. The eyes contain about 70 percent of the body's sensory receptors. Vision is accomplished through a coordination of the delicate structures of the eye and a substantial part of the brain: Nearly half of the cerebral cortex is given over to visual processing.

The adult eye is a sphere about one inch in diameter, with about one-sixth exposed. The rest is sheltered within the eye socket, cushioned by pads of fat and tugged this way and that by six sets of eye muscles. The outermost layer of the eye consists of the sclera, the white of the eye, and, front and center, the transparent area of the cornea, the window that allows light into the eye's interior. The cornea has pain receptors but no blood vessels, which would block the light. Its lack of blood vessels makes it uniquely suitable for transplants, because it does not trigger an immune reaction.

Behind the cornea is a clear fluid, the aqueous humor, which overlays the iris, the colored central part of the eye, and its round opening, the pupil, which dilates and contracts to let light enter. The transparent lens behind the pupil is flexible, changing shape to focus light on the sensory cells in retina.

The interior of the eyeball behind the lens is filled with a clear, jellylike fluid, the vitreous humor. A dark membrane, the choroid, lines most of the eyeball. Its blood vessels feed the

Rods (red) and cones (pink) line the retina
in this colorized scanning electron micrograph.

RIGHT EYE
(Horizontal Section)

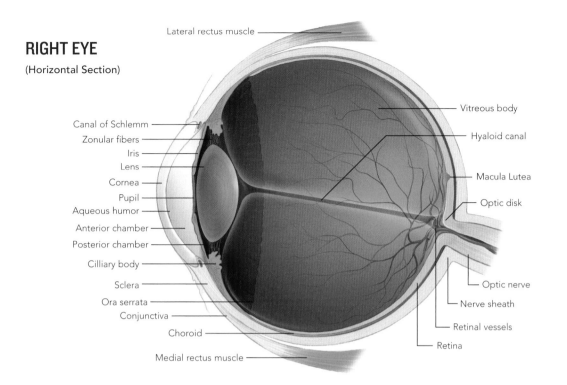

Lateral rectus muscle

Canal of Schlemm
Zonular fibers
Iris
Lens
Cornea
Pupil
Aqueous humor
Anterior chamber
Posterior chamber
Cilliary body
Sclera
Ora serrata
Conjunctiva
Choroid

Medial rectus muscle

Vitreous body
Hyaloid canal
Macula Lutea
Optic disk
Optic nerve
Nerve sheath
Retinal vessels
Retina

organ, while its dark, pigmented cells cut down on reflection within the eye.

Over the choroid is the thin, two-layered retina. The outer layer of the retina, like the choroid, prevents light from scattering within the eye. The inner, neural layer contains 125 million photoreceptors, the light-sensing cells. When light, focused by the lens, hits the retina, it reaches two kinds of photoreceptor cells: rods and cones. Rods, the most numerous, are highly sensitive to dim light but do not provide either color vision or sharp edges. Cones, fewer in number than rods, detect color and sharp details in brighter light.

Signals from the photoreceptors move forward from the retina to an overlying layer of ganglion cells. Axons from the ganglion cells make a sharp turn and run down the back of the eyeball to form the bundle that is the optic nerve. This nerve exits the back of the eyeball

BREAKTHROUGH LASIK SURGERY

Until recently, most near-sighted and farsighted people corrected by fitting additional lenses over their eyes in the form of either eyeglasses or contact lenses. A new development, surgery to reshape the corneas, began in the early 20th century. This procedure now is dominated by the LASIK technique. LASIK stands for Laser-Assisted In Situ Keratomileusis. During this procedure, eye surgeons first cut a flap in the cornea and then fold it back. Guided by a computer, they then use a special laser to vaporize an inner section of the cornea. The process flattens the cornea into a shape that can focus light accurately on the retina, correcting the misguided path of light rays that actually caused the faulty vision. If all goes well, the patient will be able to see clearly right away, without corrective lenses. The surgery can be used to upgrade vision to 20/15 or better.

at a spot called the optic disk. Because no photoreceptors exist in this area, it is known as the blind spot: Light focusing here cannot be seen, but the eye shifts back and forth so rapidly, we rarely notice this blank space in our vision.

Vision depends on a working connection to the brain and its vision centers. Optic nerves from each eye reach back into the brain and, at an X-shaped crossing called the optic chiasma, nerve fibers from the innermost halves of each eye cross over to the opposite side of the brain. This means that each hemisphere of the brain receives images from one half of the visual field—for instance, fibers from the left side of the left eye and (inner) left side of the right eye end up in the left hemisphere. People with widespread damage to one hemisphere—from a stroke, for example—may end up seeing only one-half of their visual field, despite having healthy eyes.

Signals moving along two optic tracts, holding the reorganized optic nerve fibers, then reach the thalamus, which in turn passes them on to the primary visual cortex in each of the cerebrum's occipital lobes. There, the visual information is sorted out into such categories as dark and bright edges, object orientation, color, and motion. For the conscious mind to perceive what the eyes have just seen, all these separate categories of perception recombine into an integrated image.

Taste & Smell

The special senses of taste and smell are close partners in the nervous system. Both are chemical senses, registering their sensations through chemoreceptors that detect chemicals dissolved in solution. Taste is heavily dependent on smell, the two senses usually combining to convey flavor, from the salty comfort of chicken soup to the interplay of ripe cherries, vanilla, and oak that might blend to characterize a fine glass of Cabernet.

The sensory receptors for taste are found in thousands of oval structures called taste buds, most of them located on the tongue, but a few also lining the soft palate, pharynx, inner surface of the cheeks, and epiglottis. Taste buds on the tongue sit on top of the little rounded bumps, called fungiform papillae, that give the tongue its rough appearance. Each bud holds sensory receptor cells for taste (called gustatory cells); out of each cell extends a long hairlike microvillus, or tiny projection, bathed in saliva. The variety of chemicals dissolved in saliva stimulate these hairs, which pass the signals down to sensory dendrites at their bases. Taste bud cells are frequently destroyed by friction or heat, to be replaced by new cells every seven to ten days.

A taste sensation is usually a combination of smell and one or more of five primary taste qualities: sweet, sour, salty, bitter, and umami —a fifth quality that conveys the savory taste of steak or tangy cheeses. Organic compounds such as sugars and alcohol trigger the sweet taste. Salty comes from metal ions. Bitter flavors come from alkaloids such as caffeine or strychnine. Sour tastes arise from acids. Umami comes from the amino acid glutamate. Contrary to popular opinion, the tongue has no special areas associated with particular tastes: All kinds of tastes can be detected all over the tongue. The threshold for activating a taste sensation varies. Bitter substances can be detected in minute quantities, while sweet and salty require higher concentrations. This

DID YOU KNOW?

TONGUE FACTS

+ The tongue of a young adult has **10,000 taste buds.**

+ Each taste bud contains about **50 sensory cells.**

+ The rear of the tongue contains a V-shaped row of extra-large papillae, or receptive cells, each one holding **100 to 300 taste buds,** above.

+ **Taste aversions are inborn;** even a newborn baby will grimace and turn away from a bitter taste.

+ Taste is **80 percent smell.**

configuration probably evolved because most poisons taste bitter or sour.

Once the sensory dendrites at the base of the gustatory cells have received a stimulus, they pass it on via the cranial nerves to the brain stem, then to the thalamus, and finally to the gustatory cortex in the parietal lobes. As the signals pass through the brain stem, they trigger digestive reflexes. Saliva increases in the mouth, and gastric juices flow into the stomach.

Taste may be limited to five primary qualities, but smell is a different matter. Researchers believe that the body can detect hundreds of primary smells, which combine to create at least 10,000 different odors recognizable to the human brain. Smell, like taste, is a chemical sense. After air enters the nose, it flows past a small yellowish patch of epithelium in the roof of the nasal cavity. About one inch square, the olfactory epithelium contains between 10 million and 100 million olfactory receptor cells. Dendrites extending from these cells end in a sprouting mass of little hairlike cilia that spread out in the layer of mucus that covers the olfactory epithelium. These cilia detect the chemicals from the inhaled air that dissolve in the mucus. The olfactory cells are among the few neurons that continue to regenerate; they turn over about every 60 days.

These receptors can be exquisitely sensitive, in some cases detecting an odor from just a few molecules in the air. The foul-smelling chemical methyl mercaptan, for instance, can be detected in minute concentrations, which is why it is added as a warning element to odorless natural gas. Smell receptors adapt quickly, however, and many strong smells sink out of awareness in a few minutes.

Having sensed a particular chemical, the olfactory cells flash signals along their axons through openings in the bone above the nasal cavity, where the axons connect with the olfactory bulbs. These are the ends of the olfactory tracts, which carry the neurons via the thalamus under the frontal lobes to the primary olfactory cortex of the brain. Some nerve fibers also travel from the thalamus to the amygdala and other parts of the limbic system, where the odor becomes associated with emotion and memory. Thus the sensations attached with certain smells often conjure up distinct memories, an evolutionary remnant from a time when odors carried important messages about danger.

Hearing

The delicate, curling structures of the ear are home to two disparate senses: hearing and equilibrium, or balance. As different as these senses are, they both rely on the same basic mechanism—liquid sloshing back and forth around receptor cells—to convey an amazing range of sensations to the brain.

The ear itself is usually divided into three parts—outer, middle, and inner—with the inner ear containing the most complex sensory apparatus. The outer ear, the visible part, is made of cartilage covered by skin. Its primary purpose is to funnel sound to the eardrum, or tympanic membrane, a thin partition of connective tissue that divides the outer and middle ears. The middle ear, or tympanic cavity, is a little air-filled cavity in the temporal bone. Within it, connecting the eardrum to the inner ear, are three tiny bones linked by miniature joints: the malleus (hammer), the incus (anvil), and the stapes (stirrup). The stapes is the body's smallest bone, measuring about 0.10 inch long.

The inner ear is far more complicated. Located securely in the temporal bone, behind the eye socket, it is also called the labyrinth because of its maze of spiraling canals. It is, in fact, in two parts: An outer bony canal forms a chamber and encloses an inner membranous labyrinth.

The inner ear's three main areas include a hollow vestibule, holding two little membranous sacs that contribute to equilibrium; three looping semicircular canals, also used for equilibrium; and the coiled, snail-like, fluid-filled cochlea, containing the organ of Corti, the primary organ of hearing. This organ is essentially a sheet of 16,000 hair cells projecting from one membrane, the basilar membrane, and touching another, the tectorial membrane. When vibrations roll through the fluid of the cochlea, the basilar membrane and its attached hair cells bounce against the tectorial membrane, which stimulates the hair cells and generates nerve impulses to the brain.

The pathway of sound waves to the inner ear works like this: First, the outer ear catches airborne sound and directs it to the eardrum. The eardrum vibrates (slowly for low-frequency sounds and rapidly for high-frequency sounds). Next, the eardrum's vibrations shake the attached malleus bone in the middle ear, which passes its vibrations through to the stapes bone. The stapes vibrates against the membrane of the inner ear, which in turn sends pressure waves through the fluid of the cochlea. Then the fluid's pressure waves vibrate the basilar membrane and hair cells, which bend against the upper tectorial membrane and generate nerve impulses.

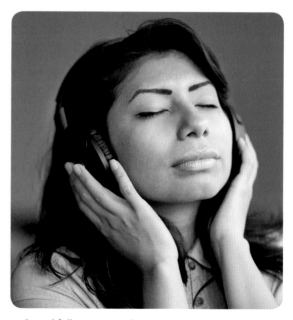

Sound follows a complicated pathway to the brain.

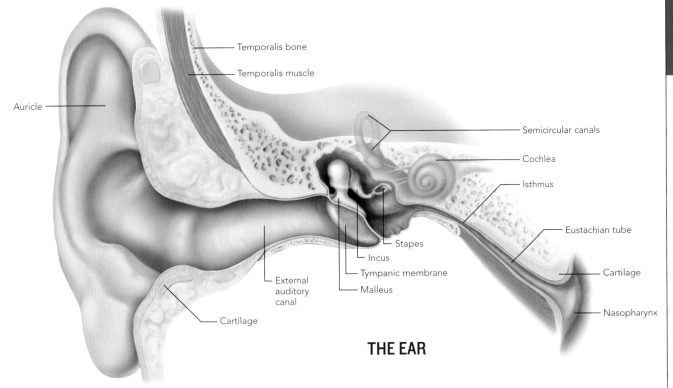

Temporalis bone
Temporalis muscle
Auricle
Semicircular canals
Cochlea
Isthmus
Eustachian tube
Stapes
Incus
Tympanic membrane
Malleus
External auditory canal
Cartilage
Cartilage
Nasopharynx

THE EAR

Nerve impulses from the cochlea travel via the cochlear nerve to the cochlear nuclei in the medulla oblongata, then onward to the thalamus, and finally to the primary auditory area of the cerebral cortex. Some of the nerve fibers for hearing cross over to the opposite brain hemisphere, while others do not, so that each hemisphere receives signals from both ears. Brain stem nuclei help to pinpoint a sound's location by comparing the subtle differences in the timing of impulses from the right and left ears.

Equilibrium is the other sense governed within the ears, also through the movement of liquid that passes over hair cells, sending signals to the brain. Two kinds of equilibrium keep our bodies balanced. Static equilibrium monitors posture, the position of the body (particularly the head) relative to gravity, as well as straight-line acceleration, such as the sudden forward movement of a car. Two little fluid-filled sacs in the inner ear, the utricle and saccule, contain hair cells that detect this kind of motion.

Dynamic equilibrium is the sense of rotational movement—spinning about, as in the sensation of an amusement park ride. Hair cells within the semicircular canals pick up this whirling motion. Arranged at right angles to one another, the semicircular canals work together to determine just where the head is moving through three-dimensional space.

Equilibrium is not just a function of the inner ear. It involves other senses as well. Vision, sense receptors in the muscles, and sensory nerves in the skin play a part in maintaining balance. Information about equilibrium needs to be processed quickly, so it travels primarily to the brain stem and cerebellum to produce quick reflex responses in the muscles and eyes. Occasionally the body needs a moment to adjust to equilibrium information, which is why you may stagger and find your eyes jerking back and forth after you have been spinning about.

Chapter 10

MESSENGERS

THE SYSTEM

The Endocrine Network

I f you wanted to set up a communications network, you could do so in more than one way. You could lay down fiber-optic cables and send light signals from one location to another. Or you could devise a system of cell phones and towers that transmit electromagnetic microwaves into the air to be picked up by a base station and relayed to another mobile device.

The human body, too, has multiple ways to communicate, body system to body system. The nervous system works more like the land-based telephone network, with neurons conducting electrical impulses through fibers, from body to brain and back. Alternatively, the endocrine system acts more like a wireless network. Its signals are hormones, chemical in nature. Unlike the brain, the organs and tissues that produce these hormones do not directly connect to their targets.

Instead, specialized endocrine cells release hormones into blood, just as mobile phones transmit microwaves into the air. Many different hormones are circulating through the bloodstream at one time, all searching for specific target cells with special receptors, designed to pick up only specific signals. Only cells that are genetically programmed for a hormone can respond to that hormone.

In general, the everyday effects of the endocrine system may go unnoticed by the casual observer. The processes and functions regulated by hormones appear seamless and

DID YOU KNOW?

PINT-SIZE PARTS

+ The organs of the endocrine system are so small that all the tissue from eight to nine adults would amount to only 2 pounds (about 1 kg) of hormone-secreting tissue.

+ The pituitary gland, which secretes nine hormones, is about the size of a pea.

+ The thyroid, above, is the largest pure endocrine gland in the body.

THE ENDOCRINE SYSTEM

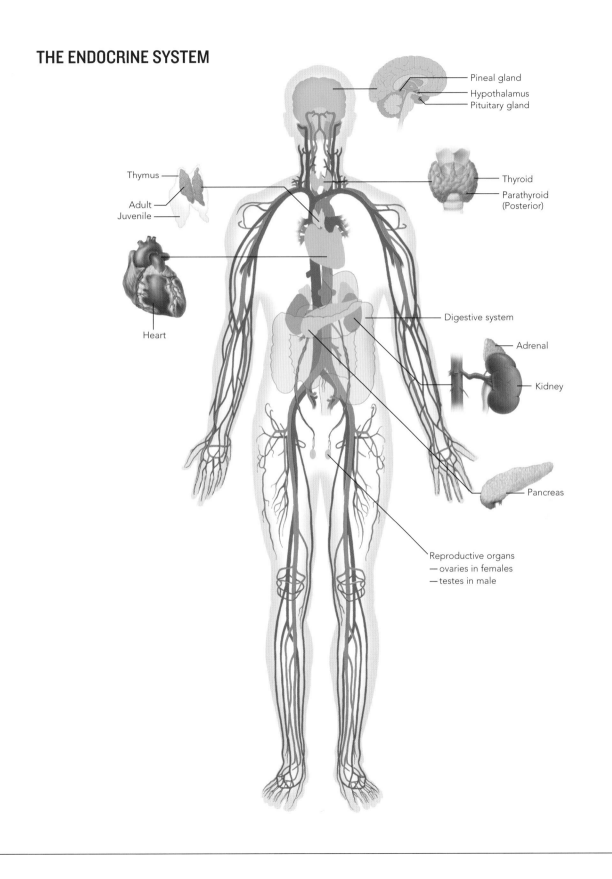

Pineal gland
Hypothalamus
Pituitary gland

Thymus

Adult
Juvenile

Thyroid
Parathyroid
(Posterior)

Heart

Digestive system

Adrenal

Kidney

Pancreas

Reproductive organs
— ovaries in females
— testes in male

smooth, making it difficult to appreciate the power of the system when everything is running smoothly. But the endocrine system's impact becomes dramatically apparent when the body enters adolescence and begins to go through puberty. Hormones (estrogen in women, testosterone in men) begin to alter the body's appearance, turning girls into women and boys into men. Puberty doesn't occur overnight; it is a series of changes and events that occurs over longer stretches of time, like many processes regulated by the endocrine system.

Timing

Because of their structural differences, the nervous and endocrine systems also differ in the amount of time it takes for their messages to travel and be received. The time between neuronal impulse transmission and reception is so short, it is often measured in milliseconds. The time between endocrine transmission and reception almost always takes longer. Some reactions, like the fight-or-flight response, can take seconds whereas others, like sleep regulation controlled by the pineal gland, can last for days or even months.

While responses in the endocrine system—ovulation, for instance—can seem relatively

The adrenal gland is made up of two separate glands.

sluggish in comparison, it means that their effects last a long time. Hence the body uses the endocrine system to regulate prolonged processes such as reproduction; maintaining the balance in electrolyte, nutritional, and water concentrations; and fine-tuning cellular metabolism and the flow of energy—all processes that require a steady state of regular rhythms. These body rhythms, guided by waves of chemical messages, represent the major forces at work in the body's endocrine system.

BREAKTHROUGH THE FIRST HORMONE

In the early 20th century, scientists believed that nervous stimulation controlled digestion. London physiologists Earnest Starling and William Bayliss, right, observed that the pancreas secreted juices in response to food entering the small intestine, but they suspected another player, too. They cut all the nerves to the rats' pancreases and watched with amazement as digestion continued. A chemical secreted from the intestinal lining, not a nerve fiber, caused food breakdown and processing. In 1905, Starling and Bayliss named the substance secretin, the first hormone to be discovered.

Organs & Glands

The organs of the endocrine system, like cell phones that do not need a physical base, are small and scattered throughout the body. Hormones in the gastrointestinal system are not even associated with specific glands, nor are endocrine cells in organs such as the lungs and the kidneys. Because their goal is mainly to secrete hormones, the equivalent of organs within the endocrine system are mainly glands. A gland may consist of a single cell or a group of cells that secrete substances into ducts, onto a surface, or directly into the blood.

There are two kinds of glands in the body: exocrine and endocrine. Exocrine glands have ducts, and they produce non-hormonal secretions—like sweat and saliva—that are released in specific places—like the skin or inside the mouth, respectively. Endocrine glands specialize in producing hormones, the messengers that perform the essential functions of this system.

This cell from the pituitary gland, an organ located at the base of the brain, produces growth hormone.

On the other hand, endocrine glands lack ducts; their secretions are released into the interstitial fluid and then diffuse into the capillaries and into the bloodstream.

Located throughout the body, the endocrine glands—namely, the pituitary, thyroid, parathyroid, adrenal, and pineal glands—have a webbed, or cordlike, appearance. This anatomical structure allows these organs maximum contact with capillaries, allowing the glands to suffuse their potent chemicals into surrounding tissue fluid. In most cases, high concentrations of hormones are not necessary to normal function; circulating hormone levels in the blood are generally low.

Other players in the body's endocrine system—such as the pancreas, the hypothalamus, and the gonads (the ovaries in women and the testes in men)—are indeed organs that contain tissue capable of secreting hormones. In the liver, cells make hormones while performing other functions as well.

Meanwhile, in various places throughout the body, isolated clumps of fat cells release a hormone called leptin, which serves the function of regulating fat storage and appetite. Patches of hormone-producing cells line other major organs like the stomach, intestines, kidneys, and heart.

Development

Because the endocrine glands are found all over the body, they do not develop in a localized fashion when we are in the womb. Instead they develop from widely separated parts of the embryo. The pituitary gland emerges from the ectoderm, about three weeks after fertilization. The thyroid gland develops in the fourth week from the ectoderm. The adrenal glands arise from the mesoderm.

HORMONES
Chemical Communicators in the Body

Anyone who has had the experience of premenstrual syndrome (PMS) or sexual desire can recognize the effects of hormones on the body. The word *hormone* comes from the Greek word meaning "to excite"—and that is exactly what hormones do in the body. They are chemicals excreted at key locations that travel through the bloodstream and stimulate certain cells to cause, control, and monitor the necessary functions that keep us alive, well, and growing.

Hormones trigger many different events in the body. They initiate and control the important processes of growth and life, including reproduction, development, defenses against disease, metabolic balance, and energy regulation. When a cell spews out a hormone, it does so in order to regulate some aspect of body metabolism. The hormone's effect often happens significantly later and in a place far from the body location where it originated.

While some hormones can appear similar in structure, such as the two sex hormones, estrogen and testosterone, hormones are distinctive from each other. They span a broad range of molecular sizes and complexities. Each hormone is as unique as the physiological message it carries.

Amino Acids
Scientists group hormones into two basic categories, one proteinlike and the other fat-based. Proteinlike hormones arise from the building blocks of proteins, called amino acids, hence their name: amino acid hormones. They tend to be the smallest ones, although they do range widely in size. For example, thyroid hormone, which regulates metabolism, is built from parts of two molecules joined together.

Strings of several amino acids join to form hormones called peptides. Longer chains, such as the hormone insulin, are called proteins. All amino acid hormones—amines, peptides, and proteins—are able to dissolve in water. They are flexible molecules: Some can bend, turn, and twist into a variety of unique shapes. Their specialized structures allows them to perform their very specific functions in target cells.

Unlike the amino acid–based hormones, steroids are fatlike. Examples include the adrenocortical and the gonadal hormones, such as cortisol, estrogen, and testosterone. These steroid hormones are synthesized from a fatty substance called cholesterol.

They are fat-soluble, but they cannot dissolve in water or in the plasma of the bloodstream. These substances are carried through

The color-enhanced crystalline structure of a hormone connected to body growth

the bloodstream on transport proteins and can slide in and out of oily cellular membranes more easily than some of their water-soluble amino acid counterparts.

Communication

Hormones deliver chemical messages to cells. The cell, in turn, has to be able to receive the message. When a cell answers the signal, its activity is often altered as well. For example, during stress, the hormone epinephrine, produced by the adrenal glands, induces a powerful fight-or-flight response, stimulating the heart and the body's metabolic system. Bio-chemically, the interaction between a hormone and its receptor means that the hormone binds to smooth muscle cells in blood vessels and unleashes a programmed cascade of chemical reactions that cause the blood vessels of the heart and muscles to open, increasing blood flow to the tissues needed for the action of fighting or taking flight.

But that same hormone, when binding to smooth muscle cells in the stomach, causes their interwoven blood vessels to contract. No one wants blood flowing to the gut when facing a life-or-death situation. Best that it should stay where it's needed, in the muscles, when it's time for a fight. So the question is, how does the same hormone perform two opposing tasks in two different places in the body?

Targeting

The answer is to be found in the receptors, the receivers that bind to the original hormone signal. Each receptor is linked up to its own biochemical network. That network of mole-cules, like relay runners, hands off the message, molecule to molecule, and eventually shuttles it to the cell's nucleus, which receives a chemical command to act.

In general, the binding of a hormone to a receptor works like a key in a lock. The shape of both molecules is critical. A cell with the wrong-shaped receptor simply won't bind to an active hormone. That cell won't react, while one with the right receptor will.

Most peptide and protein hormones cannot directly penetrate a cell. They are water-soluble, while cell membranes are fatty. Therefore these hormones enlist the help of "second messengers," molecules embedded in the target cell.

Generally, a peptide or protein hormone binds to its receptor partner on the cell mem-brane. In essence, that binding squeezes the receptor and changes its shape. This shape change activates a signal transducer, often called a "G protein." Activation is akin to turn-ing on a molecular switch. A G protein switch

HORMONE TARGETING

Hormone receptor

Hormones

Blood vessel

Hormone

Target cell

target cells. Activated hormone and receptor cells form a large complex that ambles to the nucleus. There, the complex binds to other proteins, strung along a cell's DNA.

Since DNA is the component of genes, such binding action can turn on a different kind of switch: It can turn on a gene. It will make a copy of its genetic instructions and send that copy back out to the cytoplasm. At that, the cell will make a protein, based on the genetic specifications it has received. That protein, often an enzyme, causes a specific cell action, such as the stress response of cells in reaction to the steroid hormone cortisol.

initiates the activity of a second messenger, called cyclic AMP (cAMP). This potent molecule, tiny and fast, can move almost anywhere in the cell and trigger cascades of reactions in a cornucopia of body processes.

By using the same messenger, cAMP, to broker so many diverse reactions, the endocrine system is resourceful. It even commissions second messengers to relay to third messengers, in fact. For example, oxytocin, the orgasm-generating hormone, binds to its receptor and activates a second messenger, which then releases calcium as a third messenger. As in the case of orgasm, second and third messengers amplify some reactions. Second messengers also take an ongoing, but aggressive initial reaction (say, the fight-or-flight response) and later, when the stress is gone, dim the reaction so that the body can maintain homeostasis.

Amino acid hormones are not the only players. Adding to the repertoire, steroid and thyroid hormones can diffuse directly into target cells. Testosterone, progesterone, and estrogen all bind to receptors located within the

Smart Receptors

The endocrine system controls myriad processes by doling out receptors strategically and placing them in cells where they are needed the most. Most cells of the body bear receptors for thyroid hormones, for example, which stimulate cellular metabolism. All cells need energy to work, and consequently nearly all body cells have thyroid hormone receptors.

Another tactic for control is to raise or lower the levels of a hormone circulating in the blood. Concentration can be a trigger to action. Still another move is to change the number of receptors on a target cell or the strength of hormonal triggers. Depending on circumstances, cells can actually make more receptors in a process called up-regulation. During development, for example, an embryo's primitive heart cells might up-regulate receptors that control new blood vessel growth.

At the same time, the endocrine system can also downregulate receptors. When people become obese, it has been observed that their brain cells will often downregulate the appetite-dampening hormone called leptin. When food is plenty, and fat stores increase, leptin levels in the bloodstream rise, shutting down appetite. But after too much time at such a high level, the cells of the body tend to become desensitized, and they stop overreacting to a prolonged response. Hence people who are obese may have high leptin levels in their blood, but their bodies have lost the ability to respond appropriately to the appetite-decreasing hormone.

All these control mechanisms, from hormone type to second messenger system to receptor dynamics, compose the diverse showcase of the endocrine system. Control means balance, and that is the job of hormonal flow: maintaining a continual, rhythmic balance through the body's everyday functions.

Transport

Hormones are messengers. They all must travel to reach their targets and convey their messages. But given their widely different functions, they all move at different rates and in reaction to different conditions. The full system of hormonal deployment and transport is elaborate and versatile.

For proteins and other water-soluble hormones, transport is straightforward. These hormones can dissolve in the blood plasma and therefore can float free, unattached to any other molecules. Lipid-soluble hormones pose a different challenge, however, for they do not dissolve in blood plasma. The body has devised a cadre of specific transport proteins, each designed to transport its own steroid or fatty cargo.

Transport proteins also slow the passage of small hormones through the kidney filter, which allows the tiny molecules to stay in the body longer, acting for longer periods of time in the

KEEPING HEALTHY
QUICK TREATMENT FOR ANAPHYLAXIS

Anaphylaxis can be a death sentence. By definition, it is an allergic response that hits multiple systems of the body when it comes into contact with an allergen (like bee venom, certain foods, or medications). Although normally short-term, severe anaphylaxis can become anaphylactic shock, which can lead to respiratory failure, cardiovascular collapse, and, if left untreated, organ failure. Therefore, quick treatment is potentially life-saving.

Out of this emergency, doctors learned that epinephrine, a hormone that naturally occurs as part of the body's stress response, can bring about temporary relief of anaphylaxic symptoms. Meridian Medical Technologies packaged the hormone into an ingenious device called an EpiPen, an epinephrine autoinjector, right. Other pharmaceutical companies have followed suit with their own versions (for example, Twinject, manufactured by Versus Pharmaceuticals). People with severe allergies can carry the pen anywhere to deliver a quick shot.

body's various processes. Transport proteins, in essence, sequester and store the hormones, keeping them in ready reserve, available for quick action.

When hormonal action must end, the body has a way to eliminate outdated hormones. The endocrine system has an arsenal of enzymes that chew up unnecessary hormones. The kidneys and liver step in as well, filtering out aging hormones and their breakdown products.

Cooperation

Hormones are by no means solitary players. Often they work in pairs or relay teams to encourage, build upon, or oppose each other's actions. For example, reproductive hormones spur the development of gonads and other sex-related structures, but reproductive hormones cannot do their job without the action of thyroid hormones. Thyroid hormones allow the reproductive hormones to act, but at the right time, at the advent of puberty.

In another scenario, glucagon, a hormone from the pancreas, acts with epinephrine to prod the liver to release glucose, a sugar, into the bloodstream. The effect of both, acting together, is 150 percent more potent than either hormone acting alone. This effect, a good example of the whole being greater than the sum of its parts, is called synergism. It works to create a big impact in a short time as is needed when the body requires fuel, say, during exercise.

On the other hand, insulin acts to oppose glucagon's effect of raising the body's blood sugar, the measure of glucose in the blood. Called antagonism, this opposition provides the endocrine system with its signature checks in the check-and-balance system.

Hormones are not the only substances that act as checks to the system. Ions, or nutrients, can

MAJOR HORMONES

HORMONE	ORGAN	TARGET	EFFECT
Growth hormone	Pituitary gland	Liver, Muscle, Bone, Cartilage	Stimulates growth
Thyroid hormone	Thyroid gland	Body cells	Maintains metabolism; regulates blood pressure; promotes normal development of the nervous system, muscles, and skeleton; regulates hydration of skin
Cortisol	Adrenal gland	Body cells	Assists body in resisting stress; mobilizes fats for energy; depresses inflammatory response
Insulin	Pancreas	Body cells	Lowers blood sugar levels
Gastrin	Stomach	Stomach	Secreted in response to food; stimulates production of hydrochloric acid to break down food
Erythropoietin	Kidney	Bone marrow	Stimulates production of red blood cells
Leptin	Adipose tissue	Brain	Suppresses appetite

Hormones play a role in many aspects of human physiology and behavior, even romantic relationships.

also rein in hormonal action. For example, calcium in the bloodstream bathes cells of the parathyroid gland. When calcium levels plummet, the parathyroid secretes parathyroid hormone, which works to reverse a low-calcium state in the body.

Control

Hormones can also inhibit their own release in a process known as "negative feedback regulation." In essence, the end product bathes the gland that is producing it, which causes that gland to shut down and halt any further secretion—a feedback process that can happen directly or indirectly.

The endocrine system is stocked full of layers. Glands secrete hormones that target other glands, which in turn produce hormones or other substances to perform a body process. One example of this is the three-layered system involving the hypothalamus, the pituitary gland, and the gonads (the ovaries in women and testes in men). First, the hypothalamus produces a hormone called gonadotropin-releasing hormone. It triggers the pituitary gland to release two other hormones, FSH (follicle-stimulating hormone) and LH (luteinizing hormone). Next, these two chemicals flow into the bloodstream, down toward the gonads. There, in response, the ovaries begin to produce estrogen, and the testes start to produce testosterone. Not only do the sex steroids affect tissues directly; the sex hormone level in the bloodstream also rises, and the hormone flows back to the pituitary, where it halts the production of FSH and LH.

DID YOU KNOW?

HORMONE HISTORY

The word *hormone* derives from the Greek word hormon, "to excite" or "to urge on." Hippocrates used the word to mean "a vital principle." In the early 20th century, Sir Earnest Starling, discoverer of the first hormone, established the modern medical use of the word. As the story goes, he dined with a poet the night before giving a historic speech to the Royal Academy of Physicians on body chemistry. Their dinner conversation resulted in his creation of the new term.

PRODUCERS

Organs of the Endocrine System

E very system in the body has its movers and shakers. At the top of the chain in the endocrine system sits the hypothalamus, a region of the brain that controls the main glands of the endocrine system: the pituitary, thyroid, parathyroid, adrenal, and thymus. Together these glands secret the hormones that regulate functions such as growth, reproduction, and cellular metabolism.

Hypothalamus

The hypothalamus is a region of the brain and not an organ on its own. As part of a brain's tissue, it is wired with neurons and has neural functions. But, at the same time, the hypothalamus produces and releases hormones. Therefore it is considered a neuroendocrine organ. The hypothalamus takes its name from its position in the brain, below the portion of the brain called the thalamus.

Deep within, the hypothalamus is embedded at the base of the brain stem, ready to perform its role in the body. As leader and guide, the hypothalamus produces a wide variety of hormones. Some stimulate action—such as growth—while others inhibit the action of these same targets.

Why the opposing action? The two hormones illustrate the yin-yang genius of the endocrine system. Once a body response gets going—the growth of a child's bones, for example—there will come a time when that action has to stop— once the child's body reaches adulthood. The hypothalamus is a master of balance, keeping the body in tune and on time. And, as a bonus,

the hypothalamus can work in two ways: first, by stimulating target cells directly, as in the case of oxytocin, a hormone that causes the contractions of uterus in childbirth, and second, by activating other endocrine glands, which in turn excrete more hormones that stream toward target cells.

Pituitary

One of those secondary glands sits right next to the hypothalamus, joined by a stalk of nerves and blood vessels. The pea-shaped pituitary gland, right, is a master in its own right. It is double-lobed, no accident of evolution. The lobe in the back, the posterior pituitary, contains neural tissue. This portion of the gland is therefore dubbed the neurohypophysis. It was born out of brain tissue, which explains its neural nature. This lobe connects directly to the hypothalamus via nerve appendages called axons.

Xi Shun, one of the world's tallest men, is almost seven feet tall. The interplay of glands and hormones determines height.

The front-facing lobe has a more glandular role. Called the adenohypophysis, this lobe secretes six potent hormones, each with a widely different function in the body. The adenohypophysis emerges from embryonic oral tissue.

The adenohypophysis over time adheres to the neurohypophysis, separating from its embryonic oral contact. The nervous part of the gland receives chemicals from the hypothalamus, the hormones streaming down the axons for eventual storage in the pituitary gland. The glandular part, while not connected by nerves to the hypothalamus, gets signals from hypothalamic hormones that travel through the blood vessels that bathe the cells of the lobe.

Both the hypothalamus and the pituitary gland rely on an elaborate system of checks and balances. The hypothalamus begins the job by producing stimulatory hormones such as corticotropin-releasing hormone (CRH), which acts on the pituitary gland to produce and release adrenocorticotropic hormone (ACTH, also called corticotropin).

The ACTH travels to the adrenal glands, which respond by producing cortisol, among other hormones, which acts in inflammatory and stress responses. Circulating cortisol flows back to the hypothalamus and, at the right concentration, shuts down further production of CRH and ACTH.

This loop is called "feedback inhibition." It can affect not only initial hormone production but also the signaling of neurotransmitters in the brain, as well as the manufacture of potent immune-boosting chemicals called cytokines elsewhere in the body.

Growth Hormone

Discussed and debated in the news of late is growth hormone, a molecule that stimulates body cells to grow and reproduce. Bones and muscles reap the most benefit. Cells with receptors specific for the hormone respond by enlarging and proliferating, especially during the period of childhood. Growth hormone is an anabolic or tissue-building hormone.

Besides fashioning bones, this hormone also mobilizes fats from stores in the body and breaks them down into fatty acids for fuel. The action is meant to conserve glucose as an energy source. And, in fact, growth hormone affects gluconeogenesis, the formation of glucose from amino acids, and it can have the effect of adding more sugar to the blood. The effects of too much growth hormone can mimic the symptoms of diabetes mellitus, a condition in which blood sugar levels rise and remain high.

The system needs to be in balance. Too much growth hormone causes gigantism, where a person continues to grow, sometimes up to and beyond eight feet in height. If growth plates of bones are still open and susceptible to growth hormone (GH) action, body proportions remain normal. But if a child has grown and his or her growth plates closed, the result is a condition known as acromegaly, in which a person has enlarged and thickened bones of the hands, feet, and face, especially the jaw.

Too little growth hormone causes no such extreme problems in adults, but it can be associated with loss of muscle and bone mass and unusual accumulation of fat. In children, the lack causes long bones to grow too slowly, resulting in pituitary dwarfism.

Problems with growth hormone deficiencies extend beyond lack of the hormone itself. Growth hormone often exerts its effects by a proxy: insulin-like growth factor 1 (IGF-1). IGF-1 is a protein hormone much like insulin, stimulated by growth hormone. It reaches nearly every cell in the body, and especially cells of muscle, cartilage, bone, liver, kidney, nerves, skin, and lungs.

Approximately 98 percent of IGF-1 is always bound to one of six transport proteins. If any one of these is lacking, though, IGF-1 becomes impotent, resulting in growth failure and certain kinds of dwarfism.

Gonadotropins

When it comes to reproduction, the hypothalamus and pituitary work as a winning relay team. The first player is gonadotropin-releasing hormone, produced by the hypothalamus. The hormone nudges the pituitary to secrete two so-called gonadotropins: follicle-stimulating hormone (FSH) and luteinizing hormone (LH). These, in turn, regulate the function of the ovaries and testes. Depending on the body's gender, FSH works to stir either egg or sperm production. The hormone also coaxes the maturation of eggs and follicles, the cells surrounding a developing egg, in the ovaries of females.

LH works on the ovaries or the testes and prods them to produce their own hormones, estrogen or testosterone. LH action is special because the pituitary sends it out in pulses. Although no one knows exactly why, researchers have discovered that the staccato rhythm of LH production is crucial to proper ovulation and the synthesis of ovarian hormones.

The hypothalamus and pituitary gland (both shown in green) sit squarely in the middle of this scan of the human brain.

DID YOU KNOW?

HORMONE FACTS

+ Growth hormone promotes growth of muscle and bone. It also controls protein, fat, and carbohydrate metabolism.

+ Thyroid-stimulating hormone affects metabolism rates.

+ Adrenocorticotropic hormone influences the stress response.

+ Prolactin causes the mammary glands to produce milk.

+ Antidiuretic hormone controls conservation of body water.

+ Oxytocin stimulates uterine contractions and orgasm.

Prolactin stimulates the production of milk after a woman gives birth.

Of note, prepubertal boys and girls have virtually no gonadotropins in their bloodstreams. Only when puberty dawns does the pituitary animate and begin the reproductive relay. Later, hormones such as testosterone and estrogen feed back to the hypothalamus to suppress FSH and LH release. In this way reproduction, the most crucial system to the survival of the species, remains in balance.

Prolactin

This protein hormone is known as the trigger for milk production in the breasts after birth. Despite its brief cameo role in any single woman's lifetime, it is crucial to survival of the species as a whole.

The hypothalamus makes a releasing factor to stimulate prolactin production via the pituitary. Oddly, scientists still do not know this releasing factor's true identity. They suspect that the neurotransmitter serotonin causes prolactin synthesis and release. On the other hand, scientists do know the opponent of this releasing factor: dopamine.

This inhibitory hormone prevents prolactin secretion.

Male bodies produce a prolactin-inhibitory hormone in slightly larger amounts than female, although the differences diminish once a woman has had children. The level of its production rises and falls with the tide of monthly estrogen fluctuations in a woman's bloodstream. Low estrogen means high prolactin-inhibitory hormone release; high estrogen means high prolactin-releasing hormone levels. Women experience the ebb and flow of estrogen in their bodies as breast tenderness and abdominal swelling before their menstrual periods. Prolactin levels are surging at this time, but only briefly.

With pregnancy, however, everything changes. Prolactin levels rise dramatically at the end of pregnancy, as a woman's body begins to produce milk to feed the coming newborn. Suckling triggers prolactin release further, along with encouraging ongoing milk production.

Pituitary gland tumors can bring on results that include the secretion of too much prolactin. Other symptoms include inappropriate lactation, lack of menses, infertility in females, and erectile dysfunction in males.

Oxytocin

Of all the hormones in the body, none more than oxytocin can be dubbed the "love and cuddle" hormone. It is a substance associated with explosive spasming. During childbirth, oxytocin mediates the contractions of the uterus that push the baby out of the mother's body. During breastfeeding, oxytocin targets special cells surrounding the mammary glands that squeeze out milk, commonly termed the "letdown reflex"

experienced by nursing mothers. For both men and women, oxytocin stimulates sexual arousal and orgasm.

The hormone's production begins with neural triggering of the hypothalamus. That connection in the brain also explains the link of each of these processes—childbirth, nursing, and sexual climax—to intense emotions. The hypothalamus is intricately linked to the limbic system of the brain, the site of pleasure, pain, and rage. Thus, oxytocin is a wonderful kind of Cupid inside the body, stirring feelings of satisfaction, of nurturing, and of affection as these body processes unfold.

Antidiuretics

Diuresis means urine production, hence the antidiuretic hormone (ADH) stops the process of adding water to urine. Produced directly by the hypothalamus, this hormone tempers the water balance in the body, preventing wide

BREAKTHROUGH GOITER

A disease known by abnormal swelling in the neck, called goiter, right, drew attention during World War I when soldiers drafted from the Great Lakes region started showing up for active duty with its symptoms. Doctors knew that goiter arose from enlarged thyroid glands, but they did not understand why a cluster of cases would come from one geographic region. Eventually investigators linked hypothyroidism to a lack of iodine in the diet. The iodine content of Midwestern soil was low. Saltwater fish, another natural source of iodine,

was not in ready supply. So these men lacked the iodine needed to keep their thyroids functioning properly. That breakthrough in understanding led to a new treatment for the prevention of goiter: iodide supplements. Only about 150 mg of iodide keeps the thyroid functioning properly, but people worldwide often lack even that little. More than 8 million people worldwide suffer from complications arising from iodide deficiency. In regions of South America, Southeast and Southern Asia, and Africa, where malnutrition is prevalent, the condition is

seen quite often. In the United States, goiter is generally avoided because iodide is customarily added to commercial table salt.

Electronic scanning created this 3-D depiction highlighting the hypothalamus,
pituitary gland, thyroid gland, and parathyroid glands.

swings like dehydration to bloating. A high level of ADH causes blood vessels to contract—a characteristic that gives the hormone its other name: vasopressin.

The keys to this system are the osmoreceptors in the hypothalamus: structures that can sense the level of dilution or concentration in a body fluid. The receptors' job is to monitor the level of electrolytes (solutes) and water in the blood. If solutes, such as salt, are at levels that are too high—after excessive sweating, for example—the osmoreceptors send off impulses to the hypothalamus to trigger the synthesis and release of ADH.

The pituitary stores and releases the antidiuretic hormone into the bloodstream, sending it toward its target, the kidney tubules. They remove water from the urine as it is forming and return it to the blood. On the other hand, if the solute concentration drops, the osmoreceptors stop firing, effectively ending ADH release.

Many people have experienced ADH action firsthand after a night of drinking. Alcohol inhibits ADH secretion, generating lots of urine output. The next day, the body experiences dehydration and the dry mouth of a hangover.

Drugs called diuretics actually encourage this same process of ADH inhibition. They cause water to flush from the body. This can be a necessary step in the treatment of hypertension and congestive heart failure, conditions in which excess water accumulates in the tissue and must be eliminated.

Individuals with eating disorders often take diuretics in a misguided attempt at weight loss. The practice is dangerous because it often ends in dehydration, electrolyte imbalance, and sometimes even death.

Adrenal Hormones

The adrenocorticotropic hormone (ACTH) is a polypeptide hormone that targets the adrenal gland to boost the synthesis of corticosteroids (glucocorticoids and sex steroids). One glucocorticosteroid called cortisol has been fashioned into a drug, cortisone, used in the treatment of rheumatoid arthritis because of its anti-inflammatory effects.

The entire process begins in the hypothalamus with the production of corticotropin-releasing hormone (CRH). The pituitary also secretes a related molecule called pro-opiomelanocortin (POMC), a prohormone, in that it is a precursor cut up by enzymes in the body into several other hormones. Among them are two opiates, enkephalin and beta endorphin, known for their feel-good effects. Both target the μ-opioid receptor in the brain, the same receptor that responds to chemicals derived from opium, such as morphine and codeine. All these substances provide relief from pain; the drugs, though, bring the risk of addiction along with pain relief.

One other progeny of POMC is melanocyte-stimulating hormone (MSH), which stimulates special cells in the skin and hair to produce and release the pigment melanin. Coincidentally,

MSH also acts as a neurotransmitter in the brain with profound effects on appetite and sexual arousal. That dual role helped in POMC's discovery, since mice defective in it grew orange hair and obese bodies.

Thyroid & Parathyroid

At the front of the throat, underneath the voice box, sits one of the largest and most important glands in the endocrine system: the thyroid gland. It controls the essential process of metabolism, by which the body converts food into energy. Therefore this 10- to 20-gram, butterfly-shaped gland affects every cell and most systems of the body, essential in processes ranging from the development of the fetal and neonatal brain to the regulation of many functions in the adult body: temperature, protein synthesis, growth, heart rate, fertility, and central nervous system matters.

Smaller glands known as parathyroid glands attached to the thyroid lobes help to control blood calcium and phosphate levels. Follicles within the thyroid gland encircle a protein-rich material called colloid, which absorbs and stores iodine ions from the body.

Iodine is critical to the process of synthesizing the hormones thyroxine (T4) and triiodothyronine (T3). In fact, it is so important to proper thyroid functioning, the body stores 25 percent of its iodine supplies in the follicles of the thyroid gland (see "Breakthrough," page 280).

Thyroid hormones rely on an intricate system of control that adjusts itself continually to maintain balance. Deficient production of thyroid hormones can lead to infertility, chronic fatigue, and, in severe cases, mental retardation in infancy or coma in adulthood. Other symptoms of thyroid deficiency include an intolerance for cold, muscle pain and weakness, dry skin,

DID YOU KNOW?

DISCOVERY

In 1880, Swedish medical student Ivar Sandstrom discovered the parathyroid gland—but to little fanfare. He displayed data from extensive dissections in several species and reported the gland's anatomical position, but no one believed that a mere student could find a heretofore unknown organ. Sandstrom even predicted that the gland had an endocrine function, but only later did the discovery gain acceptance.

a slow heart rate, and excessive sleepiness. Overproduction of these hormones can lead to nervousness and anxiety, insomnia, and increased heart rate.

Thyroid hormones begin their journey at their site of synthesis, the epithelial cells that line the outer edge of the follicles. Thyroid hormones then enter the bloodstream, searching for their target cells. Carrier proteins ferry the potent hormones through the body via a complex vein and artery system to target cells within the body and brain, where carriers release their hormonal cargo and target cells take it in.

Once inside a cell, these hormones bind to intercellular receptors, proteins that act as signaling towers that relay messages to the cell's DNA. These messages trigger or inhibit the expression of particular genes.

While this system works well for delivering hormones to most cells of the body, the hormones used by cells of the brain face an

KEEPING HEALTHY
THYROID LEVELS IN PREGNANCY

During the first trimester of pregnancy, a growing baby depends entirely on its mother's thyroid hormones for healthy brain development. If the woman's thyroid level is low, the fetus might not get enough hormones to grow normally. Studies have even linked deficiencies in thyroid hormones to an increased risk of autism and preterm births.

An estimated 2 to 5 percent of all pregnant women have hypothyroidism, a condition in which the thyroid does not secrete enough hormones. The condition often causes symptoms such as fatigue, brittle nails, and joint or muscle pains. But it's not always easy to detect. Not everyone displays symptoms, and sometimes the symptoms can be blamed on other conditions—including pregnancy itself.

Women with hypothyroidism, a family history of the disease, or symptoms of hypothyroidism should talk to their doctors before getting pregnant to make sure that their hormone levels are high enough to support a healthy pregnancy. Fortunately, hypothyroidism can be treated with synthetic thyroid hormone pills, which are safe to take during pregnancy.

The parathyroid glands are embedded in the thyroid.

additional challenge. They must cross the formidable blood-brain barrier, a wall of cells that acts as gatekeeper and restricts the passage of unknown or harmful substances into the brain.

One other important thyroid regulator is the hormone calcitonin, which stimulates the movement of calcium from the blood to the bones and helps prevent osteoporosis. Calcitonin also inhibits the activity of osteoclasts, cells that otherwise would deprive the bone of calcium and transfer it back into the blood. Calcitonin prevents the absorption of calcium by the tubules of the kidney as well, which causes more calcium to be excreted in the urine.

The parathyroid glands secrete an opposing hormone, parathyroid hormone (PTH), when blood calcium levels fall too low. PTH stimulates the removal of calcium from the bone and raises the blood calcium levels. It also increases the activity of osteoclasts and triggers absorption of calcium by the kidney tubules, which suck the calcium back into the blood. PTH also indirectly affects calcium absorption in the small intestine, introducing more calcium into the blood. The system works in yin-yang balance to maintain daily homeostasis.

Adrenal Glands

Resting atop the kidneys, two wedge-shaped glands, the adrenals, control the stress response, one of the body's life-or-death reactions. When faced with danger, hunger, or any other threat to survival, the adrenal glands respond, sending the body into its mode of fight-or-flight response.

Anatomically, the adrenal system seems small, almost inconsequential. The adrenal glands are dwarfed by the kidney, yet they interact with it in one of the most beautiful orchestrations of two body systems. The adrenals also coordinate

with the nervous system, which allows quick reactions to mental and physical stress.

The adrenal glands contain two parts, the cortex and the medulla. The cortex, composed of three distinct layers of cells arranged in concentric layers, produces a critical stress hormone called cortisol. The inner medulla, or the bull's eye, synthesizes and secretes epinephrine and norepinephrine, more commonly known as adrenaline and noradrenaline. These hormones work together and engender a repertoire of responses that keep the body in balance.

Cholesterol, the foundation for several adrenal hormones, grouped as "steroid hormones," is critical to adrenal functioning. It can be absorbed from the diet or synthesized within the body from precursor metabolites, small molecules produced during the metabolism of food. Once formed, cholesterol begins an odyssey toward its eventual targets: the middle and inner rings of cortex cells.

As the journey begins, cholesterol travels with transport proteins that shuttle it across the cell membrane. Because cholesterol cannot cross from blood to cell interior on its

The cells of the adrenal gland

The "adrenaline rush" from dangerous activities is actually produced by several different hormones.

own, a single defect in the transport protein can block or entirely shut down the function of the adrenal cortex.

Inside the cell, cholesterol must cross through another barrier: the membrane of the mitochondria, tiny organelles that function as power plants and provide the energy needed for all cell processes. In the mitochondria, various enzymes convert cholesterol into a variety of essential hormones, including gluco-corticoids. Of these steroid hormones, cortisol is the most renowned.

Cortisol clearly provides a demonstration of how the body can closely integrate the nervous and the endocrine systems. When either mental or physical stress occurs, the body alerts the hypothalamus, located in the brain. In response, the hypothalamus secretes a corticotropin-releasing hormone, which in turn stimulates the release of the adrenocorticotropic hormone (ACTH) from the anterior pituitary gland. ACTH

DID YOU KNOW?

ADRENALINE

A person who seeks dangerous thrills earns the title "adrenaline junkie," but the adrenaline part of that colloquialism is a misnomer. The rush is actually the complex interplay of several hormones. Adrenaline and noradrenaline are released from the adrenal medulla, while the adrenal cortex secretes cortisol, aldosterone, and corticosterone.

travels through the bloodstream to the adrenal cortex, where it binds to surface receptors and exerts an influence on its cells. In response, they unleash a stream of cortisol.

Every cell in the body has a glucocorticoid receptor, so every cell in the body has the capability of reacting to stress, as signaled by cortisol. Like cholesterol, cortisol needs a transport protein to taxi the steroid hormone to the receptors of its target cells. Inside those

In an islet of Langerhans, the red alpha cells (left) emit glucagon while the green, yellow, and brown beta cells (right) secrete insulin.

cells, special molecules transport the hormone into the nucleus, where it signals the expression of a wide variety of genes. These genes catalog information that, when expressed through molecules such as RNA and proteins, stimulates the breakdown of fat for energy, the conservation of glucose through a process called gluconeogenesis, and the release of protein building blocks called amino acids from muscle tissue. All of these processes coordinate to regulate the proper levels of blood sugar, or glucose, and the metabolism of carbohydrates.

In addition to carbohydrate metabolism, glucocorticoids can block inflammation or suppress immune system functioning. Researchers have exploited this property by formulating medications derived from cortisol, also known as hydrocortisone, to treat inflammatory diseases, such as arthritis, as well as autoimmune disorders, such as lupus, in which an overzealous immune system attacks the body's own tissue.

While these responses take place in the inner or middle cortex, cells in the outermost layer conduct yet another symphony of hormonal synthesis and body regulation. This layer synthesizes and secretes mineralocorticoids, hormones that regulate electrolytes—ions that carry an electrical charge and can act as conduits for molecules such as water, moving in and out of cells, thus regulating the flow of fluids in the body.

The most important mineralocorticoid is aldosterone. It travels through the bloodstream and targets the cells of the kidneys. Its action stimulates the absorption of water and sodium ions and the excretion of potassium ions.

The kidneys constantly filter the body's total blood supply. By regulating the concentration of the sodium and potassium ions, they can control blood pressure and volume. Attesting to this system's importance, a body devoid of mineralocorticoids cannot properly balance fluids in the circulation. The kidneys malfunction, the heart fails and the body eventually dies.

When blood pressure or blood volume drops, special cells in the kidney release a substance called renin into the blood. Renin interacts with plasma protein, ultimately forming angiotensin II, which stimulates the release of aldosterone and helps to raise blood pressure.

Finally, the adrenal medulla, the most interior part of the adrenal glands, synthesizes and secretes adrenaline and noradrenaline, the "fight or flight" hormones. Unlike the steroids, such as cortisol or aldosterone, adrenaline and noradrenaline are protein-based hormones. The body constructs them both from the amino acid tyrosine.

Cells of the medulla store the two hormones, ready and waiting for the time when the brain signals that it recognizes danger. Within milliseconds, brain cells release the neurotransmitter acetylcholine. In response, the medulla deploys a flood of adrenaline and, to a lesser extent, noradrenaline, which travel to receptors on the surface of their target cells, including those of the heart, lungs, and intestines.

As with other hormonal relay systems, triggered receptors pass their hormonal message to the DNA in the nucleus. Genes turn on and alter their expression patterns. The body experiences this change in a number of ways. The heartbeat speeds up; the bronchial tubes dilate, or enlarge, allowing more oxygen intake; the pupils dilate, absorbing more light; and fat and glycogen begin to break down, speeding metabolism and supplying a rush of energy.

The adrenaline cascade also coordinates an overall give-and-take in the body, giving to organs in need while taking from others deemed temporarily inconsequential. For example, during a rush of adrenaline, the blood vessels in the leg muscles expand while secretion of gastrointestinal juices in the digestive system halts. Adrenaline is often used to treat anaphylaxis, a severe allergic reaction. Because of its bronchodilator effects, the same hormone is used to treat asthma.

The Pancreas

Some would think that the stomach is the organ of hunger. But actually the pancreas, tucked underneath the stomach, wins the title of hunger regulator, managing its important feat with the help of blood sugar monitoring. The organ itself resembles a tapered cylinder, divided into three parts: a head, a body, and a tail. This anatomy, however, has nothing to do with distinctions in function.

Instead, function arises from two types of tissue: endocrine tissue, which secretes hormones, and exocrine tissue, which secretes digestive juices. These two types of tissue work next to and with each other. The

DID YOU KNOW?

PANCREAS FACTS

+ A small, flat organ, the pancreas is about 5–6 inches long.

+ The pancreas contains both exocrine and endocrine cells. Called acini, the exocrine cells make up 99 percent of the organ.

+ Groups of endocrine cells are clustered together in islets of Langerhans, named for the German physician who first described them in 1868.

LIVING WITH DIABETES

Diabetes, also called diabetes mellitus, occurs when the body experiences a chronic deficiency in either the secretion or the action of insulin, a hormone made by the pancreas. Those who suffer from this condition must manage chronic problems with blood sugar regulation that grow more severe over time.

TYPES 1 & 2

There are two major types of diabetes. Type 1 diabetes shows up primarily in children and young adults. Often genetic factors play a role in an individual's predisposition for this type of diabetes. Type 1 diabetes often involves damage to cells in the pancreas, almost always the result of an autoimmune attack, that is, a situation in which the body's own cells militate against its proper functioning.

The discovery of insulin changed the prognosis for people with diabetes.

When the pancreas fails to make insulin, the body experiences abnormally high glucose levels. With Type 2 diabetes, enough insulin is initially made, but cells in the body do not respond appropriately, either due to a lack of insulin receptors or to an improper functioning of the receptors. This condition is known as insulin resistance.

Doctors used to think of Type 2 diabetes as an adult disease. People do tend to develop the condition predominantly in midlife or later. With the recent upsurge in childhood obesity—cases of it have doubled in the last 30 years—younger and younger patients are now being diagnosed with Type 2 diabetes. Today, nearly 21 million Americans live with diabetes, while 41 million more are prediabetic, meaning their blood sugar levels are significantly above normal levels, putting them at risk.

SYMPTOMS

Both types of diabetes share basic features. Facing a lack of insulin, the body strives to keep blood sugar levels high all the time. Despite the effort, the sugars—essential fuel for the body—cannot get into cells without the hormone insulin, so cells in a diabetic body start to starve for glucose. In response, the body reacts by making still more sugar, breaking down glycogen and using amino acids from protein to make glucose in a process called gluconeogenesis. The blood sugar level soars even higher, spilling over into a person's urine. This is why diabetes can often be detected in the results of a urine test.

As the problem worsens, sugar cannot be used as cellular fuel. In some cases, especially among Type 1 diabetics, the body breaks down more fat.

Ketones, the product made by the body during the breakdown of fat, flood the bloodstream. When they accumulate in the blood, the blood's pH drops, acid levels rise, and the body falls into a dire state called ketoacidosis because the blood has become far too acidic. Symptoms including nausea, vomiting, and rapid breathing can ensue, leading potentially to coma or even death.

Diabetes may show itself in symptoms that include excessive urine output, thirst, and hunger. As glucose spills into urine, it forces water to follow. This process often causes such extremes of urination that dehydration can occur. A person who has no insulin function, in essence, has a body that believes it has not eaten. This person is starving in the land of plenty. Before the identification of insulin in the early 20th century, undetected diabetics lost weight precipitously. Today, physicians are sensitive to the signs, and in the developed countries, at least, diabetes rarely goes undiagnosed.

Obstetricians test pregnant women routinely for gestational diabetes, which arises for the first time in a woman during pregnancy. While this condition usually abates after giving birth, it leaves the mother more likely to develop diabetes later in life. It is also unhealthy for the developing fetus, equally sensitive to high glucose levels in the blood.

In times past, women who already suffered from diabetes were discouraged from getting pregnant. Today, with available blood sugar monitoring and management techniques and a fuller understanding of diabetes in general, diabetic mothers are simply advised to remain vigilant in watching blood sugar levels during pregnancy.

TREATMENTS

Before the discovery of insulin in 1921, doctors could do little for diabetes patients but manage their symptoms. But in the 21st century, those suffering from the disease are often able to live normal, active lives. Type 1 diabetes, caused by a lack of insulin in the body, can be treated with insulin supplements, often given as injections up to four times daily.

Even so, some people with the disease develop long-term circulatory and nervous system problems. These can be minimized by constant monitoring and management of an individual's blood sugar level. Eating right can also reduce complications. A diet high in fat and high blood cholesterol levels can cause further problems, including heart disease, stroke, kidney shutdown, and blindness.

Type 2 diabetes, in which there is insulin resistance and therefore too much insulin in the body, requires a more complicated treatment. Doctors prescribe a combination of many different categories of antidiabetic drugs. Some drugs have been designed to reduce the diabetic body's resistance to insulin; others have been engineered to increase the sensitivity of hormone receptors, so that the body's natural insulin is more likely to be put to use.

At the same time, physicians encourage a program of weight loss, exercise, and healthy diet in addition to a drug regimen in the treatment of Type 2 diabetes. Studies show that Type 2 diabetes can be significantly delayed, or even prevented, through the loss of 5 to 10 percent of body weight in persons at high risk. Overall, physicians recommend regular blood sugar screening to assess whether a person is at risk for Type 2 diabetes. Medical researchers are hoping to find methods of measuring blood sugar that are easier and do not require drawing blood.

endocrine tissue grows in small circular clusters, known as the islets of Langerhans; the exocrine tissue surrounds those clusters. Within the islets, three types of cell secrete essential hormones.

Alpha cells secrete the protein hormone glucagon. Low blood sugar, also known as hypoglycemia, triggers the production of this hormone. When blood sugar levels plummet—many hours after a meal or after exercise, for example—alpha cells go to work. They excrete glucagon, which travels to special receptors on liver cells and spurs the process of glycolysis, breaking down the molecule glycogen into glucose. Newly made glucose flows from liver cells into the bloodstream. The body now has a fuel source to continue functioning. If the sugar produced from glycolysis fails to meet body needs, however, glucagon has another trick up its sleeve. The hormone also stimulates amino acids to convert to glucose, a process called gluconeogenesis, which involves the production of a steady stream of glucose.

Meanwhile, back in the islets of the pancreas, beta cells, bathed in sugar-laden blood plasma, monitor glucose levels. When too high, the cells go to work, secreting insulin, which reverses the effects of glucagon; insulin lowers high blood sugar by binding to receptors on target cells and allowing glucose to enter cells.

Again, the liver is the workhorse. It converts glucose from the blood into glycogen, a strategy to store glucose for future needs. Insulin is such a versatile hormone that it also prevents the breakdown of fat, promotes the synthesis of fat, and inhibits the degradation of protein during its conversion into energy. The combination is breathtaking. Blood sugar drops, and the system regains its balance.

Insulin is such an important molecule, all mammals produce it. The insulin molecule is very similar, no matter whether it is found in voles or dogs or humans. Scientists have exploited this phenomenon, extracting insulin from a large mammal such as a cow and using the bovine insulin thus derived to treat insulin deficiencies in people, including those suffering from diabetes.

While the body is experiencing the balancing act between insulin and glucagon, yet a third type of pancreatic cell, the delta cell, is synthesizing and secreting a molecule called somatostatin. Somatostatin inhibits the secretion of insulin and glucagon as well as other hormones. It is the body's attempt to impose yet another layer of control on a system that extracts, stores, and uses fuel critical for life.

The pancreas shows up in pink on this abdominal scan.

The gonads create reproductive cells and are different in men and women.
Women's gonads are ovaries, left, and men's are the testes, right.

The Gonads

The endocrine system underlies the very per-
petuation of the species. As the body makes
eggs and sperm cells, and the two unite to form
a new being, the endocrine system is there,
guiding and shifting components within the
complex reproductive systems.

The main endocrine organs involved in
reproduction are the gonads: testes in males
and ovaries in females. These organs create the gametes, or reproductive cells (egg
cells in women, sperm cells in men). The
body is ingenious in how it uses the same
hormonal players to control development
in both sexes. The key is adding variations,
depending on sex.

In both males and females, the whole process
begins with the hypothalamus, which pulses
out gonadotropin-releasing hormone (GnRH).
That jars the pituitary gland to produce follicle-
stimulating hormone (FSH) and luteinizing hormone (LH).

In males, FSH and LH target the testes in what
biologists refer to as the "brain-testicular axis."
This allows a male, after puberty, to develop all
the sexual characteristics that make him male,
including the ability to manufacture sperm.
The male endocrine system then adds in a
protein called inhibin, produced by cells in the
testes, as well as the steroid hormone testosterone, which both feed back to dampen GnRH
production. In essence, testosterone keeps
the male reproductive system in check. But
because it is an anabolic steroid, it also builds
up the muscles and controls behavioral traits
such as aggression.

Meanwhile, in the female, tiny pulses of FSH
and LH stimulate growth, maturation, and estrogen secretion from the follicles in the ovaries.
Over the course of 14 days, estrogen levels
climb to a threshold that feeds back to amplify
a surge of LH and FSH from the pituitary. This
hormonal gush stimulates the developing egg
to ovulate. The ruptured egg sac becomes a

DID YOU KNOW?

PINEAL GLAND

+ Doctors used to think the pineal gland was vestigial, with no use to humans.

+ Julius Axelrod linked the pineal gland to its role in the sleep cycle and melatonin production.

+ The pineal begins to produce melatonin in humans at age three months. Production peaks at puberty.

+ Melatonin secretion may be suppressed by light. It may also figure in seasonal affective disorder (SAD).

structure called the corpus luteum, which causes the production of progesterone. The uterus thickens, and the pituitary gland receives signals that cause it to shut down its production of LH.

If fertilization does not occur, hormonal sustenance of the uterine lining dries up. Menstruation results. At the same time, the brain senses lowered LH levels. These trigger the hypothalamus to begin pulsing out GnRH once again, the beginning of another reproductive and hormonal cycle.

Pineal Gland & Thymus

Hanging like a pinecone within the brain, the pineal gland secretes melatonin, a hormone that affects sleep cycles. The actual cells that do the job are called pinealocytes. Developing in

compact clusters, the cells are surrounded by a bed of calcium salts, which show distinctly in a brain x-ray.

Melatonin honors the nocturnal cycle, waxing at midnight and waning at noon. The hormone's production is linked to the visual pathway of the brain. Light and dark flip this hormone off and on, a process at the heart of melatonin's function as a biological clock.

The hormone works by targeting a patch of tissue deep in the hypothalamus called the suprachiasmatic nucleus. It is stocked with melatonin receptors. Their exposure to bright light resets the clock and connects the night and day cycle in with other rhythmic body processes, such as body temperature, sleep and waking, and—in mammals others than humans—seasonal mating.

Another tiny but crucial endocrine structure is the thymus, located deep in the sternum area of the chest cavity. The thymus makes a family of peptide hormones thought important in the development of immune responses. Early on in human life, the thymus is relatively large. It diminishes in size in adults.

Other Producers

Other organs have cameo parts in the performance of the endocrine system. The gastrointestinal tract is well stocked with special cells that release amino acid and peptide hormones such as secretin and cholecystokinin (CCK). CCK is an appetite-regulating peptide that pours out in response to food. It triggers the gallbladder to release bile and opens the sphincter entering into the small intestine.

Enteroendocrine cells, as they are called, act a lot like neurons, earning their moniker as "the brain in the gut." Some peptide hormones of the stomach and intestines, such as serotonin and histamine, are identical to neurotransmitters in the

BREAKTHROUGH LEPTIN

Body fat contains a hormone that tracks fat (or lipid) stores and should keep them in perfect balance with energy output. That lipometer is leptin, a protein hormone that has profound effects on appetite. In 1994, endocrinologist Jeffrey Friedman, right, at Rockefeller University discovered the hormone in a strange breed of mice. The rodents were obese, three times the weight of normal mice. Friedman's team determined that the animals got this way because they carried mutations in their so-called obesity genes, which encode the protein leptin. These obese mice lacked the substance. Friedman's group and others traced the site of hormone production not to a gland but to fat cells themselves. For clues to what the hormone does, Friedman needed only to look at the mice. Missing the hormone, they ate as if in a state of perpetual starvation, unable to stay warm or grow normally. Now researchers know that leptin levels increase in the bloodstream as fat deposits grow. The hormone signals the brain to stop eating, increase activity levels, and thus raise body heat. Ironically, the bodies of people who are obese actually make too much leptin. Overweight and obese individuals

develop leptin resistance, probably due to downregulation of leptin receptors or to malfunctioning of receptor response. Since the discovery of the hormone, researchers have identified a host of other molecules that regulate appetite and fat storage.

brain and act locally on tissues nearby. There are many unique gastrointestinal peptides as well. This wealth of hormones allows precise and local regulation of hunger, satiety, and energy balance—along with a connection to brain sensations of appetite and fullness.

The top of the heart, the atria, contains special cardiac muscle cells that secrete atrial natriuretic peptide, which targets the adrenal cortex, inhibits aldosterone, and signals the kidneys to increase the salt level in the urine. Why? Removing salt and water from the bloodstream reduces blood volume, pressure, and sodium concentration. The process gives the heart a rest when the body is at rest.

Cells in the kidneys make the glycoprotein called erythropoietin. This hormone's function is to stimulate bone marrow and step up the manufacture of red blood cells. Not surprisingly, drugs related to erythropoietin are used to treat anemia.

The skin also contains special messengers that set off a cascade of reactions in the body. Skin cells produce cholecalciferol, an inactive form of vitamin D. When cells are exposed to ultraviolet light—during an afternoon outdoors, for example—the latent hormone enters the bloodstream through capillaries buried in the dermis. The liver takes the next step, modifying the prohormone, and sends it to the kidneys for ultimate activation. The final product, vitamin D or calcitriol, works as part of a transport system and helps the body take in calcium, a mineral critical for bone strength, from ingested food.

One last endocrine player is adipose tissue, or fat cells. These energy packets take in glucose and lipids during a meal and, if the body does not immediately need the fuel, convert them to fat. As this occurs, fat cells churn out leptin, which binds to its receptors on neurons in the brain. This complex circuitry controls appetite, temperature, and reproductive prowess.

Chapter 11

REPRODUCTION

MALE SYSTEM

The Masculine Body and Its Functions

I n reproduction, both the man and the woman have important parts to play. The man's main biological function is to supply sperm, the cells that combine with the woman's egg when conception occurs. For that reason, the male reproductive anatomy is relatively simple and centered on the production, storage, and delivery of sperm. The operation of those systems, however, involves a complex interaction with glands and hormones that begins even before birth.

Male Anatomy

The testes, or testicles, are two plumlike glands suspended within the scrotum, a skin sac that hangs outside the body. The testes produce and cultivate spermatozoa, immature sperm cells that contain a man's entire genetic heritage. Reproductive glands and vesicles create fluids such as semen that lubricate, feed, and propel potent sperm cells on their journey toward new life.

A color-enhanced testosterone crystal

THE MALE REPRODUCTIVE SYSTEM

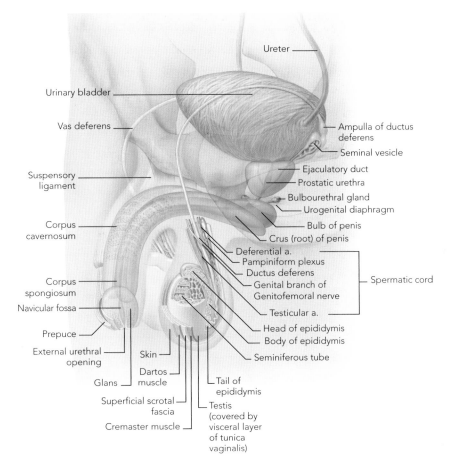

Ureter

Urinary bladder

Vas deferens

Ampulla of ductus deferens

Seminal vesicle

Suspensory ligament

Ejaculatory duct

Prostatic urethra

Bulbourethral gland

Urogenital diaphragm

Corpus cavernosum

Bulb of penis

Crus (root) of penis

Deferential a.

Pampiniform plexus

Ductus deferens

Genital branch of Genitofemoral nerve

Spermatic cord

Corpus spongiosum

Testicular a.

Navicular fossa

Head of epididymis

Body of epididymis

Prepuce

Seminiferous tube

External urethral opening

Skin

Dartos muscle

Glans

Tail of epididymis

Superficial scrotal fascia

Testis (covered by visceral layer of tunica vaginalis)

Cremaster muscle

The rest of the male reproductive anatomy is a delivery system, including the epididymis, coiled and nestled against each testis; the vas deferens, joining the epididymis to the ejaculatory duct; and the urethra, part of the urinary system and also the sperm's exit path. The anatomic pinnacle of this system is the penis, which delivers sperm cells to the female reproductive tract. The penis consists of the root, the shaft, and an enlarged tip, called the glans penis. It is cuffed by loose skin called the foreskin, which many cultures remove surgically in a procedure called circumcision.

The endocrine and nervous systems produce hormones that jump-start and run the male reproductive engine. Meanwhile, neurotransmitters imbue a sense of gender and sexual drive. All of these systems interact to create male identity.

Testosterone

Long before birth, male fetuses "masculinize" within the womb. The process hinges on the male sex hormone testosterone, manufactured by cells of the testes. Prenatal testosterone bathes the male fetus's brain and primes it for later development. This exposure is so critical

SPERMATOGENESIS

Nucleus
Acrosomal vesicle
Mitochondrial spermatid

Acrosomal vesicle
Flagellum

Mitochondrial
Acrosomal cap
Nucleus
Centriole
Mitochondrial spiral

Acrosomal cap

Shed cytoplasm
Residual cytoplasm
Flagellum

Fibrous sheath

that if a male fetus lacks the correct hormonal environment, he can grapple with infertility later in life.

Testosterone works alongside three other regulatory hormones: gonadotropin-releasing hormone (GnRH), which pulses from a brain structure called the hypothalamus; and the gonadotropins, follicle-stimulating hormone (FSH) and luteinizing hormone (LH), both produced by the pituitary gland, a bulbous structure suspended from the hypothalamus. The four hormones pulse and work as a team, one hormone triggering the next in what biologists call the "brain-testicular axis."

This interaction in a male infant produces high levels of GnRH, FSH, and LH from birth until he reaches six months of age. In fact, just after birth an infant's hormonal state equates to that of a midpubertal boy. Within months, however, a restraining system develops. It involves a protein called inhibin, produced by cells of the testes, as well as testosterone. These substances act like a molecular brake upon GnRH production.

Puberty is simply a release of this biochemical brake. Scientists speculate that the trigger is a tiny molecule secreted by the pituitary gland. Once the putative trigger begins to circulate through the body, the reproductive system, on hold for years, begins to rev up. Testosterone stimulates sperm production. The body develops sexual characteristics, including the sprouting of pubic, underarm, and facial hair. The voice deepens. The skeletal muscles increase in size and mass. From the perspective of biology, a boy becomes a man.

Creating Sperm Cells

Sperm are so important that the male body has evolved an elaborate system to safeguard them. Sperm thrive only if incubated three degrees below normal body temperature. Hence, the scrotum, which holds the testes and their vulnerable sex cells, hangs below the too-warm abdomen. The scrotum also has built-in structures to control this positioning. When the body is cold, for example, the cremaster muscles—the muscles by which the testes are suspended—contract to pull the testes up toward the belly. The dartos muscle, woven within the tissue of the scrotum, wrinkles the skin, reducing the surface area available for testicular heat loss. At other times, the cremaster muscles loosen, the dartos slackens, and a massive blood vessel system, comprising almost two thirds of the scrotum, cools the blood as it circulates around the testes.

Mature sperm are streamlined cellular rockets that shuttle genetic material to their target, an egg. Each sperm possesses a tail, which is connected to the cell body by a robust molecular engine. This allows the sperm, upon launch through ejaculation, to swim relatively great distances through the female reproductive tract. Sperm also possess special "heads" that can penetrate the formidable capsule that protects the egg.

But sperm do not start out as accomplished voyagers. They begin as immature spermatogonia: tailless, impotent, nonmotile stem cells buried deep within the testes. From a boy's infancy to puberty, spermatogonia divide to reproduce but remain immature. Then, at about age 14, a boy enters puberty. From a sperm cell's perspective, puberty means his spermatogonia are also ready to mature.

All living organisms carry the genetic instructions for life in the form of DNA. In humans, DNA is carried on 46 chromosomes in the nucleus of every cell, 23 from the mother and 23 from the father. During fertilization, the parents combine their genetic material into one embryo. It is the sperm, however, that carries the X or Y chromosome that, when added to the egg's X chromosome, determines whether the child will be a girl (XX) or a boy (XY).

DID YOU KNOW?

SPERM FACTS

+ Sperm cells (colorized and magnified, above) are the smallest in the body at 0.002 inches.

+ The male body creates 300–400 million sperm cells per day.

+ Formation of mature sperm cells takes 64–72 days.

+ A milliliter of semen contains 50–130 million sperm cells.

+ Sperm cells live only 48 hours after ejaculation.

But before that union can happen, sex cells must halve their chromosome number to 23. This reduction occurs in a process called meiosis, or "lessening." During meiosis, immature spermatogonia, now called spermatocytes, cycle through two precisely orchestrated cellular divisions to become four daughter cells, each with half as many chromosomes as a typical human cell.

At this stage, the male sex cells are called spermatids. Even though they possess the right chromosomal number for fertilization, they are still nonmotile and far too bulky to make a lengthy journey. A streamlining process must occur, in which the spermatids shed excess cell contents, grow tails, and compact DNA into their bullet-like "heads."

The human body has evolved an anatomical pipeline through which a sperm cell must travel as it develops. The entrance is one of four tightly coiled seminiferous tubules that make up the bulk of the testis. Most of the cells lining these tubules are sperm-forming cells in various stages of development. The seminiferous tubules function as virtual sperm factories, providing supporting cells that bathe, feed, and anchor the developing sex cells.

Spermatids on their way toward maturity enter the epididymis, which hugs the outside of the testis. Because these spermatids do not yet have tails, muscle contraction and pressure

NORMAL SPERM COUNT
About 113 million sperm per milliliter of ejaculate populate a normal sperm count.

LOW SPERM COUNT
Only 60–70 million sperm per milliliter of ejaculate are found in low sperm counts.

from testicular fluid propel them from the tubules into the epididymis. In essence, the epididymis is 20 feet of tightly wound tubing. Developing spermatids have 20 days to travel its length while accomplishing the many tasks of development, such as learning how to swim on their own. If sperm cells remain within the epididymis too long, they are consumed by neighboring cells and do not complete the journey.

During ejaculation, smooth muscles within the epididymis contract to expel sperm into the next duct of the system, the vas deferens. This duct squeezes the sperm forward and out the urethra.

In a sterilization procedure known as a vasectomy, surgeons cut the vas deferens and tie off the end. The body continues to make sperm even though the sperm cells cannot exit the system. The body will eventually begin to recognize these as "foreign" cells and produce antibodies to consume them.

Semen Creation

But sperm cells do not make the next step alone. They must combine with other substances to create semen, a life-giving elixir that is a mixture of DNA-carrying sperm and the fluids that support their ability to achieve fertilization. Sperm mingle with a small amount of fluid in the vas deferens, but most of the fluid in semen is secreted by the prostate gland

and fingerlike structures called the seminal vesicles that are perched on the backside of the bladder.

The seminal vesicles secrete seminal fluid and, together with the vas deferens, form the ejaculatory ducts. The thick, yellowish seminal fluid contains sugar for fuel; vitamins to regulate cellular actions; enzymes that help speed sperm velocity; and regulatory molecules (called prostaglandins) that will thin the mucus that guards the entrance to a woman's cervix, the gateway to the womb. As the sperm exit the vas deferens, they mix with seminal fluid in the seminal vesicles. During ejaculation, the semen enters the first stretch of the urethra.

There, semen flows by the prostate, a walnut-size gland that encircles the neck of the urethra. The prostate pipes in yet another secretion—a milky, slightly acidic fluid that contains citrate, enzymes, and prostate-specific antigen (PSA). These molecules will protect and activate the sperm for optimal performance as they exit the urethra on their way toward their ultimate destination—a union with an egg.

The seminal vesicles secrete fluid to nourish and transport sperm.

FEMALE SYSTEM

The Feminine Body and Its Functions

B efore the 18th century, females were viewed by physicians as incomplete males. A superficial knowledge of anatomy helped to make the case. Early physicians saw that females had a womb, with a shape on top not unlike a scrotum but with the testes, represented in the woman by the ovaries, lying outside it.

THE FEMALE REPRODUCTIVE SYSTEM

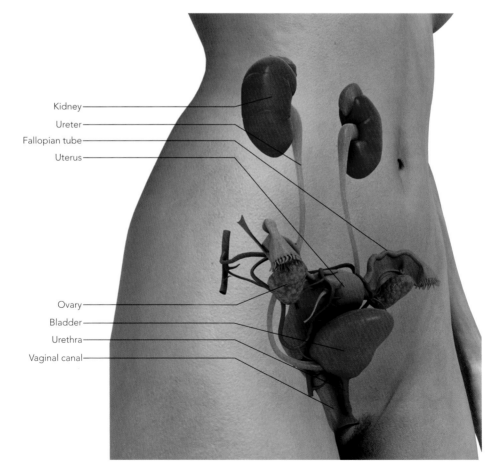

Kidney

Ureter

Fallopian tube

Uterus

Ovary

Bladder

Urethra

Vaginal canal

The womb opened into a tubular canal, something like a penis turned inside out. Thus, the notion of sexual difference put forth by the second-century Greek physician Galen prevailed for many years: Females must be males who, because of a lack of "vital heat," retained their reproductive structures inside their bodies, while those of true males emerged outside.

In fact, modern medicine has come to understand that the female reproductive system is actually more complex than its male counterpart. Females, after all, must create and store eggs for fertilization, provide an environment where those eggs can develop and thrive, and house and nurture a growing fetus.

Female Anatomy

Given this dual context, the ovaries command central focus. The shape of an almond and twice the size, each ovary is both an egg production factory and a hormonal powerhouse. The ovaries grow the eggs, or ova, and prime their development. The ovaries also make and dispatch cycles of sex hormones that steer the development of both mother and baby.

The female reproductive system has a second seat of command: the womb, or uterus. This inverted-pear-shaped structure is a muscular organ interwoven with an extensive blood-vessel system to feed menstrual flow. The top of the uterus reaches toward the ovaries, and their egg supply, with arms called oviducts, or fallopian tubes. The bottom of the uterus curves into a neck called the cervix. The cervical canal leads to a three-inch vagina, or birth canal. These structures are anything but an incomplete male tract. They are the centerpiece of "femaleness," the creator and container of a growing baby, and a locus of sexual pleasure.

KEEPING HEALTHY
PREVENTING CERVICAL CANCER

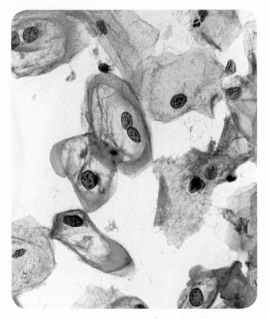

Cervical cells that are infected with HPV

Cells of the cervix constantly reproduce to replace those sloughed off by menstruation or sexual activity. Cervical cancer, the second most common cancer in women, occurs when cervical cells begin reproducing uncontrollably. Recent studies suggest that more than 90 percent of cervical cancer cases are caused by the human papillomavirus (HPV), a sexually transmitted virus that infects the cells and mucous membranes of the cervix. In 2006, the Food and Drug Administration approved the first HPV vaccine, Gardasil. While celebrated as a medical breakthrough, the vaccine is not foolproof. There are 120 known strains of HPV. The vaccine targets strains that cause about 70 percent of all cervical cancers and two other HPV strains that cause most forms of genital warts. It is effective for a minimum of four and a half years in females unexposed to those HPV strains.

Despite differences, there is a parallel between the male and female reproductive systems. Modern scientists have studied development of the external genitalia and found that many male and female structures arise from the same type of embryonic tissue.

In females, the area containing the external genitalia is called the vulva. It includes the mons pubis, the mound of fatty tissue above the pubic bone that becomes covered in hair after puberty and that contains sebaceous glands that secrete sexual attractants called pheromones. At the lower end of the mons pubis are the labia majora ("large lips"), which protect the other external genitalia from injury and guard the internal genitalia from infection; they are considered the female counterpart of the scrotum. They lie outside the labia minora, the inner lips that surround the vaginal opening. These "small lips," which swell with blood during sexual excitement and become sensitive to touch, are analogous to the ventral penis.

Between the labia minora, the projecting clitoris is composed of erectile tissue, richly innervated, and capable of engorgement during tactile stimulation, which can produce orgasm. Thus, the clitoris is

OVULATION BEGINS
In the primary stages of ovulation, the ovarian follicle bursts open.

OUT OF THE OVARY
A ripe ovum begins to emerge from the ovary wall.

PREPARATION
Smaller, protective cells encircle the ovum before its journey.

penislike. In fact, it is hooded by a skin fold called a prepuce, which is much like the male foreskin.

Unlike in the male, however, in the female the clitoris and surrounding structures are separate from the urinary tract. The urethra embeds itself in the front wall of the vagina, allowing the latter to evolve into a structure capable of great strength and flexibility. The vagina is stocked with muscles that stimulate the penis. Lubricants, secreted by Bartholin's glands alongside the vaginal opening, and elasticity allow the vagina to withstand friction during intercourse.

Women and men now celebrate and debate their similarities and differences. As the debates continue, their reproductive anatomy remains what it has always been: analogous but unique.

Puberty

For four years after the onset of puberty, a girl will experience hormonal changes that cause development of breast buds first, then pubic and underarm hair, and an overall growth spurt. Breasts are a combination of connective tissue, fat, and mammary glands. The latter are present in both males and females but only functional in women.

Mammary glands are actually modified sweat glands. Each connects to a duct which opens into the nipple. Each is stimulated to produce milk during pregnancy by the hormones progesterone and prolactin.

The hormone estrogen triggers fat deposition. Adolescent girls experience this as an increase in the size of breasts and hips. The hormone also lightens a girl's pelvis in anticipation of holding a growing fetus. Her bones experience increased calcium uptake to sustain density of her skeleton. Her body goes through an estrogen-triggered growth spurt around the age of 13. Estrogen maintains a low total cholesterol and high HDL. All these changes coalesce during adolescence to allow a girl to reach womanhood, in biological terms.

Her developing system must learn how to time itself with the appropriate hormonal levels. When she achieves this, she experiences her first menstrual period, called menarche. It heralds a 28-day cyclic pattern of hormonal interactions that will take about three more years to become regular. Producing a mature egg, called ovulation, does not usually occur with menstruation for

EXIT
A plume of follicular fluids follows the ovum's departure.

JOURNEY BEGINS
The ovum drifts towards one of the fallopian tubes.

TO THE WOMB
Smaller escort cells surround the ovum as it moves into the tube.

approximately the first year to 18 months.

The Egg

When a girl is born, her complete potential egg supply is born with her. While still in the womb, she creates about seven million egg cells, but most of these will die. At birth, about two million remain, dormant and primordial.

By puberty, only about 400,000 are left, many times more than she will need during her life. Yet only a small number of these will mature into eggs, with one egg typically released during each menstrual cycle. She has until menopause, when her menses cease—which typically occurs between ages 45 and 55—to execute her chances at perpetuating her genes.

As with developing sperm cells, immature ova must go through a development process to become fertile. The female counterpart to spermatogenesis is oogenesis. Before birth, the immature egg cells, called oogonia, proliferate by mitosis (cell division) and then, through partial meiosis, differentiate into primary oocytes. These infant eggs, in developmental terms, have two sets of chromosomes, not the single set needed to form half of fertilization's egg-sperm genetic

BREAKTHROUGH THE PILL

In the 1950s, birth control activist Margaret Sanger asked researcher Gregory Pincus to develop a pill to prevent pregnancy. Pincus teamed with Dr. John Rock, and their research led to the Food and Drug Administration's 1957 approval of Enovid for the treatment of menstrual disorders. Word spread of Enovid's contraceptive abilities; by 1959 over half a million women were taking it. In 1960, the FDA approved Enovid's use as birth control, giving women unprecedented control over their fertility. When *Time* put "The Pill" on its cover in 1967, right, more than 12.5 million women worldwide were on "The Pill."

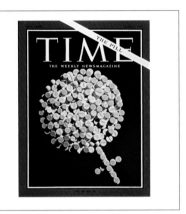

partnership. The primary oocytes remain in this state of arrested development until the girl reaches puberty. After puberty begins, a young woman's ovary will pick only one primary oocyte to complete the first phase of meiosis.

In females, this process differs in one key aspect from meiosis in male spermatogenesis. The cell division of oocytes produces two daughter cells that are unequal in size. One, called the secondary oocyte, gets all the cytoplasm and organelles of the mother cell. The other, called the polar body, gets little more than genetic material. The female body engages in this scheme because the potential egg will need nutrients for its seven-day journey to the uterus. The polar body will eventually degrade and disappear.

The secondary oocyte then starts phase two of meiosis. Again, the cell halts before completion. The oocyte is hedging its bets, waiting for ovulation and a sperm cell. Only if a sperm cell penetrates will the developing egg, now called the ovum, quickly finish meiosis. The process spits out a second polar body while the ovum keeps all the nutrients. Researchers still do not consider this ovum fertilized. Fertilization occurs only when the nuclei of ovum and sperm unite.

Backup Plan

Fewer than 500 secondary oocytes are released during a woman's lifetime. One might ask, "Why such an inefficient system—seven million oogonia cut down to 500 ova?" Perhaps the reproductive system has evolved a strong backup plan that includes producing a superabundance of egg cells, and then timing their release with precision.

"Backup" also applies to other parts of the egg-making system. For example, follicle cells cover, coax, and cater to developing oocytes throughout their life spans. Follicle cells begin as a single layer encasing primary oocytes. Later, the cell covering divides into two layers around secondary oocytes. The follicle cells do more than protect. They also secrete a hormonal precursor to estrogen. This hormone guides the development of the immature egg inside its follicular shell, as well as timing and triggering other aspects of the female menstrual cycle.

As the day of ovulation draws nearer, the follicle cells grow and shape themselves into a fluid-filled cavity around the egg. On about day 14 of a woman's menstrual cycle, one dominant, full-size follicle bulges like a boil from the outside of the ovary. The neck of the follicle stretches into a stalk and erupts from the ovary surface, releasing the oocyte into the body cavity near the oviduct that will usher it toward the uterus.

Many women experience this eruption as a pinching or cramping sensation in the abdomen, sometimes called "mittelschmerz." The follicle then releases its secondary oocyte. The rest of the follicles that had been developing in parallel simply degenerate, allowing a woman to ovulate only one eventual egg per cycle.

Meanwhile, the ruptured follicle collapses and fills with clotted blood. The transformed follicle then fashions itself into a corpus luteum, a working endocrine gland. For the next 10 days, it secretes progesterone and some estrogen as the oocyte travels down the oviduct toward the uterus. If a sperm fertilizes

DID YOU KNOW?

EGG FACTS

+ The egg is the largest cell in the human body.
+ The size of a pinhead, it is the only human cell that can be seen with the naked eye.
+ The release of eggs does not alternate between the two ovaries but seems to be random.
+ After ovulation, an egg is available to be fertilized for 12 to 48 hours.

the egg, the corpus luteum will continue its hormonal mission until the placenta takes over hormone-producing duties. Otherwise, the corpus luteum degenerates, triggering a woman's body to begin again the cycle of egg development and release.

The Menstrual Cycle

While the developing oocytes wait, formed and potent in their follicular nests, two other anatomical structures stand ready at the base

In the first phase of the menstrual cycle, the lining of the uterus (endometrium) thickens.

UNDERSTANDING THE MENSTRUAL CYCLE

Menstruation
During menstruation, cells lining the uterus (endometrium) become detached from the uterine wall and fall off. Blood and tissue from the uterine lining are discharged through the vagina, a process which normally lasts around 5 days. Menstruation marks the beginning of the menstrual cycle.

Ovulation
Approximately 14 days into the menstrual cycle, a sudden rise in the blood level of gonadatrophic hormone LH triggers the release of a single mature ovum or egg from one of the ovaries. The egg travels down the uterine tube until it reaches the uterine cavity. An unfertilized egg is discharged through the vagina.

Fertilization
The fusion of genetic material from a mature ovum and a mature sperm produces a fertilized egg (or zygote). Fertilization normally occurs in the uterine tube. As the zygote begins to divide, it travels down the uterine tube towards the uterus, eventually forming an embryo that may successfully implant into the lining of the uterus (endometrium).

of the brain to trigger the next phase. The hypothalamus, the hormonal conductor, and its target, the pituitary gland, will direct a waltz of hormones to the ovary.

Together, the hypothalamus, pituitary, and ovary form a trio that plays a wonderful song of fertility from puberty to menopause. The hormonal notes are identical to those in male fertility: gonadotropin-releasing hormone (GnRH) pulses from the hypothalamus to the pituitary gland; follicle-stimulating hormone (FSH) and luteinizing hormone (LH) ebb and flow from the pituitary.

The ultimate hormonal target differs, however. While FSH and LH stimulate the testes in males, the hormones animate the ovaries in females. And the response, a 28-day menstrual cycle, is much different from the process of sperm development for ejaculation.

In females, tiny pulses of FSH and LH stimulate growth, maturation, and estrogen secretion from the follicles in the ovaries. Over the course of 14 days, estrogen levels climb to a threshold that feeds back to induce LH and FSH release from the pituitary. This hormonal release stimulates the dominant primary oocytes to begin maturing into a secondary oocyte that bursts forth in a five-minute ovulation.

The ruptured follicle then curtails its production of estrogen, as it metamorphoses into a corpus luteum. Luteinizing hormone, which prompts this transformation, then prods the corpus luteum to produce progesterone as well as estrogen. These two hormones flow toward the uterus, which, in response to rising estrogen, has been building up a blood-enriched lining, called the endometrium.

The uterine lining thickens from day 6 to day 14 of a woman's monthly cycle. Within it endometrial cells grow receptors for the hormone progesterone, which will maintain the lining, helping it to secrete mucosa and nutrients. These would bathe, cushion, and sustain a fertilized egg, if it arrived, until it implanted. Progesterone, along with rising estrogen, also feeds back to the pituitary gland to shut down its production of LH.

If fertilization does not occur, progesterone will lose its target, the thickened lining of the endometrium. LH coming from the pituitary begins to dwindle. The hormonal supply from the corpus luteum decreases as well. Devoid of hormonal sustenance, blood vessels that feed the uterine lining kink and spasm. Cells within self-digest. The uterine blood vessels constrict and relax. Blood pours into weakened capillary beds and they fragment. All but the deepest layer of the endometrium sloughs off over the next three to five days. A woman experiences this process as menstruation, as the lining sheds. Some women experience symptoms of cramping and back pain as the uterus strongly contracts to expel its unused lining.

Meanwhile, the brain senses lowered LH levels. They trigger the hypothalamus to once again begin pulsing GnRH—the beginning of another cycle and another chance to pass on her genes. And so it repeats, month to month.

KEEPING HEALTHY
NATURAL RELIEF FOR CRAMPS

Most women are no strangers to menstrual cramps. During "that time of the month," hormones called prostaglandins cause the uterus to contract, which can result in painful menstrual cramps. Ibuprofen and birth control pills can help relieve cramps, but making different decisions about certain foods can be a more natural way to ease the pain.

+ **Hummus:** In addition to being delicious, this Middle Eastern spread is packed with nutrients that can make cramps more bearable. It contains omega-3 fatty acids, which reduce inflammation and can block the production of cramp-causing prostaglandins. It also contains calcium, which helps relieve cramps.

+ **Tea:** Herbal tea doesn't have caffeine and helps with hydration. Many nutritionists say green tea (which is low in caffeine) is particularly effective.

+ **Ginger:** According to some studies, ginger can be as effective as ibuprofen in relieving painful menstrual cramps, so ginger ale or ginger tea (made with real ginger) can go a long way.

+ **Coffee:** The high levels of caffeine in your morning cup of joe can increase tension and lead to dehydration. Plus, cream and sugar contain chemicals that can exacerbate menstrual cramps.

+ **Red meat:** Save the steak for another day. Red meat contains arachidonic acids, which have been linked to the production of the prostaglandins responsible for causing cramps.

CONCEPTION

How Babies Are Made

Nature has given humans a virile incentive to reproduce: the feeling of sexual arousal. During sexual excitement, a host of stimuli bombard the parasympathetic system, a division of the autonomic nervous system that, as one of its functions, influences sexual stimulation. Erotic touch, sounds, smells, or visual cues all can activate parasympathetic neurons. So, too, does the thought of a sexual encounter. (As therapists often note, the mind is the most powerful sexual organ of all.)

Arousal

For males, arousal centers on blood flow. A mesh of arteries supplies the erectile tissue in the penis. Normally, they are constricted, and the penis stays flaccid. When a man becomes sexually excited, nerve cells promote release of a gas called nitric oxide. It prompts penile arteries to dilate, flooding the spongy tissue with blood. The result is an erection: The penis extends outward and grows large and stiff so that it may penetrate the vaginal canal to deposit sperm inside.

Women experience arousal much as men do, but sensations are not limited to the genital area. The clitoris does stiffen, and the vagina lubricates. The breasts engorge, and the nipples become erect.

As arousal continues, impulses build until they unleash a massive discharge of nerve impulses over sympathetic nerves. The result is called an orgasm, a climactic burst of pleasure accompanied by a rise in blood pressure, an increase in heartbeat, and powerful muscular contractions throughout the body.

In men, the exploding, squeezing sensation of orgasm involves ejaculation, the propulsion of semen from the male duct system at tremendous speed. This high-velocity force during intercourse is paramount. As the penis spasms

DID YOU KNOW?

UNIQUE SYSTEM

The reproductive system differs from all others in the body because:

+ It is dormant for the first 10–15 years of a person's life. The reproductive system animates only after puberty occurs.

+ In order to fulfill its primary function, it depends on interaction with the complementary system of another person. It cannot act alone.

Joining Together

Fertilization, the next step in the union of the reproductive systems, involves a complexity that approaches the miraculous. The journey proceeds on two fronts. Sperm ejaculated into the vaginal canal swim toward the egg, which is viable only 12 to 24 hours after ovulation. The sperm themselves have only 24 to 72 hours to complete their reproductive jobs.

And so the millions of sperm launch and propel through a gauntlet of acid in the vagina and mucus in the cervix. Those conditions kill sperm, as does spill-out from the vagina after intercourse. But estrogen (secreted after ovulation from the empty follicle, the corpus luteum) thins the cervical mucus, allowing a few thousand sperm to reach the uterus.

Still, the sperm are readying themselves to penetrate the egg ahead of them. Swimming through the cervical mucus, the uterus, and the uterine cavity, the tadpole-like sperm begin to shed proteins buried in their cellular membranes, which exposes the tips of the sperm heads, called acrosomes. They hold powerful enzymes that, if launched too soon, would disintegrate the lining of ducts and tubes of the male and female tracts.

When the sperm reach the egg, they cannot immediately penetrate it. The oocyte has three

Three-dimensional scanning allows an internal view of the mechanics of sexual intercourse.

inside the vagina, it must shoot potent sperm long distances toward an ovulated egg. With female orgasm, women experience sweeping muscle tension and rhythmic contraction of the pelvic floor muscles.

Orgasm in men is followed by a refractory period, in which they cannot produce another erection for minutes to hours. By contrast, women may experience multiple orgasms in a single sexual experience. And women, unlike men, do not need an orgasm to conceive.

An electron microscope allows a close-up view of the human immunodeficiency virus (HIV), the cause of Acquired Immune Deficiency Syndrome (AIDS).

protective coverings: a cellular membrane on the inside, a tough zona pellucida in the middle, and, on the outside, a "radiating crown" of cells.

And so the sperms' enzymes, mainly hyaluronidase, have stayed dormant throughout development. Only when the sperm reach and cling to an oocyte's zona pellucida do they let loose with an explosion of acrosomal enzymes. These eat away at the fortified zona pellucida. Many sperm will spill out all their enzymes and die as they break down the layer. Finally, one sperm is able to bind to a pocket within a receptor on the surface of the secondary oocyte.

Fertilization

In a two-part process, the sperm works it way into the egg's membrane, and the egg assists by internalizing the sperm. The two unite, triggering three biochemical gates to shut, in order to prevent more sperm from entering. If two sperm get in, the oocyte will die.

But this oocyte still is not fertilized. The sperm must travel to the center of the oocyte.

SEXUALLY TRANSMITTED DISEASES

More than 12 million people in the U.S., a quarter adolescents, acquire sexually transmitted diseases (STDs) every year. Historically, the most notorious were gonorrhea and syphilis, both caused by bacteria and now treatable by penicillin and related antibiotics. More recently chlamydia, genital herpes, and AIDS have taken center stage.

CHLAMYDIA	Chlamydia is the most common STD in the U.S. Afflicting 4–5 million people annually, chlamydia is an obscure and silent epidemic. Part of the reason is that 80 percent of infected women and half of infected men show no symptoms. Chlamydia is easily detected by a body fluid or urine test and treatable by the antibiotic tetracycline.
GENITAL HERPES	Most commonly caused by the virus herpes simplex type 2, herpes infections are marked by sudden breakouts of painful sores on the reproductive organs. In pregnant women, the virus can cause severe birth defects. The anti-viral drug acyclovir can speed healing and reduce flare-ups.
AIDS	The most fatal is AIDS (Acquired Immune Deficiency Syndrome), caused by human immunodeficiency virus (HIV). More than 900,000 Americans have contracted AIDS since 1981, and nearly a million are currently infected with HIV. There is no cure, but researchers have discovered cocktails of drugs that may slow the spread of HIV in the body.

There a second polar body emerges, leaving an ovum nuclei. Both it and the sperm head expand, forming male and female pronuclei. Working together, they send out cables called mitotic spindles that connect male to female genetic material. The nuclear membranes break open. Chromosomes spill out. They unite on the mitotic spindle, like towels pinned onto a clothesline. With that, the cell, now called a zygote, is truly fertilized, ready to divide into a new life—a pre-embryo.

The zygote must then make a weeklong journey down the fallopian tube. It is assisted by cilia, waving hairlike structures that line the walls. Thirty-six hours after fertilization, the zygote cleaves into two daughter cells, called blastomeres. The daughter cells diminish in size as compared with their mother. This size reduction is important because the growing zygote, now referred to as a pre-embryo, needs cells with a high ratio of surface area to volume, in order to more easily absorb nutrients and excrete wastes through their membranes. In 72 hours, after more cell division, the dynamic ball of 16 cells is called a morula. Day five after fertilization, it proliferates to a 100-celled structure, the protective

RENDEZVOUS
An egg travels in the fallopian tube, where sperm will swim to join it.

BREAKING IN
Sperm attack the zona pellucida, a membrane encasing the egg.

UNION
A sperm plunges into the egg's plasma membrane.

zona pellucida breaks down, and a blastocyst forms. This is a fluid-filled cavity composed of cells called trophoblasts, the future placenta, and an inner cell mass that will grow into the actual embryo.

Implantation

Meanwhile, the hormones secreted by the corpus luteum have primed the uterus with a thickened endometrium. About a week after ovulation, large protrusions with sticky protein surfaces emerge.

After exiting the fallopian tube and entering the uterus, the pre-embryo looks for a docking station. If the chemical environment is ripe, the blastocyst sticks to the endometrial wall and implants high up in the uterus. If the vascular wall is not ready, the blastocyst floats freely a bit longer and attempts it again later, adhering to a lower position in the uterus.

Implantation is a process that is chemical as much as physical. Once anchored, trophoblast cells of the blastocyst secrete digestive enzymes and growth factors. They inflame the endometrium in order to weaken it enough to burrow inside. Fourteen days after fertilization, endometrial cells cover the precious embryo with a cellular blanket of protection and nutrition in which it can grow.

OUTSIDE THE WOMB

While researchers have dissected fertilization into a series of perfectly timed steps, the process still holds many mysteries. Couples trying to have a baby but who do not conceive within a year are considered to have a "fertility problem." Given the emotional costs of infertility and the desire to explore the yet-unsolved puzzles surrounding conception, researchers have long proposed means to help women conceive. In vitro fertilization (IVF) refers to techniques in which clinicians extract eggs and fertilize them outside a woman's uterus.

BEGINNINGS

The idea of assisted fertilization became the buzz in the 1930s when an anonymous editorial in the prestigious *New England Journal of Medicine* suggested that the practice, if successful, would be "a boon for the barren woman with closed tubes!" The article triggered an explosion of controversy, as the public grappled with the prospect of doctors "playing God." The archbishop of Canterbury in 1948 even recommended banning the practice.

Still, researchers persisted, coming close in Australia in 1973 with the first IVF pregnancy, which did not come to term. Meanwhile, in England, Cambridge University scientist Robert Edwards and Oldham gynecologist Patrick Steptoe had begun working together on IVF. Steptoe had pioneered the technique called laparoscopy, by which a fiber-optic light could be inserted into the abdomen to inspect blockages in fallopian tubes. Using this same technique, he developed a method of extracting eggs from the ovaries for fertilization by means of a suction needle in the laparoscope.

Edwards had traveled to Baltimore's Johns Hopkins University in the summer of 1965, where he worked on perfecting the techniques for fertilizing a human egg outside the womb. While there, he was helped by gynecological surgeon Howard Jones and his wife, Georgeanna Jones, a reproductive endocrinologist, who together later pioneered IVF in the United States.

Over the years, the British team faced hostility and skepticism, as critics claimed IVF would produce abnormal babies. Finally, ten years after Steptoe and Edwards first fertilized an egg in a petri dish, they announced the birth of Louise Joy Brown, the world's first "test-tube baby," born July 25, 1978. The healthy baby weighed 5 pounds 12 ounces. The birth led to an uproar over whether this was a one-time event and, if not, what meaning this conception method had in terms of morality and safety.

OH BABY!

Despite these concerns, other in-vitro babies soon followed, including the first success in the United States. Elizabeth Jordan Carr was born December 28, 1981, in Norfolk, Virginia, at the Jones Institute for Reproductive Medicine at Eastern Virginia Medical School, the fertility clinic founded in 1980 by Howard and Georgeanna Jones. Eventually, nearly 3,000 children would be born through the efforts of the Jones Institute.

In 1984 French researchers developed a new technique for retrieving eggs. Steptoe's laparoscopic procedure, though minor surgery, still required the patient to be anesthetized and hospitalized. In the French method a long aspiration needle is inserted through the wall

One-year-old Louise Brown, the first "test-tube" baby, smiles as she appears on television in 1979.

of the vagina into the ovary, where it extracts fluid containing eggs. The procedure, which is guided by ultrasound, can be done in an outpatient setting. It is simpler, cheaper, and reduces risk to the patient; for those reasons it can increase success because more patients are willing to repeat the procedure multiple times.

NEW TREATMENTS

Today, IVF has evolved into a popular practice, a chance at parenthood for the estimated 7 percent of U.S. couples who are infertile and others like them worldwide. Today, couples may also try assisted reproductive technologies (ART), fertility treatments in which both eggs and sperm are involved. In general, ART procedures begin as gynecologists surgically remove eggs from a woman's ovaries, sometimes after hormonal treatment. Technicians combine her eggs with the father's sperm in a culture dish for about 18 hours. If sperm counts are low, technicians can even inject a single activated sperm directly into the egg in a technique called intracytoplasmic sperm injection, or ICSI. After about 48 hours, when the pre-embryos reach the six- to eight-cell stage, a doctor returns several to the woman's uterus, where—if successful—they will implant and develop. While the success rates for ART vary from clinic to clinic, according to the Centers for Disease Control and Prevention, fertility experts performed a reported 122,872 ART cycles in 2003, leading to the birth of 48,756 infants, including some twins and triplets. That same year, Louise Brown celebrated her 25th birthday. Some 5,000 test-tube babies were invited to the party, hosted by Robert Edwards, who had helped pioneer the IVF techniques that brought Louise into the world.

CONTINUING DEBATES

Researchers continue to push the frontiers of assisted fertilization, sometimes provoking new controversies. The American Society for Reproductive Medicine, among others, has drafted policies to regulate procedures. Still, ethical and religious dilemmas abound. Most recently, debate has erupted about pregnancy after menopause, the use of genetic diagnosis before implantation to screen embryos, and the possibility that embryos might be created to be stem cell sources.

Fears that the children born of these scientific techniques would be abnormal has long since been laid to rest. Louise Brown's parents went on to have another daughter, Natalie, also by means of IVF. In 1999, Natalie became the first test-tube baby to give birth. Then, in December 2006, Louise Brown, the world's first test-tube baby, bore a son, Cameron Mullinder, conceived naturally.

PREGNANCY

A Body Changes As a New Life Grows Within

As a pregnancy develops, it also changes a mother-to-be's body. Hormones wax and wane, resetting her body chemistry in profound ways. Some women will experience numerous symptoms as a result; others, relatively few as the baby grows over the next nine months.

Early Stages

Pregnancy is divided into three-month stages called trimesters. In the earliest stages of the first trimester, a common symptom is "morning sickness." Despite the name, bouts of nausea and vomiting can occur at any time of the day. In rare cases, nausea is constant, and vomiting occurs several times a day. The condition, called hyperemesis gravidarum (HG), typically subsides by the 20th week of pregnancy but may require brief hospitalization to correct fluid and metabolic imbalances.

Surging estrogen and progesterone cause blood vessels to sprout throughout a woman's reproductive organs, giving her genitals a purplish cast and increasing their sensitivity. Blood supply to the breasts increases, sometimes appearing as a network of bluish lines. The breasts swell and become more tender as the number of milk glands increase.

Beyond hormonal changes, the first trimester holds many surprises. A growing embryo, measuring about three inches in length and weighing about half an ounce by the end of 12 weeks, fills the pelvic cavity. Weight gain is slight, only about

DID YOU KNOW?

PREGNANCY TEST

+ An Egyptian papyrus from 1350 B.C. describes a pregnancy test in which a woman would urinate on wheat and barley seeds. If the seeds grew, she was pregnant.

+ In the early 20th century, a woman's urine was injected into rabbits, rats, and frogs to detect human chorionic gonadotrophin (hCG), a hormone produced only by a developing embryo.

THREE TRIMESTERS OF PREGNANCY

| 1ST TRIMESTER | 2ND TRIMESTER | 3RD TRIMESTER |

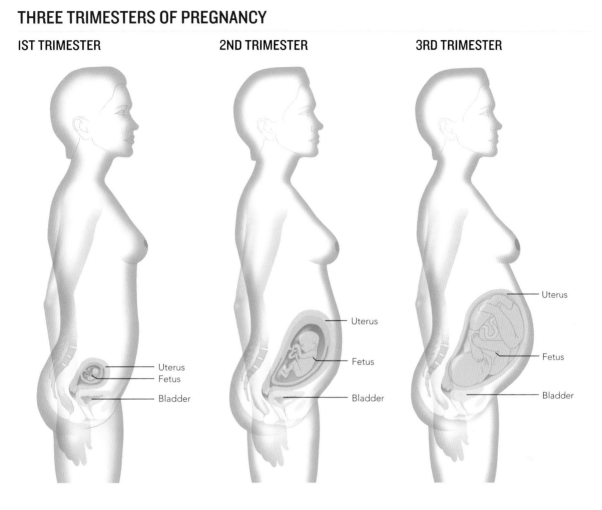

a pound per month. The abdomen may protrude slightly by the end of the trimester. Other symptoms include backaches and more frequent urination as the embryo, in its fluid-filled amniotic sac, begins to press upon the bladder. The heartbeat speeds up as it pumps increasingly more blood to the uterus. Feelings of fatigue are typical as the body adjusts.

Later Trimesters

During the second trimester, the placenta provides the embryo with food, energy, and hormones through two arteries in the umbilical cord while removing waste through a single vein. It also takes over the role of secreting estrogen and progesterone, a task formerly performed by the corpus luteum and the trophoblast cells of the embryo, buried in the uterine lining. (If this transfer fails, the embryo will be naturally aborted. One-third of pregnancies end during the first trimester because of failures in implantation or placental development.)

In the second trimester symptoms of nausea and fatigue fade, and hormone levels begin to stabilize. The mother's metabolism begins to process more fatty acids and less sugar (glucose) to preserve as energy for the baby. In some women, this change can cause gestational diabetes mellitus. The condition normally ends at childbirth, but about half of the women in whom it occurs will develop Type 2 diabetes later in life. During this trimester, weight gain increases to a rate of about a pound per week. Maximizing nutritional value to provide for the growing fetus is vital, as is keeping weight gain within healthy limits.

During this second trimester, the mother feels the baby's first movements. The abdomen swells as the uterus expands upward, toward the diaphragm, and down, toward the bladder. The breasts continue to activate more milk glands, and the areolas, the pigmented areas around the nipples, may darken and enlarge.

In the third trimester, the continued growth of the baby causes increased pressure on the digestive and other internal organs. The developing baby needs more nutrition than ever, causing the mother's total volume of body water to rise. Blood volume increases 25 to 40 percent to carry more nutrients where they are needed; the ankles, fingers, and face may swell as a result. Pressure from the swollen uterus pushing on the pelvic blood vessels can cause varicose veins. In the last couple of weeks of pregnancy, estrogen levels have climaxed, causing irregular ripplings of the uterine muscles—called Braxton Hicks contractions, or false labor. Estrogen also ripens the endometrium, causing it to sprout receptors for the potent pregnancy hormone oxytocin, which will play a large role in childbirth.

Birth

Signs of labor may be a series of fluttering, false contractions, as a rupture of the amniotic sac (or "water breaking"), or as full labor. As it begins, fetal cells excrete oxytocin, which causes the uterus to begin contracting. Placental cells make prostaglandins, while the mother's

DID YOU KNOW?

C-SECTION

Was the cesarean section named for Julius Caesar's surgical delivery from the womb? It's unlikely. In Roman times, cesareans were typically performed if a mother was dead or dying, and Caesar's mother lived to see him to adulthood. Another possibility: It comes from the Latin *caesones*, the term for male infants born by a postmortem cesarean.

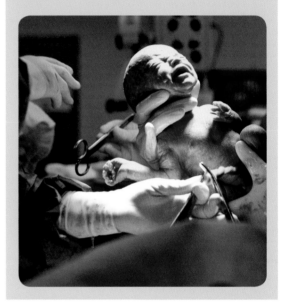

hypothalamus churns out more oxytocin. The result is rhythmic, expulsive contractions, which squeeze the upper uterus at first in intervals about 15 to 30 minutes apart.

The contractions gain force and propagate downward, reinforced by muscles of the lower uterus. The cervix thins and softens, prompting the birth canal to open. The cervix dilates, and as the baby moves into position the mother feels an intense urge to bear down. Lasting from six to twelve hours, this first phase of labor is generally the longest.

Eventually the cervix fully dilates to about ten centimeters. With the mother's pushing, the uterine muscles compress until the largest part of the baby's head reaches the vulva, an event called "crowning." The baby's head is delivered, followed, much more easily, by the neck, shoulders, and body. After the baby is delivered, the umbilical cord is cut. In the last act of birth, the placenta is shed and expelled.

Before and after the birth, hormonal interactions prepare the mother's body for lactation, or milk production by the mammary glands. Hormones, including lactogen, trigger the hypothalamus to excrete prolactin-releasing hormone (PRH), which stimulates the pituitary to make prolactin, the milk-triggering hormone.

This process takes about three days, during which the mammary glands produce colostrum, a low-fat, vitamin-stocked substance that, like milk, helps sustain a nursing baby and prevent infections. Hormones continue to exert influence in breast-feeding. When an infant latches onto its mother's nipple, touch sensors there trigger the production of PRH and prolactin. Milk supplies accumulate in the sinuses of the mammary glands. Then oxytocin causes the "let-down reflex" in which milk is ejected from the glands and the mother feels a sense of well-being, in parallel with her baby's comfort at being fed. Oxytocin also stimulates contractions of the uterus and that organ's return to its normal size. During this phase, mother and baby form a strong emotional bond. And so the species lives on.

Hormonal interactions help foster an emotional bond between mother and child during the first few days after birth.

Chapter 12

PROTECTION

IMMUNE SYSTEM

Keeping the Body Healthy

Unlike some of the body's other systems (such as the digestive and circulatory systems), the immune system is a functional system that operates throughout the human body, its work carried out by the body's trillions of immune cells and specialized molecules.

A protective macrophage (blue) prepares to attack an advancing tuberculosis bacterium (purple).

The first line of defense is the skin. Our body's covering of tissue protects us from all kinds of dangers large and small. One obvious invasion of this first defense is a bump, bruise, burn, or cut; nerves in the skin register pain and tell us to avoid the rock or the hot stove that caused the problem. A bump resulting in a bruise means that blood vessels have broken beneath the skin, and the escaped blood has formed little clots called hematomas. A cut allows microorganisms to breach the first wall—the skin—and trigger internal defenses, which we will examine in greater detail later in the chapter.

Injuries are generally the least of the immune system's problems. Armies of germs—viruses, bacteria, parasites, and fungi—make daily sorties against our body, whether we are injured or not. Through the mouth, nose, eyes, ears, genitals, and skin pores, these saboteurs infiltrate the body. Since skin can't seal out every possible opening (or else we could not breathe or eat), the first line of defense has a backup—the mucosal lining. Pathogens entering the gastro-intestinal tract, genitourinary tract, and lungs are generally stopped by the sticky, acid-secreting mucous membranes lining the organs.

A second line of defense comes into play against those persistent invaders not blocked by the skin and mucosal lining. This second line is composed of cells, which include a broad category called phagocytes, or engulfing cells whose basic job is to eat the invaders. Among the phagocytes

MAST CELL
This kind of white blood cell is activated by allergic reactions and physical injuries.

BASOPHIL
Basophils are the smallest and least common white blood cell at play in the immune system.

ANTIBODIES
These specialized proteins combat specific microorganisms.

are white blood cells called neutrophils and eosinophils, which help initiate inflammation. Another broad category of generalized cells, the natural killer cells, are unleashed immediately against viruses, cancer cells, and other foreign cells. In addition to these immune cells, numerous chemical compounds respond to infection and injury, move in to destroy pathogens, and begin repairing tissue. While these cells and chemicals are at work, the body often makes another generalized response to foreign invasion—it heats up, or produces a fever.

The body's third line of defense—the adaptive defense system—is often slower to react, because the body must determine what kind of attack is underway and organize this final, more specific response. When the first two lines of defense fail, specialized warriors then go into action against the most perfidious of pathogens, whether they be microbes from the outside or internal cells in revolt such as cancer cells. These elite fighting units are not just lying around in wait; instead, they are trained on the job—that is, they are created in response to a virus, for instance, that the body has not seen before. The adaptive defense system is thus an improvised response to a specific invader.

Chief among the immune cells of the adaptive defense system are white blood cells called lymphocytes that come from bone marrow and thymus and circulate through the body. The body also produces antigen-presenting cells that swallow pathogens or chemical constituents of the pathogens called antigens and cause lymphocytes to start reacting against the invaders. Two remarkable qualities of the adaptive defense system are that it works systemically—that is, once activated in one part of the body, it functions throughout; and it memorizes the antigens so that the next time they come along the body hits back quicker and harder.

INNATE
Bumps, bruises, and swelling are all signs that the innate defenses are at work.

ACQUIRED
The body's adaptive defenses can learn to target pathogens, like chicken pox, after exposure.

The specific immune system is sometimes called the adaptive, or acquired, immune system, because unlike the innate system it does learn to distinguish between particular pathogens and antigens. People afflicted with the chicken pox virus are sick for about a week; after this they are usually immune to the disease for life. Why? Specific antibodies and T lymphocytes recognize the virus upon re-exposure and eliminate it before it causes any harm. Mumps, measles, and most other infectious diseases can also be memorized by the body's immune system so that when the pathogens appear, the body quickly starts making the appropriate antibodies; each kind of antibody is a specific key that works to unlock and block only that specific pathogen's antigens. Immunizations for these

Reactions

The nonspecific, or innate (natural), reaction requires no previous exposure to the antigen being encountered. Something potentially harmful enters the body, and its defenses go up. The nonspecific immune system can kill or limit the spread of pathogens, but it does not increase its efficacy against any specific invader—it doesn't learn to recognize one particular antigen as more harmful than any other. It simply does the best job it can and leaves the rest up to the body's specific immune response.

DID YOU KNOW?

COMPARISONS

+ The innate response is broad-based; the adaptive response targets specific pathogens.

+ The innate response is found in all organisms; the adaptive response is found only in jawed vertebrates.

+ The innate response has no "memory"; the adaptive response memorizes individual pathogens and increases its effectiveness on re-exposure.

T lymphocyte cells (shown in purple) are white blood cells that attach to and destroy cancer cells.

diseases are a refined shortcut: They train the body to recognize specific antigens without your having to get sick.

Adaptive immunity can be divided into two basic kinds—humoral and cell-mediated. The humoral immune response is carried out by antibodies. Also known as immunoglobulins, these antibodies are molecules of protein that derive from B lymphocytes and their progeny, plasma cells. Antibodies circulate in the blood and other bodily fluids—hence the name "humoral," from the ancient concept of body humors. The cell-mediated immune response is an immunity granted specifically by activated T lymphocytes, which destroy (or lyse) infected cells, cancerous cells, or foreign cells in transplanted organs.

There are a number of other organs and tissues involved in the body's immune response. Bone marrow and thymus generate the important white blood cells known as lymphocytes. They migrate to the body's lymph nodes, which

are found in the neck, armpits, groin, abdomen, and at the base of the lungs.

The body's largest lymph organ is the spleen, a five-inch-long organ located on the left side beneath the diaphragm and behind the stomach, tucked safely underneath the rib cage. It produces antibodies, is the site of activation of T lymphocytes, destroys aging and damaged blood cells, and detects and destroys certain pathogens.

The thymus, located between the top of the lungs at the base of the neck, is a small organ involved in developing the T-cell part of the immune system. Another aggregate of lymphoid tissue, the tonsils are situated in the throat. In addition to these lymphatic masses, the body has a network of lymphatic vessels that transport lymph—a protein-containing fluid. Lymph is a fluid, while lymphocytes are white blood cells found in both blood and lymphatic tissues.

INNATE DEFENSES

A Broad Response

Every person from birth is protected against a wide variety of pathogens and foreign substances by his or her innate defense mechanisms. The skin and mucosae form a first line of defense. A second line of defense, also part of the innate system, is composed of phagocytes and other destroying cells. Other nonspecific defenses of the second line include antimicrobial proteins, inflammation, and fever. Both these lines of defense—sometimes called external and internal—are broad-based, relatively unsophisticated ramparts against bodily invasion.

Although in a sense they are less powerful than the body's adaptive immune system, they are able to keep out a large number of potentially harmful agents and thus save the adaptive system for only the most troublesome of pathogens.

Barriers

Let's take a closer look at how the skin keeps out invaders. Composed of tough squamous epithelial cells, the epidermis or outside layer of the skin forms a strongly protective sheath against microbes and environmental toxins, while keeping vital moisture and nutrients in. A keratinized epithelial membrane, the skin, like hair and nails, is heavily endowed with the protein keratin, which is impervious to weak acids and bases, as well as bacterial enzymes. Sebaceous glands in the skin produce an oily substance called sebum, which films the skin; the unsaturated fatty acids in sebum make

Powerful magnification zooms in on cells from the lining of the nose. Inside the nasal cavity, they secrete mucus to trap foreign particles.

DID YOU KNOW?

A clear saline fluid, tears are secreted by lacrimal glands located above each eyeball and spread over the eye when the eyelid blinks. Tears drain via the lacrimal canal into the nose. They flow as a reflex response to irritants, bright lights, spicy foods, and heightened emotions. The act of blinking is both voluntary and reflexive. In addition to washing away foreign substances, tears fight infection with lysozyme, an enzyme that dissolves the coating of many bacteria.

the skin partially water-resistant and restrict the growth of some bacteria and fungi.

Fatty acids and other skin secretions such as lactic acid are acidic enough to prevent bacterial proliferation. Sweat not only helps keep the body cool, but also washes the skin and contains the enzyme lysozyme, which breaks down the cell walls of some bacteria. Regular shedding of dead skin cells helps cleanse the body of bacteria and other pathogens. These invaders get no farther than the surface of the skin, unless they can find an entry point, as through a cut, burn, or body orifice, for example.

Body cavities opening to the outside—the respiratory, digestive, urinary, and reproductive tracts—are not lined with skin, but with mucous membranes. These mucosal layers secrete

EXTERNAL DEFENSES

DEFENSE	METHODS
Skin	Physical barrier against microbes; secretes perspiration and sebum, which contain anti-bacterial chemicals; keratin in skin cells guards against acids, bases, and bacterial enzymes
Mucous membranes	Though not as impervious as skin, still prevent entry of various microbes
Mucus	Secreted by mucous membranes; traps pathogens and foreign particles in respiratory, digestive, and urogenital tract
Nasal hairs	Coated with mucus; help trap dust and other foreign substances
Cilia	Lining the respiratory tract, sweep out microbes and dust
Tears	Dilute and wash out eye irritants; contain lysozyme
Saliva	Cleanses teeth and mouth of microbes; contains lysozyme
Urine	Flowing action clears urethra of microbes
Vaginal secretions	Acidic fluid inhibits bacterial and fungal growth
Gastric juice	Contains hydrochloric acid and enzymes to kill pathogens in the stomach
Vomiting and defecation	Expel pathogens from gastrointestinal tract

mucus, a viscous fluid that lubricates and moistens the body cavity lining. Since it is sticky, it traps a large number of germs. The respiratory tract is further protected by hairs in the nose that screen out microorganisms, dust, and many other airborne pollutants. Those invaders that make it past the nose face a gauntlet of tiny cilia in the cells of the respiratory tract. These microscopic hairlike projections continually sweep mucus-trapped dust and microbes up to the throat where they can be coughed or sneezed out. Another trap the body secretes is earwax, which prevents microbes, insects, and other foreign bodies from invading the ears.

The mouth is protected by secretions of saliva, which wash away bacteria from the teeth and oral cavity, preventing them from colonizing and gaining a foothold. Similarly, tears cleanse the eyes, whether you cry or not—blinking does the job. Both saliva and lacrimal fluid (tears) contain the bacteria-destroying enzyme lysozyme.

Likewise, the stomach protects itself by secreting concentrated hydrochloric acid and protein-digesting enzymes, which are toxic to many harmful microbes. While these agents are at work, a larger, mechanical process often takes over in cases of stomach bugs or food poisoning. Reverse peristalsis, or vomiting, physically rids the body of pathogens, as does diarrhea or normal defecation. Urination has a cleansing function in addition to its obvious job of ridding the body of waste—it scours the urethra, expelling any bacteria attempting to lodge there. Likewise, secretions from the vagina make for a cleansing tide, and the

White blood cells—rounded with small projections on the surface—play an important role in the body's immune response.

BREAKTHROUGH ANTISEPTICS

Before the introduction of anti-sepsis, up to 50 percent of surgery patients died from infection. Knowing French chemist Louis Pasteur had proved that disease was spread by microorganisms, English surgeon Sir Joseph Lister began using carbolic acid as a wound dressing in 1865. The acid, a sewer cleaner, reduced infections so dramatically that in four years mortality in his ward dropped from 45 to 15 percent. Realizing that germs were not just airborne, he used antiseptics to sterilize surgeon's hands and instruments. For his pioneering work, he was made a baron, the first doctor elevated to peerage.

acidity of vaginal secretions is yet another shield against bacterial growth.

Internal Defenses

The pathogens that get past the skin and mucosae encounter a second line of defense, the internal innate defense. This system includes cells, antimicrobial proteins, inflammation, and fever. Here we will examine the two kinds of cells of the innate system involved: phagocytes and natural killer cells (or NK cells).

The phagocytes are cells that eat microbes and other particles. The main phagocytes are macrophages, the big guns of the innate immune system. Literally "big eaters," macrophages develop from monocytes, white blood cells that exit the bloodstream and grow into macrophages in the body's tissues. Wandering macrophages move about in search of infections, where they feast on microbes and debris. Fixed macrophages live in such organs as the liver and the brain and stand ready to swallow up invaders. Each of these fixed macrophages has a name—Kupffer cells in the liver, microglia in the nervous system, histiocytes in the skin, tissue macrophages in the spleen, and so on.

After macrophages, the most important kind of phagocytes are neutrophils, which rank as the most abundant type of white blood cell.

Another kind of white blood cell, the eosinophil is not a heavy eater as phagocytes go, but it is important in killing worms and other parasitic invaders.

Phagocytes kill their prey in a five-step process called phagocytosis. In the first step, chemotaxis, phagocytes sniff out their prey, lured by chemical signals from the offending microbes, damaged tissues, or white blood cells. Then, in adherence, the phagocyte attaches itself to the microbe or other particle. This can only happen if the phagocyte can read the surface structure of the pathogen. A pathogen with a complicated structure or protective capsule, such as the pneumococcus bacterium, is not easy for phagocytes to latch onto. During ingestion, the third phase, the phagocyte surrounds the microbe by extending projections called pseudopods that engulf the microbe in a sac called a phagosome.

The fourth step is digestion. In this key stage, organelles called lysosomes within the phagocyte's cytoplasm join with the phagosome and induce metabolic production of hydrogen peroxide and other highly toxic oxidants, produce a bleach-like chemical with chloride ions, and release lysozyme and other digestive proteins destructive to pathogens. Lysozyme is an enzyme that breaks down the invader's cell walls. Other digestive enzymes within the lysosome act on proteins, carbohydrates, lipids, and nucleic acids.

In some cases, such as with the tuberculosis bacillus, the pathogen inhibits the joining of the lysosomes with phagosomes and simply multiplies within the phagocyte. This is when the cavalry shows up—cells of the adaptive immune system come and stimulate the release of more powerful chemical mediators that release cell-killing free radicals such as nitric oxide. In the final step in phagocytosis—the killing—the enzymes and oxidants take the microbe apart; residual bodies of indigestible material leave the cell through exocytosis. Neutrophils are such thorough killers that they take themselves out in the process.

A minor cut on a finger invokes the inflammatory response, which begins the healing process.

DID YOU KNOW?

INFECTED WOUNDS

A few days after the body's response to an infected wound, an aggregate of dead neutrophils, tissue debris, and live or dead microbes form in the wound. This creamy yellow fluid is called pus. Normally the pus either oozes out onto the skin or drains into an internal cavity. When it is sealed into a pocket of collagen fibers, it becomes an abscess that may need to be surgically drained.

The other family of cells involved in the innate immune response are the natural killer cells, which compose some 5 to 15 percent of all lymphocytes. Like the phagocytes, they also reside in the spleen, lymph nodes, and red bone marrow. These helpful assassins can terminate a number of infectious microbes and cancer cells, zeroing in on cells that have inadequate or abnormal plasma membrane proteins called major histocompatibility complex (MHC) antigens. We will take a closer look at MHC molecules later; for now suffice it to say that each person has a unique MHC signature, allowing for "self-recognition." Foreign cells without proper identification are thus subject to destruction by killer cells.

Part of a small cadre of large granular lymphocytes, natural killers are indiscriminate, or nonspecific, in their killing, earning them the adjective "natural." Instead of engulfing microbes as do phagocytes, natural killers shoot their targets full of holes: They attack the plasma membrane of a target microbe with chemicals called perforins, which make the microbe leak to death. They can also kill by releasing into target cells molecules that lead to apoptosis—programmed cell death in which the membrane stays intact but the nucleus and cytoplasm shrink, the DNA breaks apart, and the mitochondria quit functioning.

Inflammation

When you injure yourself—say you sprain your ankle, or cut or burn your hand—what happens? The affected area reacts with an inflammatory response, just the same as it does when invaded by microbes. Inflammation is yet another weapon in the body's innate defense arsenal, and there is more to it than meets the eye.

Any tissue damage, no matter what the cause, triggers the body's inflammatory response. Physical trauma, heat or chemical burns, infection by bacteria, parasites, viruses, and fungi—all these elicit inflammation. The four signs that signal inflammation are pain, redness, heat, and swelling. Loss of function is sometimes considered a fifth sign—the immobilized area (a joint, for instance) is forced to rest and thus can heal more quickly. Inflammation is not simply a reaction. It serves three main purposes: It rids the injured area of harmful pathogens and cell debris; it prevents pathogens and toxins from spreading; and it prepares the area for tissue repair.

Inflammation occurs in three stages—vasodilation and increased permeability of blood vessels, emigration of phagocytes, and tissue repair. The first thing that happens at the site of an injury is vasodilation of arterioles: Vessels dilate so that they can carry more blood to the injured area. At the same time, capillaries become more permeable, allowing antibodies, oxygen, nutrients, and clotting proteins to swarm the inflamed area. The extra flow of blood helps wash out microbes and dead cells; it's the body's first way of cleaning itself. Think of this stage as the

KEEPING HEALTHY
HEALING AFTER SPRAINS AND STRAINS

Serious athletes and weekend warriors know the danger of sprained joints and strained muscles. While inflammation occurs when you pull or tear a muscle, you can speed up healing by following the first-aid procedure of R.I.C.E: rest, ice, compression, and elevation. Rest to prevent further injury and allow the injured area to begin healing. Apply ice packs for up to 20 minutes to reduce swelling, pain, and muscle spasms. Compress by lightly wrapping the affected area with an elastic bandage. Finally, elevate the injury above the heart to further reduce swelling. Swelling helps by diluting toxic substances, but since muscle strains do not introduce foreign microbes, swelling can be counterproductive by slowing down the healing process, causing pain and spasms. If you suffer a serious sprain or any injury that causes pain for more than 48 hours, see your physician or visit the emergency room.

emergency medical team roaring down the highway to the accident site.

What causes vasodilation and increased permeability? How do the vessels know when they need to dilate and become leaky?

DID YOU KNOW?

WARNING SIGNS

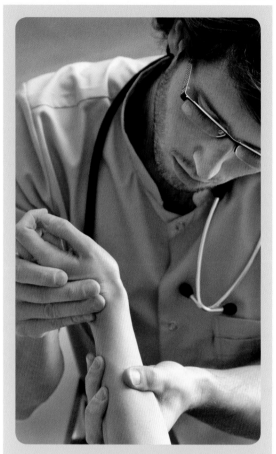

+ **Pain**: from neuron injury, toxic chemicals from microbes, vasodilating chemicals, and edema

+ **Heat**: from increased blood flow in affected area

+ **Redness**: also from increased blood flow in affected area

+ **Swelling**: from greater permeability of local blood vessels

Several chemicals are released after an injury, stimulating vasodilation. Let's take a closer look at the major ones. When an injury occurs, mast cells, platelets, and white blood cells sound an "alarm" by releasing histamine, causing vasodilation and leaky blood vessel walls. In addition, kinins stimulate chemotaxis (the attraction of leukocytes to microbes), and they make neutrophils release lysosomal enzymes, which help produce even more kinins. Kinins also induce pain. Another group of chemicals, the prostaglandins heighten the effects of histamine and kinins, and they produce free radicals, which themselves can trigger the inflammatory response. Prostaglandins may also prompt the second stage of inflammation—emigration of phagocytes from the blood into interstitial fluid. Leukotrienes stimulate increased vascular permeability, and they attract phagocytes and help them adhere to pathogens. Complement protein, a group of about 20 plasma proteins, enhance phagocytosis, lyse microorganisms, stimulate the release of histamine, and attract neutrophils.

Besides increasing the flow of blood and associated antibodies, vasodilation and increased vascular permeability cause three of the four symptoms of inflammation—heat, redness, and swelling. The localized increase of blood and metabolic activity raises the temperature at the injury site and accounts for the redness (erythema) beneath or within the epidermis. The swelling (edema) occurs when the more permeable capillaries leak fluid into the interstitial spaces and surrounding tissues. The fourth symptom, pain, results from damaged nerve endings and the toxic chemicals of microbes. Kinins can produce inflammatory pain, and prostaglandins can intensify and prolong it. Pressure from swelling also induces pain.

About an hour after the inflammatory response has begun, phagocytes arrive and start emigrating from the blood vessels into the damaged tissues. This emigration process is caused by chemotaxis. Once out of the vessels, neutrophils (a kind of phagocyte) go to work eating up microbes. The bone marrow keeps a steady supply of neutrophils coming, increasing their numbers in the bloodstream four to five times during times of bacterial and certain fungal infections and with severe tissue injury. The neutrophils are followed shortly by monocytes, which within 8 to 12 hours develop into macrophages, the serious microbe-devouring professionals. After a few days the phagocytes die off, and they and the dead tissue cells clump together to form pus (see "Did You Know?," page 330). Blood-clotting agents have in the meantime been leaking from the capillaries and weaving a mesh of gel-like fibrin threads, which prevents the spread of microbes. This mesh, along with the work of the busy phagocytes gobbling up microbes and debris, sets the stage for tissue repair.

While inflammation is a localized innate response, a more general response is fever. During a microbe invasion, leukocytes and macrophages release pyrogens, chemicals that reset the body's thermostat above the normal temperature of 98.6°F (37°C). A moderate fever helps inhibit microbe multiplication, but the mechanism for this is still not well known. By speeding up metabolic activity, fever can also promote healing.

In the immune response, neutrophils like this one (shown in purple) trap and destroy bacteria (yellow).

ADAPTATION

Acquiring Specialized Defenses

Greeks in the fifth century B.C. noted that people who survived the plague never suffered from it again. Somehow their bodies remembered the disease and were resistant to it in the future. What the Greeks observed was the body's acquired or adaptive immune system in action. Unlike the broad attack of the innate defense system, the adaptive system attacks pathogens with specific, made-to-order antibodies and cells.

The phagocytes of the innate system don't discriminate between harmful agents; the lymphocytes (white blood cells) and their derivative antibodies do. In fact, the antibodies and lymphocytes of the adaptive system can recognize millions of different pathogens, some not even found in nature, and they can distinguish infected, cancerous, or foreign cells in an organ from normal cells of the same type.

DID YOU KNOW?

DISEASE VECTORS

Broadly speaking, a disease vector is a vehicle or pathway through which an infectious disease spreads; specifically, it is an organism that transmits a pathogen. Malaria is spread through the mosquito vector—mosquitoes can carry the malaria parasite and transmit it to the blood of a human host via a bite. Similarly, lice carry the typhus parasite. Control of these diseases is a matter of eliminating the middle man—the vector.

Each antibody and lymphocyte recognizes only one kind of antigen (a substance that provokes an immune system response). The receptors on the surface of each lymphocyte recognize the specific chemical structure of the antigen and fit the antigen like a key sliding into a lock. The body must be primed by an introduction to these antigens. Then the old adage applies: That which does not kill you makes you stronger. The adaptive immune system manufactures the necessary antibodies and cells so that upon re-exposure you can fight back and win. Thus, unlike the innate system, the adaptive system is a highly specialized fighting force with a "memory."

From stem cells within the bone marrow, some one million lymphocytes are created every second. At any given time, billions of lymphocytes are moving through your bloodstream and tissues, each of them seeking out the antigen that fits its receptor. Since there are millions of possible antigens, the body may hold only a few of each type of lymphocyte programmed to attack each type of antigen. Upon finding a match, the lymphocyte rapidly multiplies. Immunity in the adaptive system is granted system-wide, throughout the body, instead of just at the site of infection.

The two kinds of adaptive immunity are humoral and cell-mediated. Humoral immunity is that provided by antibodies, which circulate in bodily fluids or "humors," especially blood and lymph. These antibodies, also called immunoglobulins, are protein molecules produced by B lymphocytes and their progeny, plasma cells. Antibodies seek out bacteria, viruses, and other pathogens, deactivate them, and tag them for destruction by phagocytes. In cell-mediated immunity, the lymphocytes themselves go after cells infected with viruses, cancerous cells, or foreign cells; after that they either lyse the cells or summon macrophages or other killers.

Plasma cells are a type of white blood cell. They produce antibodies that seek out and deactivate bacteria, viruses, and other pathogens.

Fighting Cells

The three kinds of cells of the adaptive immune system are the T and B lymphocytes and the antigen-presenting cells. T lymphocytes, or T cells, are the engines of the cell-mediated wing of the adaptive immune system. They comprise 65 to 85 percent of all bloodborne lymphocytes. Unlike cells of the humoral wing, T cells produce no antibodies but directly destroy microbes. All blood cells, lymphocytes included, come from stem cells in red bone marrow. All lymphocytes start life practically identical; as they mature, they differentiate into T and B cells. The earliest T cells multiply in the thymus (hence the name T, for thymus-derived), and then mature over a two- to three-day span. T cells that poorly recognize antigens do not survive, nor do those which strongly bind to self-antigens (surface proteins of body cells). The latter process, negative selection, prevents the immune system from attacking the body.

The B cells (from "bone-marrow derived") mature in the bone marrow. Comprising the humoral wing of the adaptive immune system,

CREATION OF T, B, AND NK CELLS

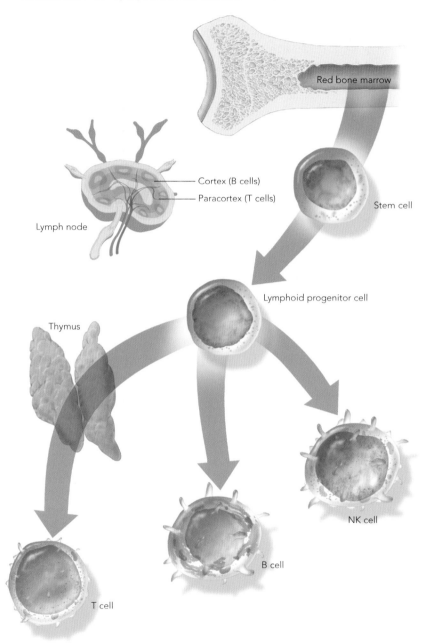

Red bone marrow

Cortex (B cells)

Paracortex (T cells)

Lymph node

Stem cell

Lymphoid progenitor cell

Thymus

NK cell

B cell

T cell

After the T and B cells become immunocompetent they have one of 10,000 to 100,000 possible chemical receptors on the surface of each of their cells. The receptors of one cell are capable of binding to only one specific antigen. For instance, the hepatitis A virus and each of the more than 200 kinds of viruses that cause common colds are recognized by only one receptor combination on a lymphocyte.

Though the body makes the necessary lymphocyte force to fight a particular antigen, scientists believe that the body cannot create a new immune cell receptor upon meeting an antigen. Rather, the body is genetically programmed to produce all the possible receptors it might need. Thus genes, not antigens, dictate what your body can recognize and resist. Many of the kinds of lymphocytes you make will never be needed in the environment you inhabit; they simply stand on reserve for life.

Upon reaching immunocompetence, T and B cells move to the lymph nodes, spleen, and other secondary lymphoid organs. At this point, the T and B cells are not fully mature because they have not

B cells produce antibodies—protein molecules also called immunoglobulins—that provide defense against extracellular microbes and other foreign substances.

Dendritic cells, above, help match
T cells with antigens.

matched with antigens. In the secondary lymphoid organs they can bind with antigens and become activated, fully functional T and B cells.

The antigen-presenting cells (APCs) swallow microbes and other antigens and then display a piece of them like signal flags to T cells. Thus they present the antigen to the lymphocyte. The major APCs are dendritic cells, macrophages, and activated B cells. B cells can be either antibody producers or antigen presenters. It is also important to understand that APCs don't simply signal T cells; they help activate them. In turn, T cells then crank up the maturation and mobilization of APCs.

Lymphocytes circulate throughout the body and thus increase the likelihood they will encounter antigens. Once they are needed in a specific area, homing signals call them into the appropriate tissues. Some lymphocytes hang out continuously in lymph nodes, waiting for lymph capillaries called lymphatics to bring pathogens by. Either strategy—circulating or standing in one key place—works well for encounters. Bloodborne antigens are often encountered in the spleen; invaders of the mouth and nose are often met by lymphocytes and APCs in the tonsils.

Adaptive Responses

After maturing in the bone marrow or thymus, lymphocytes travel to all parts of the body via the bloodstream. Certain sites, though, have a special strategic importance. Within the skin and internal mucosa, there are lymphocytes lying in wait for any microbes that attempt to invade. Then the lymph tissue itself houses vast numbers of lymphocytes. Within the tissues of lymph organs, including lymph nodes, lymphocytes multiply and stand ready for activation or transport to other sites.

Lymph nodes filter lymph as it travels back to the bloodstream. Located in the neck, under the arms, in the groin, and within the chest and abdominal cavities, these half-inch-long, bean-shaped organs lie along the network of lymphatic vessels. They keep microorganisms from entering the blood and help activate the immune system by spotting antigens. Lymph nodes can become swollen and tender when they detect invaders (like bacteria or viruses) and become activated. Once the nodes have cleaned the lymph, the fluid is carried through

DID YOU KNOW?

AGING CELLS

Before birth, the human body produces huge numbers of T cells and B cells. But as we age, the production slows down, and the thymus begins to atrophy. It is unclear why the immune system breaks down as we age, but because it does we become more susceptible to diseases such as cancer. Even though the thymus diminishes, T cells are replaced by cell division outside the thymus and thus cell-mediated immunity continues.

the lymph vessels, a one-way system that returns fluid to the heart; up to three quarts a day of interstitial fluid is thus returned to the bloodstream.

The tonsils are lymph organs positioned around the tongue and throat that entrap microbes from food, water, and air and begin activation of B and T cells to initiate an immune response. Some cells in the tonsil also produce signaling chemicals, called cytokines, that help orchestrate other parts of the immune system. The tonsils and adenoids form a ring of lymphatic tissue. The tonsils in effect "invite" infection by trapping bacteria, but for this risk the body gains a memory of harmful antigens.

The largest lymphoid organ is the fist-size spleen. As with the lymph nodes, the spleen acts as a site for lymphocyte proliferation and immune surveillance and response. Here also, old and damaged blood cells are destroyed. The spleen saves the iron from destroyed red blood cells for recycling in the bone marrow, stores blood platelets, and produces blood cells in fetuses. If the spleen ruptures (from infection or injury) or is removed surgically, the bone marrow and liver pick up some of its duties, though a person with no spleen is more vulnerable to some bacterial infections.

A humoral immune response begins with an antigen challenge. In the lymphatic tissue—usually the spleen or lymph nodes—the

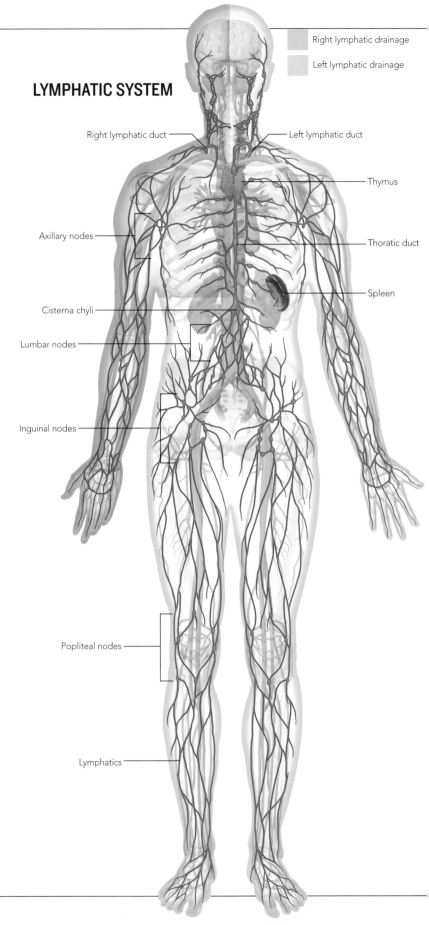

LYMPHATIC SYSTEM

Right lymphatic drainage

Left lymphatic drainage

Right lymphatic duct

Left lymphatic duct

Thymus

Axillary nodes

Thoratic duct

Cisterna chyli

Spleen

Lumbar nodes

Inguinal nodes

Popliteal nodes

Lymphatics

invading pathogen or other antigen encounters an immunocompetent but inactivated B lymphocyte. The antigen is detected by antibodies found on the surface of B lymphocytes. Antigen-presenting cells similarly ingest the pathogen or antigen, process it, and present the processed antigen to helper T lymphocytes programmed to recognize the antigen using a customized surface receptor. At this point the antigen-bearing B lymphocytes and antigen-presenting cells induce the helper T lymphocytes to produce cytokines, which stimulate the antigen-bearing B cells to proliferate and evolve into antibody-secreting plasma cells, each capable of producing 2,000 antibody molecules per second. The antibodies seek out antigens and tag them for destruction. Within five days the plasma cells die.

Instead of evolving into plasma cells, some of the antigen-specific B cells become memory cells. These important players can launch a humoral immune attack upon recognizing an antigen in the future. The memory cells thus provide immunological memory. While a primary humoral response can take three to six days to build up enough antibodies to begin fighting an invasion, a secondary response—one that relies on immunological memory—is much quicker and stronger. Within hours an army of plasma cells is mobilized; in two or three days the antibody level is higher than it gets at the ten-day peak of a primary response. Antibody levels then remain elevated in the blood for weeks or months.

Cell Mediation

The antibodies of the humoral immune system provide only a simple, limited defense against pathogens. What they leave unfinished, the cell-mediated division picks up. Whereas antibodies only deactivate antigens or tag them

DID YOU KNOW?

THE SPLEEN

Largest of the lymphoid organs, the fist-size spleen lies on the left side of the abdominal cavity beneath the diaphragm. The spleen serves as a base for lymphocyte proliferation, as well as immune surveillance and response. In addition, it helps clean the blood of worn-out and damaged red blood cells. At the same time, the spleen saves the iron from destroyed red blood cells for future use by the bone marrow; it stores blood platelets; and it produces blood cells in fetuses.

for destruction, the activated T cells (along with cells from the innate system) actually do the killing. T cells may also take aim on body cells, either cancerous or foreign cells in a transplanted organ. Since T cells cannot distinguish free-floating antigens, they rely on antigen-presenting cells to process antigens first and then present them to the T cells to begin the cell-mediated response.

Several kinds of T lymphocytes make up the cell-mediated wing of the adaptive immune system. The helper T cells are crucial links in the immune response. Once it is activated by the antigen on an antigen-presenting cell, a helper T cell then begins to stimulate the production of more T cells and B cells. The helper T cells are the directors of the cellular immune response, a job that they perform by both direct contact with other immune cells and by the release of chemicals called lymphokines or cytokines, proteins that enhance the immune and inflammatory responses. Lymphokines mobilize a formidable army of cells into action against the antigens—more T cells, B cells, macrophages, and other white blood cells heed the call and muster in overwhelming numbers to the battlefield.

DID YOU KNOW?

VIRUSES

Neither plant or animal, a virus is something between a simple microorganism and a complex molecule. As parasites, viruses, below, cannot reproduce or live outside a host cell. Viruses consist of nucleic acid (either DNA or RNA) wrapped in a protein sheath. They are pieces of bad genetic code in a protective protein wrapper, causing disease by entering a host cell, taking over the cell's machinery, replicating, destroying the host cell, and spreading to other cells.

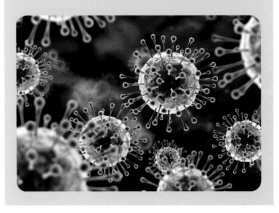

Cytotoxic T cells, or killer T cells, are the only T cells that attack and directly destroy other cells. Once activated by an antigen, a cytotoxic T cell is recruited and pumped up by a helper T cell. It then begins stalking the blood and lymph systems and lymphatic organs, hunting down cells with the antigen it is programmed to destroy. These are usually virally infected cells, cancer cells, or cells from transplanted organs. They may also hunt down tissue cells infected by certain bacteria and parasites. In this way they stop the invasion of microbes that are hidden from antibody detection within host cells.

To kill their prey, the cytotoxic T cells first have to dock onto the target cell. All body cells have a surface antigen ID unique for each person (except for identical twins). Called the major histocompatibility complex (MHC) antigen, this ID flashes its "self" identity. An infected or cancerous cell flashes a fake ID or an inadequate self ID. This fake or inadequate ID is what the cytotoxic T cell binds to. The mechanism by which it kills the bad cell is not completely understood, but in many cases the T cell releases a chemical called perforin into the target cell. Perforin helps destroy the target cell by creating pores in its membrane, allowing granzymes to enter the cell and begin the process of cell death.

Once the intense battle against the antigens is drawing to a close, something has to declare victory and call off the winning force. That's where the suppressor T cells come in. The suppressors slow down B-cell and T-cell activity. Without these specialized T cells, the immune system would keep turning out more and more lymphocytes, like a Hydra that cannot stop itself. It is likely that suppressor T cells play an important role in prohibiting the autoimmune response—the production of lymphocytes that attack a person's own healthy tissues.

As with B cells, there are also memory T cells. Clones of T cells remaining after a cell-mediated immune response carry immunological memory and are thus termed memory T cells. At any given time there will be thousands of memory T cells available the next time the same pathogen invades the body. These cells will then react swiftly and efficiently to halt the invasion, generally before the infection has had time to produce symptoms.

In summary, the various kinds of T cells work with each other and with other immune cells to launch attacks on invading pathogens. With helper T cells directing the battle, the cytotoxic T cells kill cells hosting viruses and other microorganisms, as well as cancerous or transplanted foreign cells. The helpers also enlist B cells, and together they stimulate the body's innate defenses against invaders that are hidden away in infected cells or free-floating in blood and tissue spaces. The suppressor T cells call the action off when the antigen threat is neutralized. Meanwhile, memory T cells are spawned as cell clones, ready for the next attack of the same antigen.

Immunization

When your B cells have encountered and acted against antigens, your immune system is showing active humoral immunity. This immunity is exhibited whether the infection is naturally acquired through a microbe invasion or artificially acquired through a vaccination. In both cases, antibodies are produced against a specific antigen; helper T cells then organize an attack on the antigen. Vaccination with dead or weak pathogens, or more recently with genetically engineered fragments of microbial antigens, simply saves us the suffering that we might otherwise go through to achieve active humoral immunity. With immunity from natural infection, memory B and T cells are created during the primary response—the first exposure to the pathogen. While the primary response may take days to build up to its maximum strength, the secondary response generally takes only a few

To protect the body, T lymphocytes (shown in pink) attach to the surface of this invading cancer cell.

DEFENSE DESTROYER

One of today's most serious diseases, acquired immune deficiency syndrome (AIDS) targets the body's immune system and cripples its ability to defend itself from infection. First reported in 1981, the disease is thought to have originated in Africa, possibly first in monkeys.

Since then, the disease has spread around the world, through sexual contact, sharing of contaminated needles by intravenous drug users, and, before effective blood-screening methods were developed, through transfusion of contaminated blood and blood products. First recognized when young homosexual men reported strange cases of pneumonia and skin cancer, the disease is now primarily spread through heterosexual contact.

Currently, more than 35 million people worldwide have HIV/AIDS. Some 2.3 million new cases of HIV infection occur each year. Since 1981 more than 36 million have died from AIDS, and nearly 2 million more die each year. About 70 percent of all AIDS cases are in sub-Saharan Africa. In the United States, upwards of 1.1 million people are infected with the virus that causes AIDS.

CAUSES

HIV (human immunodeficiency virus) is the virus that causes AIDS. It is transmitted through blood and body fluids, primarily semen and vaginal fluid. During unprotected sex contaminated fluids make contact with mucous membranes lining the vagina, penis, rectum, or mouth, but in order to enter the bloodstream, the virus has to gain access through an abrasion or cut in the membrane. The virus can also be passed along by a contaminated needle or syringe, such as

Covering the National Mall in Washington, D.C., the AIDS Memorial Quilt commemorates the lives lost to the disease.

those shared by intravenous drug users. Before routine and effective blood screening for HIV was initiated in the mid-1980s, people receiving transfusions were at risk. The virus can also be spread through the placenta or breast milk of a mother. It is not spread by coughing or shaking hands, and although saliva, sweat, and tears can carry the virus in low levels, it is not believed that transmission occurs through them.

ATTACKS

Once inside the body, HIV attacks the immune system's helper T cells, also known as CD4+ cells because of the surface protein CD4. The helper T cells orchestrate the body's entire immune response to pathogens, activating B cells, macrophages, and other T cells. HIV binds to the CD4 molecule on the surface of the helper T cell, which allows the virus entry into the cell's cytoplasm. Once inside the cell, the virus sheds its protective protein coat and hijacks the cell's engines to make a DNA copy of its own ribonucleic acid (RNA), which is then incorporated into the host cell's DNA. From then on, every time the cell divides, it adds a copy of the viral DNA. Furthermore, the virus can force the host to make millions of copies of viral RNA, as well as protein coats. In effect, one virus can enslave a cell to make an army of viruses. In a devious twist, the production of HIV is often imperfect. Thus, mutations crop up frequently as the virus replicates, allowing it to evolve rapidly and keep ahead of antiviral immune defenses, drugs, and potential vaccines.

It takes about a day for an infected T cell to die. The immune system responds by helping to kill infected cells, but many of the virus particles move into other T cells. The body continues to produce more T cells, but it is a losing battle: Over time more are lost than replaced.

The first stage of HIV infection generally results in a short flu-like illness. Symptoms include fever, fatigue, swollen lymph nodes, rash, sore throat, and joint pain. At this point the levels of antibodies against HIV are still so low that routine blood tests may not yet show evidence of the infection. But within a few weeks, the immune system has created enough antibodies that a test can reveal HIV-positive status. For several months to years the HIV virus will then continue to replicate, and the afflicted person will show no symptoms.

After the helper T cells decline to about 20 percent of normal levels, opportunistic infections begin occurring that the body cannot fight. The patient now has AIDS. Among the more common infections are a lung disease caused by the parasite *Pneumocystis carinii,* tuberculosis, herpes simplex, and toxoplasmosis (which affects the brain). Additionally, cancers such as lymphoma and Kaposi's sarcoma (causing skin and other lesions) are frequent, as is dementia. One or more of these infections or cancers is likely to kill the patient.

TREATMENTS

There is no cure or widely effective vaccine yet for HIV infection. But there are drugs that can slow its progress, and researchers continue to develop new treatments. Some drugs, including AZT (zidovudine), inhibit the action of the reverse transcriptase enzyme that enables the viral RNA to be copied. Another group of drugs interferes with the enzyme protease, which helps make protein coatings for the HIV viruses.

Since 1996, many physicians have been treating HIV cases with a "triple therapy" called HAART (highly active antiretroviral therapy), which calls for a combination of two or more reverse transcriptase inhibitors and one protease inhibitor. This triple cocktail has met with success in drastically reducing the number of viral particles in an infected person. The downside is that it cannot completely eradicate the virus, and some people cannot withstand its toxicity. Also, at a cost of more than $10,000 per year, the therapy is often beyond the reach of many people in the world who need it.

Education about the dangers of unprotected sex has often proved to be the best hedge against AIDS. In Uganda, awareness campaigns begin in schools as early as kindergarten, with songs and posters aimed at getting children to realize their ability to stop the AIDS epidemic. In Senegal, prostitutes must be licensed and routinely tested for HIV. Other West African campaigns warn men during soccer matches. National campaigns in other countries still face an uphill climb in battling superstitions about the disease.

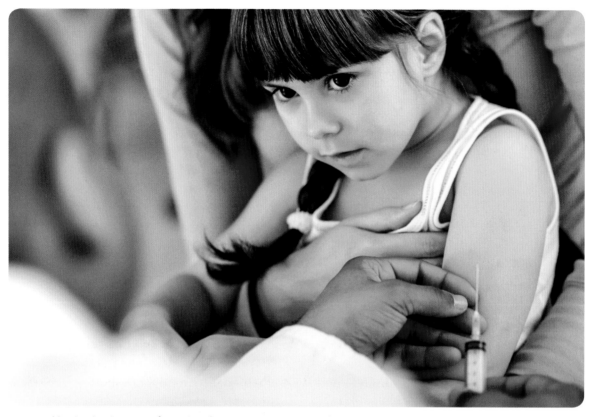

Vaccination is a way of spurring the immune response and protecting the body from certain diseases.

hours. A secondary response is any re-exposure to the pathogen after the primary response. In this case, the humoral and cell-mediated arms of the immune system are already primed by the existence of thousands of memory B and T cells. When they recognize the same old antigen, they go into immediate action, literally dividing and conquering—that is, they multiply and differentiate into plasma cells (antibody-producing B cell progeny), cytotoxic T cells, and so on.

Secondary responses are quicker and more potent: The antibodies produced on reexposure to a pathogen can be hundreds of times greater than during the primary response. These antibodies exhibit more affinity with the antigens—they bind more readily and tightly to quickly overcome them.

Memory cells can remain in a person for decades. Of course, the original ones die, but their progeny maintain a steady population, a vigilant unit of trained guards. Often with a secondary infection with the same pathogen the response is so fast and overwhelming that you show no symptoms of the infection.

Vaccinations rely on the body's immunological memory. A small amount of killed or attenuated microbes, or portions of microbes' antigens, are introduced into your body. The vaccination is enough to activate your B cells and T cells and produce memory cells. The body can then handle natural infections without ill effects. Booster shots are often given after a period of several years to reinforce the immunological memory gained by vaccination.

Early efforts at immunization were often desperate attempts to stop epidemics, such as smallpox. Crude preparations of live attenuated microbes or microbe components closely mimic a real infection and thus lead to long-lasting protection. They may cause mild symptoms; immunocompromised patients may even suffer severe infections. In recent years, recombinant DNA technology, or genetic engineering, has helped refine vaccinations by producing purer forms of microbial antigens.

Passive Immunization

Like active immunity, passive humoral immunity can also occur in two ways. First it can be acquired naturally, when a mother's antibodies pass through the placenta to a fetus and in breast milk to her baby. Or it can be acquired artificially by injection with an "antiserum" containing gamma globulin. Such borrowed protection is short-lived. Antibodies are simply introduced; since there is no antigen challenge, there is also no active immunological response, and once the passively injected antibodies are lost, no new ones are produced. Travelers to foreign countries often get a shot of gamma globulin

A white blood cell (red) engulfs a tuberculosis vaccine (blue). The bacteria in the vaccine will prime the immune system without causing sickness.

to boost their immunity to a variety of infectious diseases. Other antiserums are used for emergency treatment of rabies, snake bites, tetanus, and botulism.

Let's take a look at two serious diseases, diphtheria and tetanus. A vaccine for diphtheria came out in 1923, one for tetanus in 1927. Most children in the United States today receive a series of five immunizations for these two diseases, in combination with a pertussis (whooping cough) vaccine: The DPT vaccine is administered at 2, 4, 6, and 18 months and before entering school. Then, the general recommendation is that people receive boosters for diphtheria and tetanus every 10 years. Clearly the adaptive immune system takes a lot of priming before it is fully immune to these two diseases.

Vaccines, of course, are active immunizations. But what happens if you have not been vaccinated or are not up on your boosters? Passive immunization can come to the rescue. If you step on a dirty nail, there is a chance that

DID YOU KNOW?

FIRST VACCINES

1798 Smallpox	**1945** Influenza
1885 Plague	**1955** Polio
1917 Cholera	**1963** Measles
1917 Typhoid	**1967** Mumps
1923 Diphtheria	**1969** Rubella
1927 Tuberculosis	**1982** Hepatitis B
1927 Tetanus	**1995** Hepatitis A

C. diphtheriae bacteria caused serious illness and death in children before the development of a vaccine.

tetanus bacteria from the ground could get deep enough into the wound that, deprived of oxygen, they could germinate and produce the tetanospasmin toxin. As its name suggests, this toxin causes muscle spasms, particularly in the neck and jaw. What starts as a minor stiffness could, if left untreated, develop into actual lockjaw and painful convulsions within a week or two. If the convulsions begin early, the bacteria have propagated rapidly, and your chances for recovery are not good.

For any possibly contaminated wound, you should consult your doctor. In the meantime, clean the wound as thoroughly as you can. If you have had a recent tetanus shot, you are probably in the clear. If not, you will probably be given a booster. If you have had no tetanus immunizations, your doctor may give you a shot of tetanus immune globulin along with a standard shot of tetanus vaccine. This immune globulin is made from blood that contains antibodies for tetanus. A form of passive immunization, the antiserum is not the same as that of your body producing its own antibodies, and thus it provides protection for only a few weeks. But that is usually enough to fight off the bacteria and get you past the dangerous stage.

Another bacterial disease, diphtheria is now very uncommon in the United States. The bacteria usually are inhaled in airborne water particles, coughed or sneezed out by someone infected with or carrying the disease. The bacteria incubate for up to a week. Then you develop a mild sore throat, fever, and malaise. After this, a thick gray membrane coats the throat and tonsils, and your breathing can become labored. More serious symptoms include vomiting, heart infection, and paralysis of the respiratory muscles. Death may even occur. Those with proper vaccinations will probably develop no symptoms. Without vaccinations you must rely on passive immunization by a diphtheria antitoxin. This emergency measure usually destroys the infection. During a diphtheria outbreak in Nome, Alaska, in 1925, antitoxin serum was delivered in a heroic down-to-the-wire dog-sled relay that took five and half days.

A more deadly disease, rabies (from the Latin for "madness") is caused by a virus; untreated, it is almost always fatal. The rabies virus is usually transmitted through the saliva of an animal during a bite. The virus generally incubates for three to seven weeks before symptoms appear, but they can appear as early as ten days or as late as two years later. The symptoms begin with tingling around the bite area; the skin becomes very sensitive, and the patient begins producing excessive saliva. Swallowing becomes difficult. Periods of intense anxiety and agitation ensue, followed by uncontrollable muscle spasms, convulsions, paralysis, and death. Since the disease is so dire, bites from wild animals are often treated as though for rabies. A combination of passive antibody and a vaccine is given in five separate doses, which can be very painful. The first rabies vaccine was invented in 1885 (see "Breakthrough," below).

BREAKTHROUGH RABIES VACCINE ⊢

After Jenner's landmark small-pox vaccine in 1798, another 87 years would elapse before the breakthrough in rabies immunizations. The scientist was Louis Pasteur, founder of microbiology, born in eastern France in 1822. From early on, he began applying his skills as a chemist to problems in real-life situations, including the wine, beer, and silk industries. Using his microscope, Pasteur discovered bacteria that were spoiling wine; mildly heating it killed the bacteria. Named in his honor, the process of pasteurization of milk and other perishable liquids soon became a standard precaution. After years of groundbreaking work with various animal diseases, Pasteur, right, turned his attention to the deadly disease rabies. During experiments with dogs, he found that by using dried tissue from infected animals he was able to isolate weakened viruses. In 1885, he used a solution of these viruses to save the life of a boy, nine years old, who had been bitten by a rabid dog. The boy developed no serious symptoms. With donations from

around the globe, the Pasteur Institute was founded in 1888. It evolved into one of the world's outstanding biological research institutions. Pasteur directed it until his death in 1895.

BODY ATTACKS

Allergies and Autoimmunity

As wonderful and nearly flawless as the immune system usually is, in some people it can be too sensitive. Pollen, animal dander, or other substances that are not intrinsically harmful to the human body are "read" by some immune systems as harmful antigens. The result is an allergic reaction, which might include sneezing, tissue inflammation, mucus secretion, and more serious symptoms.

In short, the body reacts against the antigen, this time called an allergen, the substance it perceives as a foreign invader. Allergies or hypersensitivities—though usually not life threatening—may cause tissue damage and thus need to be monitored and controlled.

Allergic Reactions

There are four main kinds of allergic reactions: anaphylactic, or immediate; cytotoxic; immune complex–related; and delayed. The first three are antibody-mediated, the last is a cell-mediated response.

In immediate hypersensitivities, the reaction begins within seconds of the allergen introduction and lasts for about a half hour. During the first encounter with an allergen, antigen-presenting cells eat the innocuous "invader" and present its pieces to T cells. The T cells then secrete chemical mediators or cytokines that tell B cells to grow into antibody-secreting plasma cells. Special antibodies called IgE bind to mast cells within tissues and to white blood cells known as basophils. The immune system is now primed and ready.

At all subsequent exposures to the same allergen, the antibodies bind with the allergen, and the mast cells and basophils release chemicals: histamine, prostaglandins, leukotrienes, and kinins. All of these combine to cause an inflammatory response: vasodilation, increased capillary permeability, bronchial muscle contraction within the lungs, and excessive mucus secretion. The affected person may have itching, watery eyes, a runny nose, and troubled

DID YOU KNOW?

ALLERGY

The term "allergy" was coined in the early 1900s from the Greek *allos* meaning "other," by Austrian physician Clemens Baron von Pirquet. In 1905 he observed that patients receiving a second injection of smallpox vaccine or horse serum had a quicker reaction than that produced by the first. He theorized that primed antibodies were reacting to antigens in the serum. His findings led to a skin test for tuberculosis: A bump indicated a positive test.

ALLERGIC RHINITIS

Immediate Reaction (in minutes)

What is rhinitis?

Rhinitis is the inflammation of the mucous membranes that line the nose. The name rhinitis stems from the Greek word *rhinos* meaning "of the nose." This inflammation increases the production of nasal mucus and typically makes breathing through the nose difficult. Rhinitis can result from infections, such as the common cold and viral infections, allergic reactions, and unknown causes. Other cases of rhinitis result from common outdoor allergens such as airborne tree, grass or weed pollens. This is known as "hay fever." Certain chronic (long lasting) forms of rhinitis may cause the mucous membranes to thicken or to wear away.

Antigen attaches to antibodies

Mast cell releases mediators

Immediate reaction
(in minutes)

After the immediate allergic reaction, the chemicals released by the mast cells also may produce a late (delayed) reaction in the airways. Inflammatory cells are attracted to the area and cause the airways to again narrow and swell. This delayed reaction can last much longer than the initial reaction.

Antigen

Mucus

Cilia

Antibodies attached to mast cell

Epithelial cells

Goblet cells

KEEPING HEALTHY

FOOD ALLERGIES

Although many people believe themselves allergic to certain foods, fewer than 2 percent of Americans are affected. An allergic reaction to a food causes the immune system to produce histamines and other chemicals in response to the allergens. Food sensitivities, such as lactose intolerance, sometimes result in symptoms similar to allergic reactions but are not caused by an antibody response, and there is no release of histamine. Children are more likely than adults to have food allergies, but they often grow out of them. About 90 percent of food allergies are caused by cow's milk, peanuts, egg whites, wheat, or soybeans. Some people may be allergic to shellfish, berries, tree nuts, or corn. Most food allergies are type I, or immediate, reactions, occurring within seconds of eating. Symptoms include swollen lips, tongue, and throat; nausea and vomiting; hives; and nasal congestion. It can be a life-threatening emergency.

breathing. Hay fever, or allergic rhinitis, is one of the most common examples: Millions of people are somewhat allergic to pollen and suffer allergic rhinitis (hay or fever have nothing to do with it). If the allergen is ingested rather than inhaled, the person may experience cramping, vomiting, or diarrhea.

The anaphylactic ("against protection") reaction is a bodywide, instead of local, immediate reaction. Usually this occurs when the allergen gets into the bloodstream right away and travels throughout the body. Some bee stings, spider bites, and injections such as penicillin may act this way, causing anaphylactic shock.

Since the reaction occurs all over the body, the results can be quite serious: Breathing may be impaired by constriction of the bronchioles in the lungs and swelling of the larynx; widespread vasodilation can lead to hypotensive shock; and death can occur in minutes. The hormone epinephrine, a vasoconstrictor, is often used as a life-saving antihistamine during episodes of anaphylactic shock.

Some immediate hypersensitivity reactions require no previous sensitization to the allergen. The tendency to have localized immediate reactions to allergens, called atopy, is hereditary. About 10 percent of the U.S. population has such reactions.

Cytotoxic, or type II, allergic reactions take up to three hours to develop; the effects may last for several more hours. Again, antibodies are reacting against antigens, but this time on the surface of cells. These antigens may be located on microorganisms, red blood cells, lymphocytes, platelets, or transplanted tissue cells. When the antibodies have attached to the antigens, the plasma proteins called complement are activated, and cells bearing the antigen are destroyed by lysis. These reactions happen when incompatible transfused blood is rejected by the body.

Immune-complex, type III, reactions occur when soluble antigens and antibodies form insoluble complexes. The antibody-antigen complexes can persist deep within walls of blood vessels and other tissues where they can cause inflammation and eventual tissue damage. Rheumatoid arthritis and some types of glomerulonephritis are examples of type III allergic reactions.

Finally, cell-mediated, or type IV, allergic reactions are also known as delayed hypersensitivity reactions. Symptoms appear 12 to 72 hours after exposure. Antigen-presenting cells (such as those in the skin) bind to the allergen and carry it to lymph nodes, where they present it to T cells. The T cells then multiply, and some of them return to the allergen's point of entry. They activate macrophages, tumor necrosis factor, and other agents that then create an inflammatory response.

Allergic contact dermatitis is one of the most commonly encountered type IV reactions, caused by exposure to poison ivy, some cosmetics, heavy metals, and other seemingly innocuous materials. When skin brushes against poison ivy, the plant resin urushiol works into the skin and binds with self-proteins. Then these are bound by antigen-presenting cells located there and attacked as invaders by the immune system. Within a couple of days an itchy rash develops.

Autoimmunity

In some people the immune system can work against itself. Instead of recognizing a "self" cell as friendly, it attacks as if the cell were an invader, producing antibodies and cytotoxic T cells that destroy normal body tissues. Called autoimmunity, this condition affects 5 percent of adults in North America and Europe, with women having twice the incidence of men.

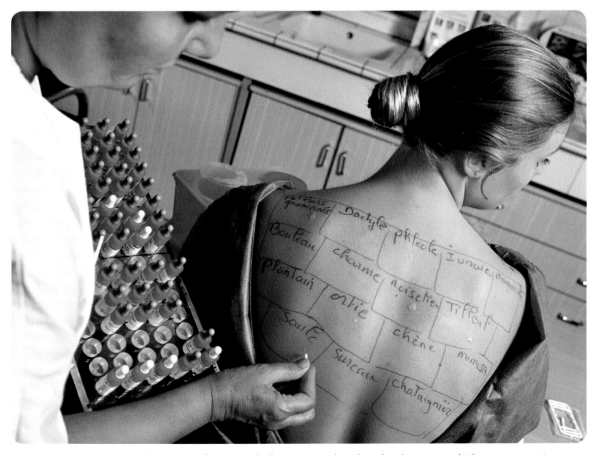

In allergy testing, small amounts of suspected allergens are placed on the skin to see which cause a reaction.

In an autoimmune disorder, T cells (yellow) mistakenly attack the glial cells that support nerve cells (blue).

Normally, B and T cells learn to recognize one's own MHC (major histocompatibility complex) proteins, a personal ID attached to all body cells. The immune system's ability to recognize and ignore its own cells is called self-tolerance. Early in life, immune cells that do not recognize your MHC proteins are eliminated or deactivated. Sometimes this safeguard doesn't work as intended. Environmental cues or faulty genes produce clones of self-reactive B and T cells that become activated and fuel antibody- or cell-mediated immune responses against "self" cells.

Besides renegade immune cells, autoimmunity can be caused by the appearance of new self-antigens, or self-proteins not previously exposed to the immune system as a result of a genetic mutation or a chemical alteration of existing self-antigens, possibly during infections. There are also foreign antigens that so resemble self-antigens that the self-antigens become targets. In rheumatic fever, for example, antibodies attacking streptococcus bacteria cross-react with heart antigens and damage heart muscles, heart valves, and joints. Another example of an autoimmune disease, multiple sclerosis

DID YOU KNOW?

LUPUS

Named for its characteristic skin lesions that resemble the bite of a wolf, lupus is an autoimmune disorder. Its most common form, systemic lupus erythematosus, is more prevalent in women than men, affecting people aged 15 to 50; the hormone estrogen may play a role in its development. Symptoms include redness or rash on the cheeks, sensitivity to light, inflammation of the lining of the heart or lungs or joints, damage to glomeruli of the kidneys, and the production of anti-DNA antibodies.

(MS) causes degeneration of tissues in the brain or spinal cord and can lead to partial or complete paralysis. Cytotoxic T cells become activated and destroy the myelin sheaths around nerve axons.

In the case of Type 1 (juvenile) diabetes mellitus, the T cells attack pancreatic beta cells, which produce insulin. Without insulin, carbohydrates cannot be normally processed. There is a period of several years in which the T cells weed out beta cells, possibly because their proteins resemble those of a virus the immune system is primed to attack. After this period symptoms may appear, usually before the person is 15 years old. Symptoms include increased thirst and urination, weight loss, fatigue, and ketoacidosis (increased ketones in the blood). Today some 750,000 Americans have type 1 diabetes.

Autoantibodies can cause other autoimmune diseases. In Graves' disease, they mimic a thyroid-stimulating hormone that leads to hyperthyroidism. Autoantibodies may block receptors for the neurotransmitter acetylcholine; nerves thus cannot properly communicate with weakened skeletal muscles, resulting in the disease myasthenia gravis. Some autoimmune diseases stem from wrongly inactivated helper T cells or from the excess output of the antiviral protein gamma interferon. Other autoimmune diseases include lupus (see "Did You Know?," left), rheumatoid arthritis, glomerulonephritis, pernicious and certain types of hemolytic anemia, Addison's disease (adrenal gland failure), Hashimoto's thyroiditis, and ulcerative colitis.

Treatments for autoimmune diseases seek to suppress parts of the immune response without interfering with the whole system. Injections of genetically altered antibodies help some MS patients.

The bright spots in this brain image represent brain lesions that are characteristic of multiple sclerosis.

BIOGRAPHIES

BOARD OF ADVISERS

WILLIAM R. AYERS, M.D. is emeritus professor of medicine and consultant in educational affairs to the academic dean at Georgetown University School of Medicine in Washington, D.C. He is a fellow of the American College of Physicians and was a Robert Wood Johnson fellow in health policy.

ROBERT M. BRIGGS, M.D. graduated from Yale University School of Medicine and completed his residency at Stanford University. Before retiring, Dr. Briggs was chair of the Department of Plastic and Reconstructive Surgery at the St. Barnabas Medical Center in Livingston, New Jersey. He served as clinical assistant professor of surgery at Robert Wood Johnson Medical School.

ELIZABETH BROWNELL, M.D. is a 1988 graduate of St. George's University School of Medicine, Grenada, West Indies, and a 1991 graduate of the Medical College of Wisconsin Waukesha Family Practice Residency Program where she served as chief resident. She served as assistant professor of clinical medicine at the Medical College of Wisconsin from 1991–1994 and vice chair of the Department of Family and Emergency Medicine at Waukesha Memorial Hospital from 2000–2002. She has been in private practice since 1994. Dr. Brownell was also a consultant on *National Geographic Investigates: Medical Mysteries.*

TED M. BURNS, M.D. is a professor of neurology at the University of Virginia. He graduated from the Kansas University School of Medicine and did a residency in neurology and a fellowship in electrodiagnostic and neuromuscular medicine at the University of Virginia, followed by a fellowship in peripheral nerve disease at the Mayo Clinic in Rochester, Minnesota.

ROBERT A. CHASE. M.D. is the Emile Holman Professor of Surgery and professor of anatomy, emeritus, at the Stanford University School of Medicine. After joining the faculty at Yale in 1947, Dr. Chase established the Section of Plastic Surgery there. He later joined the faculty at Stanford as professor and chairman of surgery and was named the first Holman professor.

D. ROBERT DUFOUR, M.D. is emeritus professor of pathology at George Washington University Medical Center School of Medicine and Health Sciences. He is also consultant pathologist at the Veterans Affairs Medical Center in Washington, D.C., where he served as chief of Pathology and Laboratory Medicine Service.

DAVID E. FLEISCHER, M.D. is a gastroenterologist at the Mayo Clinic, Arizona, and a professor of medicine at the Mayo Clinic College of Medicine. He is past president of the American Society for Gastrointestinal Endoscopy and a previous Robert Wood Johnson health policy fellow.

JEAN L. FOURCROY, M.D., PH.D., M.P.H. is a retired Navy captain who now serves as staff urologist at Walter Reed Army Medical Center in Washington, D.C. She is a former medical officer with the Food and Drug Administration. She currently serves on the board of the U.S. Anti-Doping Agency.

DONALD S. KARCHER, M.D. is chair of the Department of Pathology at the George Washington University Medical Center, where he also serves as director of laboratories and chief of the Flow Cytometry Service.

JOHN E. MORLEY, M.B., B.CH. is Dammert Professor of Gerontology and director, Division of Geriatric Medicine and Endocrinology at

St. Louis University Medical Center. He is also director of the Geriatric Research, Education, and Clinical Center at the St. Louis Veterans Affairs Medical Center.

POLLY E. PARSONS, M.D. is professor of medicine, chief of pulmonary and critical care medicine, and chair of the Department of Medicine at the University of Vermont College of Medicine. She graduated from the University of Arizona College of Medicine and completed a residency in internal medicine and a fellowship in pulmonary medicine at the University of Colorado.

ROBERT S. ZOHLMAN, M.D. retired as clinical professor of medicine at the University of California, San Francisco. He graduated from the Albert Einstein College of Medicine and completed his residency at Montefiore Hospital in New York. He completed a fellowship in nephrology at the University of California, San Francisco.

AUTHORS

PATRICIA DANIELS has written extensively on science and health. Among her publications are *The New Solar System, Human Body, Encyclopedia of Space, Childhood Medical Guide,* and *The Medical Advisor.*

TRISHA GURA, PH.D. is a molecular biologist and former journalist for the *Chicago Tribune.* She earned a Knight science journalism fellowship at MIT and Harvard. Today, she is a resident scholar at Brandeis University, a freelance writer, and author of *Lying in Weight: The Hidden Epidemic of Eating Disorders.*

SUSAN TYLER HITCHCOCK has written more than a dozen books, including National Geographic's *Geography of Religion* and

Frankenstein: A Cultural History. She is a senior editor with the National Geographic Books Division, responsible for nature, science, and health titles.

LISA STEIN is an online news editor with *Scientific American.* She was former deputy national editor at *U.S. News & World Report.* In addition, she has written and edited for many media outlets, including *TV Guide* (as Washington bureau chief) and *National Journal.*

RICHARD RESTAK, M.D. is a neurologist and best-selling author of 20 books, including his newest book, *Think Smart.* He is clinical professor of neurology at the George Washington University Medical Center School of Medicine and Health Sciences, has served on many national advisory councils for brain research, and has been a consultant to NBC's *Today* show.

JOHN THOMPSON has a background in English and biology. He is author of ten National Geographic books and two novels, *The Reservoir* and *Love and Lament.*

CAPTIONS FOR CHAPTER OPENERS

(All images are identified from left to right.)

Chapter 1, "Beginnings" (p. 11): Named for a cell theory pioneer, a Schwann cell (green) supports a nerve cell (blue); multigenerational family; chromosome; a fetus at 16 weeks. Chapter 2, "Body Armor" (p. 41): a woman's hair; feet; a magnified image of blood clotting; woman with bare skin. Chapter 3, "Structure" (p. 65): skull; baby feet; a 3-D illustration of a male skeleton; teeth. Chapter 4, "Body Movers" (p. 103): a muscular man lifting weights; a woman exercising; a 3-D illustration of male musculature; runners. Chapter 5, "Circulation" (p. 123): blood cells; a human heart; vials of blood; blood vessels. Chapter 6, "Vital Exchange" (p. 153): a digital illustration of alveoli, delicate air cavities in the lungs; a boy blowing bubbles; a 3-D illustration of lungs with bronchi in red; a man holding his breath underwater. Chapter 7, "Body Fuel" (p. 177): a woman with abdominal pains; an x-ray of a woman's intestines and stomach; nutritious berries; a scan of a human liver. Chapter 8, "Body Cleansing" (p. 207): an x-ray of a urinary bladder; a young woman hydrating; kidneys; a color-enhanced microscopic image of a cross-section of a ureter, a muscular duct that sends urine from the kidneys to the bladder. Chapter 9, "Control Center" (p. 227): neurons and nuclei; children using their senses to read and listen; a microscopic view of a human brain; a child tasting. Chapter 10, "Messengers" (p. 265): a microscopic image of a human thyroid; physical attraction; a color-enhanced microscopic image of the female hormone oxytocin, which causes contractions during labor; skydiving, which induces a rush caused by several different hormones. Chapter 11, "Reproduction" (p. 295): ovulation; a colorized 3-D image of sperm entering an egg; pregnancy; a newborn baby girl. Chapter 12, "Protection" (p. 321): *E. coli* bacteria and microphages, cells that fight off infection; a child with chicken pox dotted with blue anti-itch ointment; white blood cells; a child receiving a vaccination.

ILLUSTRATIONS CREDITS

INDEX

NATIONAL GEOGRAPHIC
THE BODY: A COMPLETE USER'S GUIDE

Published by the National Geographic Society

Gary E. Knell, *President and Chief Executive Officer*
John M. Fahey, *Chairman of the Board*
Declan Moore, *Executive Vice President; President, Publishing and Travel*
Melina Gerosa Bellows, *Executive Vice President; Chief Creative Officer, Books, Kids, and Family*

Prepared by the Book Division

Hector Sierra, *Senior Vice President and General Manager*
Janet Goldstein, *Senior Vice President and Editorial Director*
Jonathan Halling, *Creative Director*
Marianne R. Koszorus, *Design Director*
R. Gary Colbert, *Production Director*
Jennifer A. Thornton, *Director of Managing Editorial*
Susan S. Blair, *Director of Photography*
Meredith C. Wilcox, *Director, Administration and Rights Clearance*

Staff for This Book

Amy Briggs, *Editor*
Robin Terry Brown, *Project Editor*
Suzanne Crawford, Karin Kinney, *Text Editors*
Sanaa Akkach, Elisa Gibson, *Art Directors*
Kris Hanneman, Kristin Sladen, *Illustrations Editors*
Grassroots Graphics, *Design and Production*
Michelle Cassidy, *Editorial Assistant*
Erin Greenhalgh, *Editorial Intern*
Dan O'Toole, Suzanne Poole, *Researchers*
Emily Paulsen, *Contributing Writer*
Marshall Kiker, *Associate Managing Editor*
Judith Klein, *Production Editor*
Lisa A. Walker, *Production Manager*
Galen Young, *Rights Clearance Specialist*
Katie Olsen, *Production Design Assistant*

Production Services

Phillip L. Schlosser, *Senior Vice President*
Chris Brown, *Vice President, NG Book Manufacturing*
Nicole Elliott, *Director of Production*
George Bounelis, *Senior Production Manager*
Rachel Faulise, *Manager*
Robert L. Barr, *Manager*

The National Geographic Society is one of the world's largest nonprofit scientific and educational organizations. Its mission is to inspire people to care about the planet. Founded in 1888, the Society is member supported and offers a community for members to get closer to explorers, connect with other members, and help make a difference. The Society reaches more than 450 million people worldwide each month through *National Geographic* and other magazines; National Geographic Channel; television documentaries; music; radio; films; books; DVDs; maps; exhibitions; live events; school publishing programs; interactive media; and merchandise. National Geographic has funded more than 10,000 scientific research, conservation, and exploration projects and supports an education program promoting geographic literacy. For more information, visit www.nationalgeographic.com.

National Geographic Society
1145 17th Street N.W.
Washington, D.C. 20036-4688 U.S.A.

For information about special discounts for bulk purchases, please contact National Geographic Books Special Sales: ngspecsales@ngs.org

For rights or permissions inquiries, please contact National Geographic Books Subsidiary Rights: ngbookrights@ngs.org

The Library of Congress has cataloged the 2007 edition as follows:

Body : the complete human / foreword by Richard Restak ; text by Patricia Daniels ... [et al.].
    p. cm.
   Includes index.
   ISBN 978-1-4262-0128-8 (hc) -- ISBN 978-1-4262-0185-1 (deluxe)
   1. Human physiology--Popular works. 2. Body, Human--Popular works. I. Daniels, Patricia, 1955-
   QP38.B597 2007
   612--dc22
                                                    2007026161
ISBN: 978-1-4262-1414-1 (regular)
ISBN: 978-1-4262-1415-8 (deluxe)

Printed in the United States of America
14/QGT-CML/1

SHAPE YOUR FUTURE

with *National Geographic Books*

Written by medical experts, these essential resources provide the tips, advice, remedies, and recipes you need to be and stay healthy.

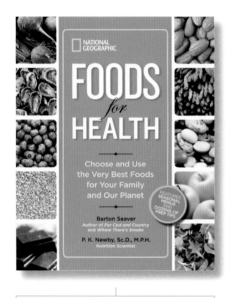

The brain weakens as it ages. This book shows you how to fight back. Full of tips, exercises, and more, it will help anyone's brain stay sharp for years to come.

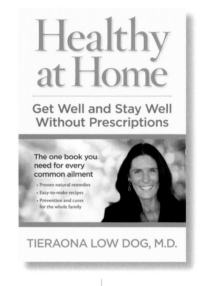

Filled with your favorite menu items and new cooking ideas, this reference book helps you make the right choices for you, your family, and the planet.

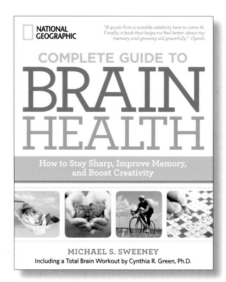

Tieraona Low Dog, M.D., integrative physician and expert in natural medicine, shares her favorite remedies and recipes to help you find milder, more natural ways to get well.

AVAILABLE WHEREVER BOOKS ARE SOLD

nationalgeographic.com/books

 Like us on Facebook: Nat Geo Books

 Follow us on Twitter: @NatGeoBooks